AF621546

Software-Defined Radio for Engineers

For a listing of recent titles in the *Artech House Mobile Communications*, turn to the back of this book.

Software-Defined Radio for Engineers

Travis F. Collins
Robin Getz
Di Pu
Alexander M. Wyglinski

Library of Congress Cataloging-in-Publication Data
A catalog record for this book is available from the U.S. Library of Congress.

British Library Cataloguing in Publication Data
A catalog record for this book is available from the British Library.

ISBN-13: 978-1-63081-457-1

Cover design by John Gomes

© 2018 Travis F. Collins, Robin Getz, Di Pu, Alexander M. Wyglinski

All rights reserved. No part of this book may be reproduced or utilized in any form or by any means, elec-tronic or mechanical, including photocopying, recording, or by any information storage and retrieval system, without permission in writing from the publisher.

All terms mentioned in this book that are known to be trademarks or service marks have been appropriately capitalized. Artech House cannot attest to the accuracy of this information. Use of a term in this book should not be regarded as affecting the validity of any trademark or service mark.

10 9 8 7 6 5 4 3 2 1

Dedication

To my wife Lauren
—Travis Collins

To my wonderful children, Matthew, Lauren, and Isaac, and my patient wife, Michelle—sorry I have been hiding in the basement working on this book. To all my fantastic colleagues at Analog Devices: Dave, Michael, Lars-Peter, Andrei, Mihai, Travis, Wyatt and many more, without whom Pluto SDR and IIO would not exist.
—Robin Getz

To my lovely son Aidi, my husband Di, and my parents Lingzhen and Xuexun
—Di Pu

To my wife Jen
—Alexander Wyglinski

Contents

Preface

Every sector of today's society is entirely dependent on connectivity. As a result, communication systems engineering has evolved into an essential profession where practitioners are required to master a diverse set of skills and tools into order to solve the latest technical challenges. This was not the case several decades ago, where individuals on a design team each possessed complementary yet different backgrounds needed in order to construct a communication system. For example, in order to design a communication system it would take several engineers, each of whom would be individually handling tasks such as algorithmic development for the transceiver, theoretical performance analysis of the end-to-end communication system, implementation of the platform in digital hardware, design and implementation of the radio frequency front-end (RFFE), and so on. These tasks were intentionally defined to be handled in silos such that each engineer would not be responsible or extensively knowledgeable about the other concurrent tasks. Now let us fast forward to today, where effectively all of these silos are extensively overlapping with each other. An engineer is expected to know how to provide solutions across several of these silos on the same project, such as designing a communication system algorithm and masterfully implementing it on a field programmable gate array (FPGA) or an embedded processing device. This is mainly due to a new technology that has evolved over the past several decades and matured into a mainstream communication system solution: *software-defined radio* (SDR). The days of working in task silos when designing a communication system are quickly coming to an end.

The objective of this book is to provide a hands-on learning experience using SDR for engineering students and industry practitioners who are interested in *mastering the design, implementation, and experimentation of a communication system*. Building on the success of *Digital Communication Systems Engineering Using Software Defined Radio* by Pu and Wyglinski (Artech House, 2013), this book provides a fresh perspective on understanding and creating new communication systems from scratch. Until now, there have been very few books and other publications available to the community that provide an opportunity not only to learn about the theoretical elements of a communication system but also provide practical real-world experiments that help synthesize these important lessons that would otherwise be omitted in a traditional communication systems book or course. There is so much more that goes into the design of a communication system than just drawing up a block diagram consisting of different functions and deriving its performance characteristics. Communication system engineers need to understand the impact of the hardware on the performance of the communication algorithms being used and how well the overall system operates in terms of successfully recovering the intercepted signal. What makes this hands-on learning experience is the SDR platform itself. Several years ago, it was a significant financial and technical investment to utilize SDR in order to prototype and experiment with actual communication systems. The hardware was at least a couple of thousand dollars and mastering the software was challenging to say the least. Nevertheless, the last decade has witnessed significant advances in SDR technology such that these platforms now cost on the order of a hundred dollars and the software is relatively straightforward to use and reliable.

This book is ideally suited for individuals who possess a fundamental understanding of continuous-time and discrete-time signals and systems, as well as possess a solid understanding of computer engineering. Additionally, individuals who already possess some basic knowledge about communication systems, for example, amplitude modulation, frequency modulation, and phase shift keying, would be beneficial. This book is written for both industry practitioners who are seeking to enhance their skill set by learning about the design and implementation of communication systems using SDR technology, as well as both undergraduate and graduate students who would like to learn about and master communication systems technology in order to become the next generation of industry practitioners and academic researchers. The book contains theoretical explanations about the various elements forming a communication system, practical hands-on examples and lessons that help synthesize these concepts, and a wealth of important facts and details to take into consideration when building a real-world communication system.

The book is organized in such a way that it can be used in either a 7-week academic term or a 14-week academic semester, as a guide for self-study on this subject, or as a technical reference about communication systems engineering. The book is structured in the following manner:

- Establishing a solid foundation
  - *Chapter 1 – Introduction to Software-Defined Radio*: Provides a brief overview of communication systems engineering and the evolution of SDR technology.
  - *Chapter 2 – Signals and Systems*: A condensed review of discrete-time signal and systems, digital signal processing, filter design, and their application to communication systems.
  - *Chapter 3 – Probability in Communications*: An overview of the must-know topics in probability theory required in designing and analyzing communication systems.
  - *Chapter 4 – Digital Communications Fundamentals*: A relatively brief treatment of various digital communication principles, including modulation, digital transmission, and receiver structures.
- Fundamental of SDR-based communication systems engineering
  - *Chapter 5 – Understanding SDR Hardware*: A tutorial regarding SDR technology with emphasis on the student targeted ADALM-PLUTO SDR, which will be used in the hands-on elements of this book.
  - *Chapter 6 – Timing Synchronization*: A detailed explanation of how to obtain timing information from an intercepted signal.

  - *Chapter 7 – Carrier Synchronization*: An extensive description of methodologies used to obtain the carrier frequency of intercepted signals, which also includes hands-on examples.
  - *Chapter 8 – Frame Synchronization and Channel Coding*: An extensive introduction to the theoretical and practical considerations when performing frame synchronization.
- Advanced topics in communications design and implementation
  - *Chapter 9 – Channel Estimation and Equalization*: Both theoretical and experimental details on the design and implementation of several equalization methodologies are provided.
  - *Chapter 10 – Orthogonal Frequency Division Multiplexing*: A detailed study of orthogonal frequency division multiplexing (OFDM) communication systems, including their implementation in SDR.
  - *Chapter 11 – Applications for Software-Defined Radio*: A brief introduction into various application areas involving SDR technology, including 5G communications and deep space satellite communications.

For a 7-week course, it is expected that the class would cover Chapters 4–8 throughout the term, while for a 14-week course it is expected that the class would cover Chapters 4–9. As for industry practitioners seeking advanced knowledge in SDR implementations, they can readily proceed with approaching Chapters 9–11. Although this book covers a substantial amount of material, it is not designed to cover topics in information theory, signal estimation and detection, RFFE design, computer networking theory, or wireless channel modeling. We believe that there are numerous other books available that handle these topics rigorously and efficiently.

This book was made possible by the extensive support of numerous individuals and organizations throughout the duration of this project. In particular, we are deeply indebted to Walt Kester for extensively reviewing many drafts of this book and providing us with valuable feedback and comments. Furthermore, we would like to gratefully acknowledge the support of the folks at Artech House, including Aileen Storry, through this entire project and helping us make this book a reality. We would like to sincerely thank our collaborators at MathWorks for their support and assistance with respect to many of the software elements of this book, especially Darel Linebarger and Mike McLernon. The generous contributions of Analog Devices with respect to providing us with the material support needed to create the hands-on elements of this publication is duly acknowledged. We would like to sincerely thank the entire class of ECE4305 *Software-Defined Radio Systems and Analysis* at Worcester Polytechnic Institute for providing extensive feedback about this book prior to its publication. Finally, we would like to thank our families for their love, support, and constant encouragement.

CHAPTER 1

Introduction to Software-Defined Radio

Various forms of communication have evolved over the millennia. The spoken word can be transmitted from one person, and heard or received by another. In modern times town criers hold an annual contest to discover who can shout a comprehensible message over the greatest distance [1]. However, while the world record is for loudest crier is 112.8 decibels, it can only be understood at less than 100 meters. The desire to communicate more effectively than shouting, is old as speech itself.

With modern advances in computing technologies, digital signal processing and digital communication algorithms, artificial intelligence, radio frequency (RF) hardware design, networking topologies, and many other elements have evolved modern communication systems into complex, intelligent, high-performance platforms that can adapt to operational environments and deliver large amounts of information in real-time, error-free. The latest step in communication systems technology is the software-defined radio, or SDR, which adopts the most recent advances in all fields to yield the ultimate transmitter and receiver.

1.1 Brief History

Given the exciting history associated with advances that directly impact SDR technology, Figure 1.1 provides a timeline describing several significant milestones over the past four centuries. This history is dominated by various people investigating ideas or concepts, publishing the results, then allowing their peers and colleagues to build on their work. Many turned their work into commercial products and became famous and rich; some became neither. For an exhaustive list of major milestones relevant to SDR technology, the interested reader is referred to Appendix A.

1.2 What is a Software-Defined Radio?

Every professional organization attempts to define a common framework of terms and definitions to allow easy communication between professionals who are working on similar areas of research or product development. Wireless communications and SDR is no different. The Institute of Electrical and Electronic Engineers (IEEE) P1900.1 Working Group has created the following definitions to

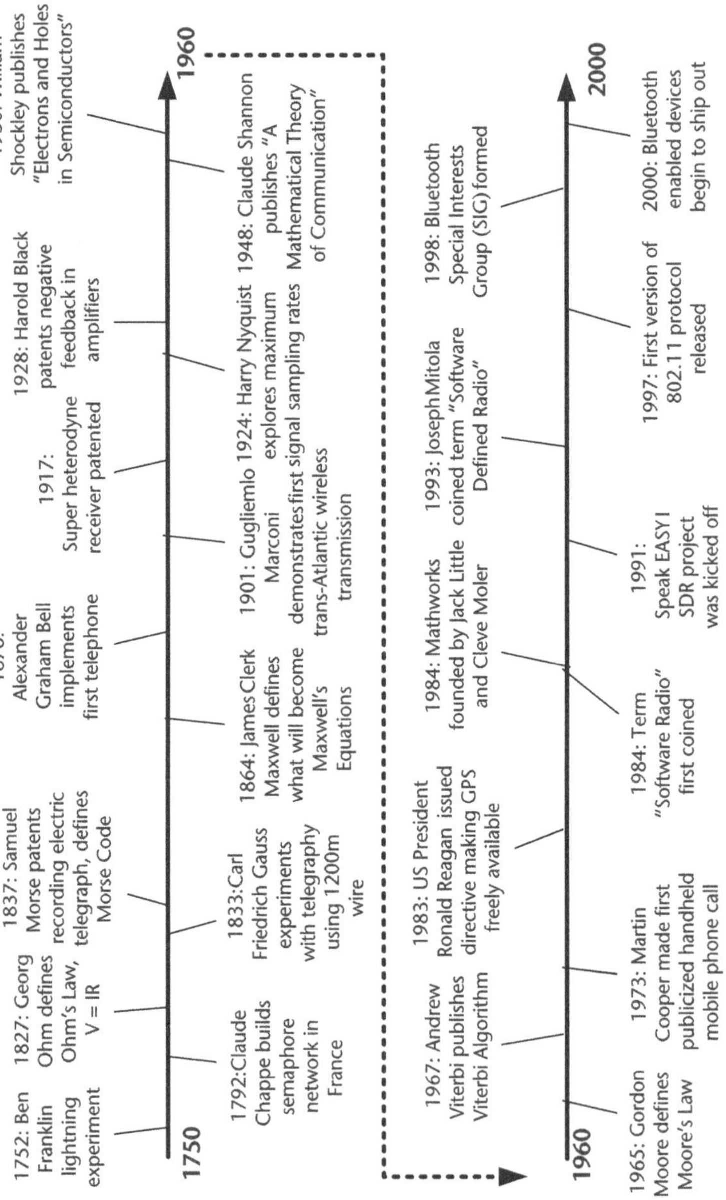

Figure 1.1 Timeline of several key milestones in communications.

ensure that everyone in the field has common terminology [2]:

Radio

1. Technology for wirelessly transmitting or receiving electromagnetic radiation to facilitate transfer of information.
2. System or device incorporating technology as defined in (1).
3. A general term applied to the use of radio waves.

Radio Node

A radio point of presence incorporating a radio transmitter or receiver.

Software

Modifiable instructions executed by a programmable processing device.

Physical Layer

The layer within the wireless protocol in which processing of RF, IF, or baseband signals including channel coding occurs. It is the lowest layer of the ISO 7-layer model as adapted for wireless transmission and reception.

Data Link Layer

The protocol responsible for reliable frame transmission over a wireless link through the employment of proper error detection and control procedures and medium access control.

Software Controlled

Software controlled refers to the use of software processing within the radio system or device to select the parameters of operation.

Software Defined

Software defined refers to the use of software processing within the radio system or device to implement operating (but not control) functions.

Software Controlled Radio

Radio in which some or all of the physical layer functions are software controlled.

Software-Defined Radio (SDR)

Radio in which some or all of the physical layer functions are software defined.

Transmitter

Apparatus producing radio frequency energy for the purpose of radio communication.

Receiver

A device that accepts a radio signal and delivers information extracted from it.

Air Interface

The subset of waveform functions designed to establish communication between two radio terminals. This is the waveform equivalent of the wireless physical layer and the wireless data link layer.

Waveform

1. The set of transformations applied to information to be transmitted and the corresponding set of transformations to convert received signals back to their information content.
2. Representation of a signal in space.
3. The representation of transmitted RF signal plus optional additional radio functions up to and including all network layers.

The combination of digital processing and analog RF has always made up communication systems. In today's modern systems signal processing has progressed to such an extent that a majority of baseband functionality is being implemented in software. The flexibility of the RF hardware to be re purposed and reconfigured has led to one radio front-end handling most RF systems. Normally the RF front-end is software controlled rather than software defined. This modern combination of flexible RF front-ends and signal processing in software has lead the birth of software-defined radio.

This can been seen in devices like Analog Devices's AD7030, shown in Figure 1.2. The ADF7030 is a low-power, high-performance, integrated radio transceiver supporting narrow band operation in the 169.4-MHz to 169.6-MHz ISM band. It supports transmit and receive operation at 2.4 kbps and 4.8 kbps using 2GFSK modulation and transmit operation at 6.4 kbps using 4GFSK modulation. It includes an on-chip ARM Cortex-M0 processor that performs radio control and calibration as well as packet management. That and a sensor is all that is needed for smart metering or active tag asset tracking applications. This is just a side effect of Moore's law—system-level integration.

An SDR system is a complex device that performs several complicated tasks simultaneously in order to enable the seamless transmission and reception of data. In general, a digital communications system consists of an interdependent sequence of operations responsible for taking some type of information, whether it is human speech, music, or video images, and transmits it over-the-air to a receiver for processing and decoding into a reconstructed version of the original information signal. If the original information is analog (like audio), it must first be digitized using techniques such as quantization in order for us to obtain a binary representation of this information. Once in a binary format, the transmitter digitally

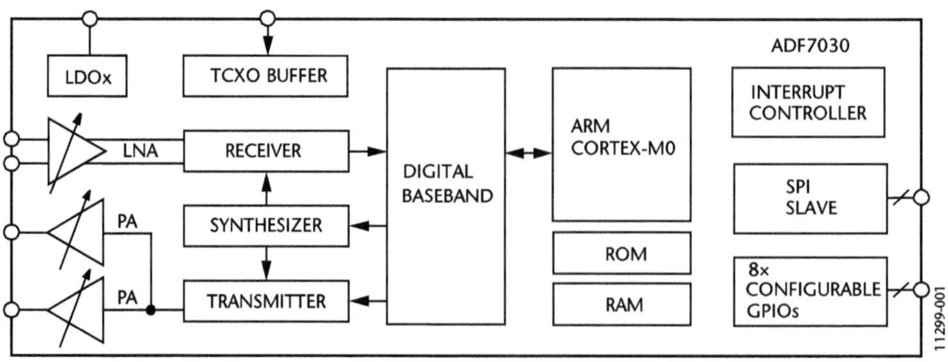

Figure 1.2 ADF7030 block diagram [3].

processes this information and converts it into an electromagnetic sinusoidal waveform that is uniquely defined by its physical characteristics, such as its signal amplitude, carrier frequency, and phase. At the other end of the communications link, the receiver is tasked with correctly identifying the physical characteristics of the intercepted modulated waveform transmitted across a potentially noisy and distortion-filled channel, and ultimately returning the intercepted signal back into the correct binary representation. The basic building blocks of a digital communication system is shown in Figure 1.3.

Figure 1.3 shows that the input to the transmitter and output of the receiver originate from or are fed into a *digital source* and *digital sink*, respectively. These two blocks represent the source and destination of the digital information to be communicated between the transmitter and receiver. Once the binary information is introduced to the transmitter, the first task performed is to remove all redundant/repeating binary patterns from the information in order to increase the efficiency of the transmission. This is accomplished using the *source encoder* block, which is designed to strip out all redundancy from the information. Note that at the receiver, the *source decoder* re-introduces the redundancy in order to return the binary information back to its original form. Once the redundancy has been removed from the binary information at the transmitter, a *channel encoder* is employed to introduced a controlled amount of redundancy to the information stream in order to protect it from potential errors introduced during the transmission process across a noisy channel. A *channel decoder* is used to remove this controlled redundancy and return the binary information back to its original form. The next step at the transmitter is to convert the binary information into unique electromagnetic waveform properties such as amplitude, carrier frequency, and phase. This is accomplished using a mapping process called *modulation*. Similarly, at the receiver the *demodulation* process converts the electromagnetic waveform back into its respective binary representation. Finally, the discrete samples outputted

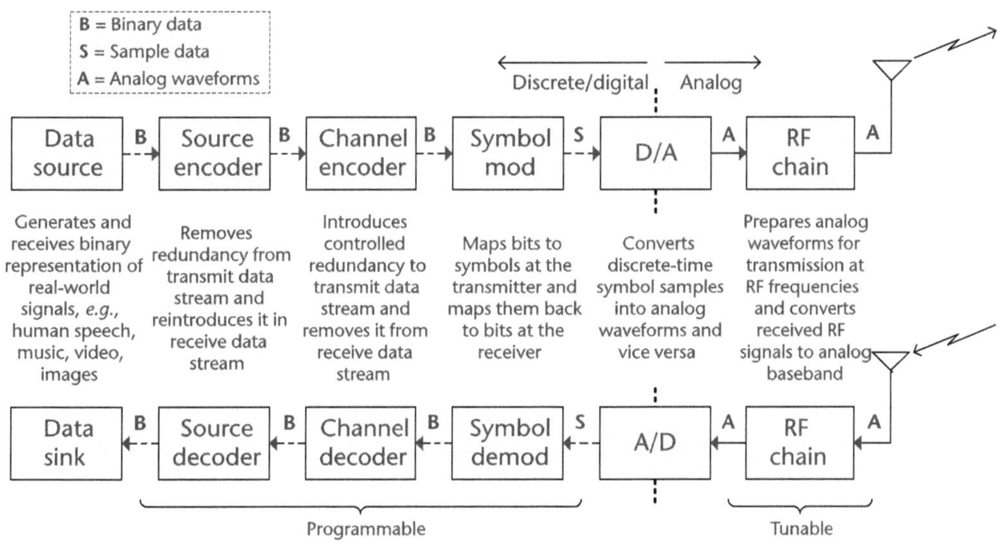

Figure 1.3 An illustration describing some of the important components that constitute a modern digital communications system. Note that for a SDR-based implementation, those components indicated as programmable can be realized in either programmable logic or software.

by the modulation block are resampled and converted into a baseband analog waveform using a *digital-to-analog converter* (DAC) before being processed by the radio frequency (RF) front-end of the communication system and upconverted to an RF carrier frequency. At the receiver, the reverse operation is performed, where the intercepted analog signal is downconverted by the RFFE to a baseband frequency before being sampled and processed by an *analog-to-digital converter* (ADC).

Given the complexity of an SDR platform and its respective components, it is important to understand the limitations of a specific SDR platform and how various design decisions may impact the performance of the resulting prototype. For instance, it is very desirable to have real-time baseband processing for spectrum sensing and agile transmission operations with high computational throughput and low latency. However, if the microprocessor being employed by the SDR platform is not sufficiently powerful enough in order to support the computational operations of the digital communication system, one needs to reconsider either the overall transceiver design or the requirements for low latency and high throughput. Otherwise, the SDR implementation will fail to operate properly, yielding transmission errors and poor communication performance.

Design considerations to think about when devising digital communication systems based on an SDR platform include.

- The integration of the physical and network layers via a real-time protocol implementation on an embedded processor. Note that most communication systems are divided into logically separated layers in order to more readily facilitate the design of the communication system (see Section 1.3). However, it is imperative that each layer is properly designed due to the strong interdependence between all the layers.
- Ensuring that a sufficiently wide bandwidth radio front-end exists with agility over multiple subchannels and scalable number of antennas for spatial processing. Given how many of the advanced communication system designs involve the use of multiple antennas and wideband transmissions, it is important to know what the SDR hardware is capable of doing with respect to these physical attributes.
- Many networks employing digital communication systems possess a centralize architecture for controlling the operations of the overall network (e.g., control channel implementation). Knowing the radio network architecture is important since it will dictate what sort of operations are essential for one digital transceiver to communicate with another.
- The ability to perform controlled experiments in different environments (e.g., shadowing and multipath, indoor and outdoor environments) is important for the sake of demonstrating the reliability of a particular SDR implementation. In other words, if an experiment involving an SDR prototype system is conducted twice in a row in the exact same environment and using the exact same operating parameters, it is expected that the resulting output and performance should be the same. Consequently, being able to perform controlled experiments provides the SDR designer with a sanity check capability.

- Reconfigurability and fast prototyping through a software design flow for algorithm and protocol description.

Instead of using fixed analog processing and fixed circuits, many of the communication systems in use every day are being implemented using microelectronic-based flexible IF, digital signal processors, programmable digital logic, accelerators, and other types of computing engines. To take advantage of new advances in processing engines, high-level languages such as MATLAB® and Simulink are being used rather than C or assembly. This transition of computing technology had the impact of enabling new communication functionalities and capabilities, such as advanced satellite communications, mobile communications, data modems, and digital television broadcasts.

1.3 Networking and SDR

With the evolution of digital communication system into highly complex devices, it became apparent that a divide-and-conquer strategy was needed in order to make the design and implementation of such systems feasible and manageable. Consequently, researchers divided a digital communication system into a collection of complementary layers, with each layer performing a specific function as part of the overall process of transmitting and receiving information. As a result of this divide-and-conquer strategy, communication systems rapidly evolved into highly capable platforms performing a wide range of operations, such as Web surfing and email to streaming multimedia content. In fact, this strategy of dividing up the operations of a communication system into layers was so successful that there are entire research communities that only focus on one of the layers and none of the others; they take for granted the information coming from the layers above and below their layer.

In general, there are two models for dividing up a communication system into layers: the Open System Interconnection (OSI) 7-layer model and the Transmission Control Protocol (TCP)/Internet Protocol (IP) 5-layer model, as shown in Figure 1.4. Both models possess approximately the same functionality, but the TCP/IP model amalgamates the top several layers into a single one. Focusing on the TCP/IP 5-layer model, this consists of the following layers, from top to bottom:

- *Application Layer*: Interfaces user with the data from the communication system. For instance, the application layer would include data originating from or intended for software running Web browsers, email clients, and streaming media interfaces. These applications are usually addressed via designated socket.
- *Transport Layer*: Responsible for transporting application layer messages between the client application and server application. This layer ensures reliable data transmission.
- *Network Layer*: Responsible for moving network layer packets from one host to another host. Defines format of datagrams and how end systems and routers act on datagram, as well as determine routes that datagrams take between sources and destinations.

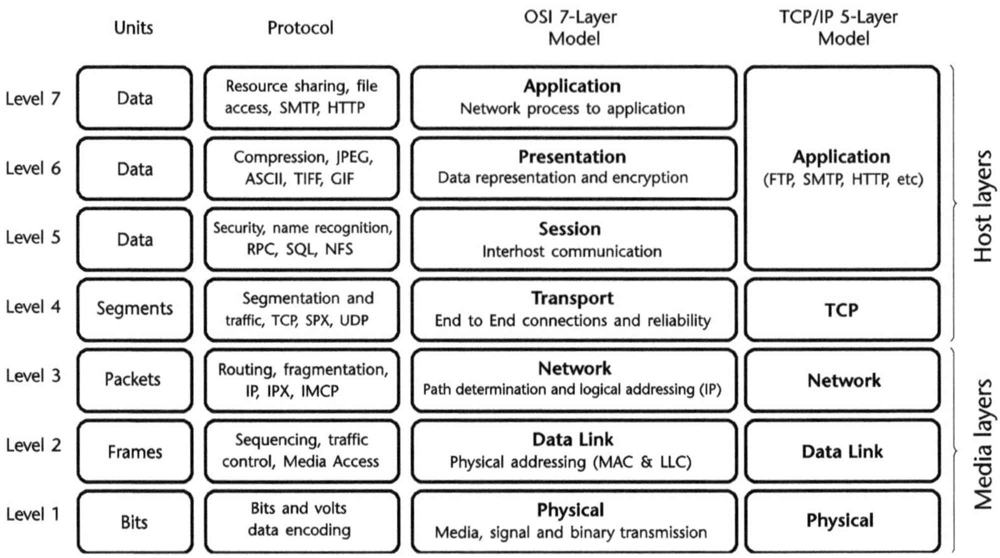

Figure 1.4 Seven-layer OSI model compared to five-layer TCP/IP model.

- *Link Layer*: Handles problem of exchanging data between two or more directly connected devices. Reliability: This includes error detection and error correction as well as addressing of different communication systems.
- *Physical Layer*: Sends individual bits from one communication system directly to another communication system. It also covers the physical interface between data transmission device and transmission medium.

From the perspective of a radio system and its implementation, much of the system design will focus on the physical layer (as does this text), but it can't be forgotten how the link layer may affect the physical layer. Nevertheless, given that the baseband processing is all conducted in software, it is possible for the communications system is to implement the higher layers of the stack in software as well. Many communication standards have adopted this scheme, where the entire communication system across all the layers are implemented in software, although depending on data rate requirements, this can require significant computational capabilities on the part of the system to achieve real-time operation. All software implementations enable functional updates without hardware replacement. In practice, this is normally only done on emerging standards or where data rates are relatively low. However, imagine applying a software upgrade to a Wi-Fi router and being able to implement the next standard without having to replace the hardware. This software upgradeable system would be more complex, and might cost more than a fixed hardware system, but would consumers be willing to pay more? History indicates no. For those types of high-volume consumer applications, many times price point is the most critical item to product success. Most end consumers do not think about long-term maintenance and total cost of ownership while looking at the variety of products on Amazon. The trade-offs of which function or layer is done in fixed hardware versus flexible software is an engineering decision based on volume, cost, power, complexity, and many other factors.

There has been a growing amount of interest with respect to combining SDR technology with software-defined networks (SDNs), where the latter focuses on adapting the higher communication layers to the prevailing operational environment. This enables things like modification of the routing to be tied to heuristics provided by the physical layers. Self-healing mesh networks are an implementation of this.

The link layer will also affect the physical (PHY) layer of a wireless communication system as shown in Figure 1.5. For example, in 802.11 (Wi-Fi), the PHY layer (layer 1) is actually broken further down into the Physical Layer Convergence Protocol (PLCP) and the Physical Medium Dependent (PMD) sublayer. The PMD sublayer provides transmission and reception of physical layer data units between two stations via the wireless medium, and passes this to the PLCP, which interfaces to the upper MAC layers, various management layer entities, and generic management primitives to maximize data rates.

At the PHY layer the unit denoted in Figure 1.4 is bits; however, across the wireless and wired links this data will be encoded in more analog-friendly forms designed for transmission called symbols. The preamble, denoted in Layer 1 in Figure 1.5 will most likely never be demodulated at the receiver from a symbol form. Such sequences are only used by the PHY layer to compensate for nonidealities in a link, and have little to no meaning to the above layers. However, for those implementing with SDRs these simple sections are the main focus, and the remaining portions of the frame are arbitrary data.

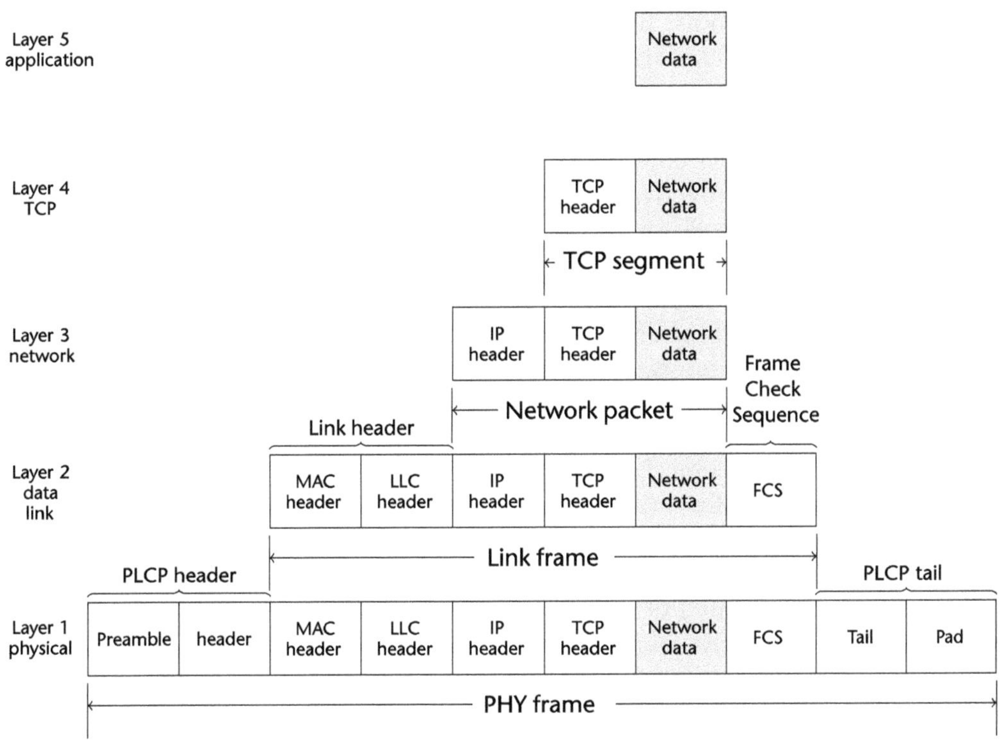

Figure 1.5 Packet structure effects SDR, PLCP = Physical Layer Convergence Protocol.

1.4 RF architectures for SDR

Next-generation communications systems introduce new challenges that require solutions beyond what can be achieved through individual device optimization. Integrating more software control and cognitive abilities to the radio demands a more frequency- and bandwidth-flexible RF design. To achieve this goal static filters need to be removed and replaced with tunable filters. Similarly, the concept of a common platform would allow for shorter development times, reduced manufacturing costs, and provide greater interoperability between systems. The common platform demands that the RF system be capable of providing full performance for applications that traditionally had very different architectures. Finally, future platforms are pushing size and power demands to a new extreme.

Handheld radios are becoming more capable and complex, but simultaneously requiring improved battery efficiency. Small UAVs lack the power generation of large aircraft and every milliwatt that the RF system consumes directly translates to payload battery weight, and thus, reduced flight time. To overcome these challenges and create the next generation of aerospace and defense solutions, a new radio architectures are being developed.

Since its inception, the superheterodyne architecture has been the backbone of radio design. Whether it is a handheld radio, unmanned aerial vehicle (UAV) data link, or a signal intelligence receiver, the single or dual mixing stage superheterodyne architecture is the common choice (see Figure 1.6). The benefits of this design are clear: proper frequency planning can allow for very low spurious emissions, the channel bandwidth and selectivity can be set by the intermediate frequency (IF) filters, and the gain distribution across the stages allows for a trade-off between optimizing the noise figure and linearity.

For over 100 years of use (see the appendix for more information), there have been significant gains in performance for the superheterodyne across the entire signal chain. Microwave and RF devices have improved their performance while decreasing power consumption. ADCs and DACs have increased the sample rate, linearity, and effective number of bits (ENOB). Processing capability in FPGAs and DSPs has followed Moore's law and increased with time, allowing for more efficient algorithms, digital correction, and further integration. Package technology has shrunk device pin density while simultaneously improving thermal handling.

However, these device-specific improvements are beginning to reach the point of diminishing returns. While the RF components have followed a reduced size, weight, and power (SWaP) trend, high-performance filters remain physically large and are often custom designs, adding to overall system cost. Additionally, the IF filters set the analog channel bandwidth of the platform, making it difficult to create a common platform design that can be reused across a wide range of systems. For package technology, most manufacturing lines will not go below a 0.65-mm or 0.8-mm ball pitch, meaning there is a limit on how physically small a complex device with many I/O requirements can become.

An alternative to the superheterodyne architecture, which has reemerged as a potential solution in recent years, is the zero-IF (ZIF) architecture. A ZIF receiver (see Figure 1.7) utilizes a single frequency mixing stage with the local oscillator (LO) set directly to the frequency band of interest, translating the received signal down

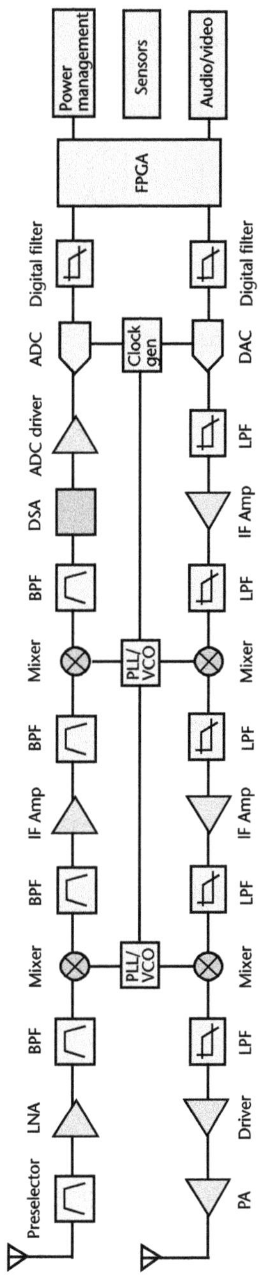

Figure 1.6 Multistage superheterodyne receive and transmit signal chains [4].

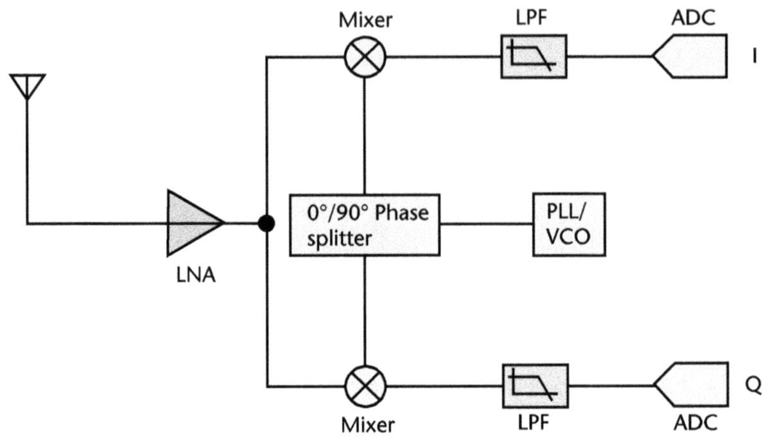

Figure 1.7 Zero IF architecture [4].

to baseband in phase (I) and quadrature (Q) signals. This architecture alleviates the stringent filtering requirements of the superheterodyne since all analog filtering takes place at baseband, where filters are much easier to design and less expensive than custom RF/IF filters. The ADC and DAC are now operating on I/Q data at baseband, so the sample rate relative to the converted bandwidth can be reduced, saving significant power. For many design aspects, ZIF transceivers provide significant SWaP reduction as a result of reduced analog front-end complexity and component count.

This direct frequency conversion to baseband can introduce the possibility of carrier leakage and an image frequency component. Due to real-world factors, such as process variation and temperature deltas in the signal chain, it is impossible to maintain a perfect 90° phase offset between the I and Q signals, resulting in degraded image rejection. Additionally, imperfect LO isolation in the mixing stage introduces carrier leakage components. When left uncorrected, the image and carrier leakage can degrade a receivers sensitivity and create undesirable transmit spectral emissions.

Historically, the I/Q imbalance has limited the range of applications that were appropriate for the ZIF architecture. This was due to two reasons: first, a discrete implementation of the ZIF architecture will suffer from mismatches both in the monolithic devices and also in the printed circuit board (PCB). In addition to this, the monolithic devices could pull from different fabrication lots, making exact matching very difficult due to native process variation. A discrete implementation will also have the processor physically separated from the RF components, making a quadrature correction algorithm very difficult to implement across frequency, temperature, and bandwidth.

Moore's law, or integration of the ZIF architecture into a monolithic transceiver device provides the path forward for next-generation systems. By having the analog and RF signal chain on a single piece of silicon, process variation will be kept to a minimum. Digital signal processing (DSP) blocks can be incorporated into the transceiver, removing the boundary between the quadrature calibration algorithm and the signal chain. This approach provides both unparalleled improvements in SWaP and can also match the superheterodyne architecture for performance specifications.

Devices like the Pluto SDR shown in Figure 1.8 integrate the full RF, analog, and digital signal chain onto a single CMOS device, and include digital processing to run quadrature and carrier leakage correction in real time across all process, frequency, and temperature variations. Devices like the AD9361 focuses on medium-performance specifications and very low power, such as UAV data links, handheld communication systems, and small form factor SDR applications. The AD9371 is optimized for high-performance specifications and medium power. Additionally, this device has refined calibration control, as well as an observation receiver for power amplifier (PA) linearization and a sniffer receiver for white space detection. This opens up new design potential for a different suite of applications. Communication platforms using wideband waveforms or occupying noncontiguous spectrum can now be implemented in a much smaller form factor.

1.5 Processing architectures for SDR

The microelectronic industry has rapidly evolved over the past six decades, resulting in numerous advances in microprocessor systems that have enabled many of the applications we take for granted every day. The rate at which this evolution has progressed over time has been characterized by the well-known Moore's Law, which defines the long-term trend of the number of transistors that can be accommodated on an integrated circuit. In particular, Moore's law dictates that the number of transistors per integrated circuit approximately doubles every 2 years, which subsequently affects the performance of microprocessor systems such as

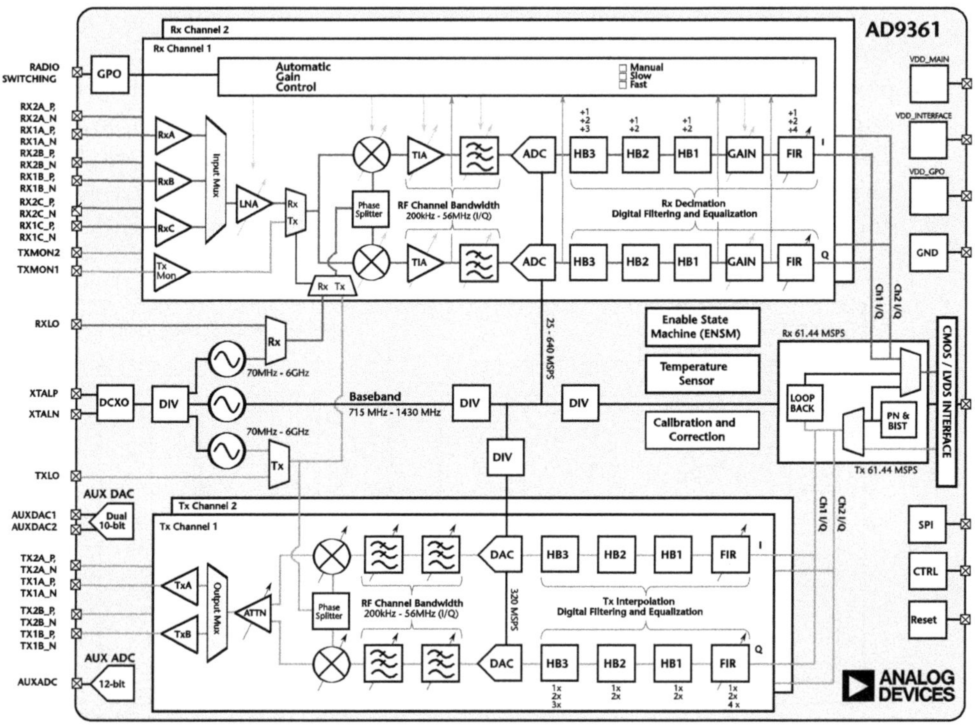

Figure 1.8 Integrated ZIF architecture used in the Pluto SDR.

processing speed and memory. One area that the microelectronics industry has significantly influenced over the past half century is the digital communication systems sector, where microprocessor systems have been increasingly employed in the implementation of digital transceiver, yielding more versatile, powerful, and portable communication system platforms capable of performing a growing number of advance operations and functions. With the latest advances in microelectronics and microprocessor systems, this has given rise to software-defined radio (SDR) technology, where baseband radio functionality can be entirely implemented in digital logic and software, as illustrated in Figure 1.3. There are several different types of microprocessor systems for SDR implementations, including.

- *General-purpose microprocessors* are often used in SDR implementations and prototypes due to their high level of flexibility with respect to reconfigurability, as well as due to their ease of implementation regarding new designs. On the other hand, general-purpose microprocessors are not specialized for mathematical computations and they can be potentially power inefficient.
- *Digital signal processors* (DSPs) are specialized for performing mathematical computations, implementation of new digital communication modules can be performed with relative ease, and the processor is relatively power efficient (e.g., DSPs are used in cellular telephones). On the other hand, DSPs are not well suited for computationally intensive processes and can be rather slow.
- *Field programmable gate arrays* (FPGAs) are efficient for custom digital signal processing applications because they can implement custom, fully parallel algorithms. DSP applications use many binary multipliers and accumulators that can be implemented in dedicated DSP slices, as shown in Figure 1.9. This includes 25 × 18 twos-complement multiplier, a 48-bit accumulator, a power-saving preadder, single-instruction, multiple data (SIMD) arithmetic unit, which includes a dual 24-bit or quad 12-bit add/subtract/accumulate. Tools like MathWorks HDL Coder are making creating new modules and targeting FPGAs easier, as it can generate portable, synthesizable Verilog and VHDL code from MATLAB functions, Simulink models, and Stateflow

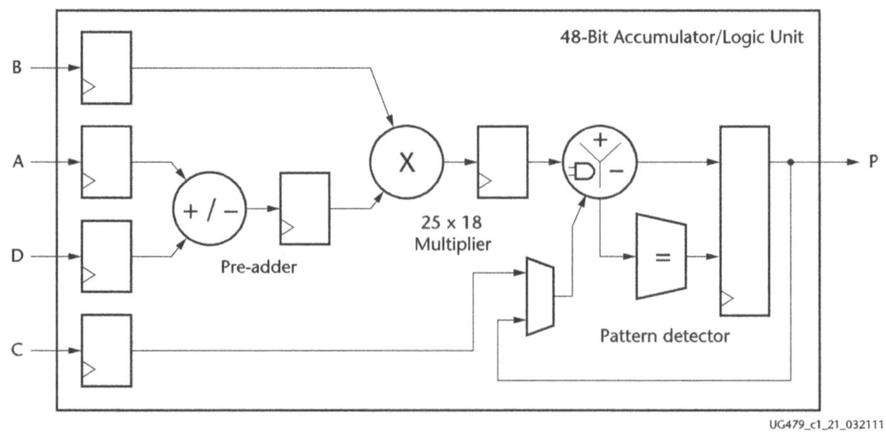

Figure 1.9 Basic DSP48E1 slice functionality [5].

charts, and is well suited for taking signal processing algorithms from concept to production.

- *Graphics processing units* (GPUs) are extremely powerful computationally. These processors have been driven to very high levels of performance and low price points by the need for real-time computer graphics in mass market gaming. Over the past 10 years, they have evolved into a general-purpose programmable architecture and supporting ecosystem that makes it possible to use them for a wide range of nongraphics applications [6]. GPU-accelerated libraries provided by manufactures like Nvidea, provide highly optimized functions that perform 2x to 10x faster than CPU-only alternatives. GPU-accelerated libraries for linear algebra, signal processing, and image and video processing lay the foundation for future software-defined radio applications to run on these types of architectures [7].
- *Advanced RISC Machines* (ARMs) have received significant attention in recent years fo their low cost, small size, low power consumption, and computational capabilities. Such processors combined with a capable RFFE make them suitable platforms for mobile communications and computing. Additions of new SIMD instructions for the Arm Cortex-A series and Cortex-R52 processors, known an NEON [8] are accelerate signal processing algorithms and functions to speed up software-defined radio applications.

It is an exciting time for algorithm developers; there are many new and advanced methods of implementing signal processing applications on hardware. The difficulty is to ensure that no matter which hardware is chosen to run algorithms on, the hardware and development methodology will be supported in 5 years.

1.6 Software Environments for SDR

As described in Section 1.2, at their most fundamental level, most commercially available SDR platforms convert live RF signals to samples at digital baseband, and use a software-defined mechanism for modulation and demodulation techniques to transfer real-world data. Referring back to Figure 1.3, the boundary between the analog and digital worlds for a communication system is located at the analog-to-digital converter (ADC) and the digital-to-analog converter (DAC), where signal information is translated between a continuous signal and a discrete set of signal sample values. Typically, the radio can be configured to select center frequency, sampling rate, bandwidth, and other parameters to transmit and receive signals of interest. This leaves the modulation and demodulation techniques, which are developed using a two-step development process.

1. Develop, tune, and optimize the modulation and demodulation algorithms for a specific sample rate, bandwidth, and environment. This is normally done on a host PC, where debugging and visualization is much easier. At this phase of development, the modulation and demodulation of the RFFEs are performed on a host, providing great flexibility to experiment and test algorithm ideas.

2. Take the above algorithm, which may be implemented in a high-level language in floating point, and code it in a production worthy environment, making production trade-offs of a product's size, weight, power and cost (SWaP-C) in mind. These platforms become truly software-defined when the onboard hardware and embedded processor are programmed to perform application-specific digital communications and signal processing functions.

While this text focuses exclusively on the first algorithmic step of the SDR development process, the second production step cannot be excluded when looking at a development flow. Unless your goal is to publish a paper and never actually have a path for a working prototype, a complete development process must be kept in mind.

The first step requires a convenient mechanism to capture data for signal analysis and development of algorithms that process those signals. This makes it vitally important to have efficient and reliable PC-based software to develop and test the data transmission and digital signal processing functions in a wireless communications system.

One software environment that meets this requirement is MATLAB from MathWorks. MATLAB is a technical computing environment and programming language, allowing ease of use development and excellent visualization mechanisms. An additional product, Communications Systems Toolbox, adds physical layer algorithms, channel models, reference models, and connectivity to SDR hardware to transmit and receive live signals. MATLAB is cross platform (Windows, Linux, MAC) offering support for many of the popular commercial radio front-ends. Using MATLAB enables an incremental and iterative development workflow for SDR consisting of:

- Algorithm development and design validation with link-level simulations;
- Algorithm validation with live signals using connection to commercially available SDR hardware.

MathWorks also offers Simulink, which is an environment for real-world system simulation and automatic code generation for hardware and software implementation. It allows the radio developer to continue to the second stage of production development. These capabilities of Simulink provide a path to production:

- Development and validation of a hardware-accurate model;
- Implementation of a prototype on SDR hardware using automatic HDL and C code generation;
- Verification of the prototype versus the validated model;
- Deployment of the implementation to production SDR hardware.

Although Simulink will largely be ignored in this text, being able to have a single environment from concept to production is very powerful and should not be overlooked for those who are trying to make a real production radio.

Another SDR software architecture is the popular open-source GNU Radio software [9], which is a free software (as in freedom) development toolkit that provides signal processing blocks to implement software-defined radios and signal processing systems. It can be used with external RF hardware to create software-defined radios, or without hardware in a simulation-like environment. It is widely used in hobbyist, academic, and commercial environments to support both wireless communications research and real-world radio systems.

In GNU Radio, a variety of C++ libraries modeling different digital communications and digital signal processing algorithms are integrated together using Python and SWIG (a software development tool that connects programs written in C and C++ with a variety of high-level programming languages including Python). These libraries are produced by the open-source community and freely shared with everyone.

The authors have used a variety of tools including MATLAB, Simulink, and GNU Radio in research, product development, and teaching undergraduate and graduate classes. Each tool has its advantages and disadvantages and can be used at different places of the research or development cycle. It was believed by the authors that all the various software environments can be used correctly or incorrectly to teach wireless physical layer fundamentals, although the prerequisites for each tool is different. For those who choose the GNU Radio path, the requirements to have a working knowledge of Linux, Python, C++, and SWIG is very high. While this is very common for a computer science student, it is not for most students in communications, and asking someone to learn the tool at the same time as the communications theory can prove difficult. One can use preexisting blocks in GNU Radio and bypass the requirements of understanding Python and C++, but then some opportunities to demonstrate and experiment with fundamental communications theory are lost, as the student just uses a block that someone else wrote, with little understanding of what it is doing. The same can be said for Simulink; it is also a very powerful tool, with many preexisting blocks for timing recovery and carrier synchronization. However, using these blocks does not allow many students to understand what is happening inside the blocks, and therefore the students have difficulty in understanding how to tune the blocks for their situation.

This is why MATLAB was chosen for this book. It is a cross-platform environment, allowing students to use what they are familiar with, and all the blocks presented are MATLAB scripts, with nothing to hide. If a student wants to better understand something, the entire algorithm is defined in the MATLAB code, with nothing to obfuscate the communications theory.

1.7 Additional readings

Although this chapter gave a brief introduction to the expanding area of SDR technology, there are several books available in the open literature that can provide a more detailed viewpoint of this topic. For instance, the book by Reed extensively covers many of the issues associated with the software architecture of an SDR platform [10], while many of the design considerations and approaches used to construct SDR hardware prototype and their RFFE are covered in the book by Kensington [11]. Another excellent reference regarding the hardware implementation of SDR systems is by Grayver [12]. Furthermore, understanding the importance of the analog-digital divide and how SDR systems bridge that divide is the subject of the paper by Machado and Wyglinski [13]. Finally, an excellent series of papers covering the latest advances in SDR technology and providing some perspective on its evolution from 20 years ago are presented in IEEE Communications Magazine [14, 15].

References

[1] American Guild Of Town Criers Website, 1997 http://www.americantowncriers.com/.

[2] IEEE Project *1900.1 - Standard Definitions and Concepts for Dynamic Spectrum Access: Terminology Relating to Emerging Wireless Networks, System Functionality, and Spectrum Management* https://standards.ieee.org/develop/project/1900.1.html.

[3] Analog Devices ADF7030 http://www.analog.com/ADF7030

[4] Hall, B., and W. Taylor, *X- and Ku-Band Small Form Factor Radio Design* http://www.analog.com/en/technical-articles/x-and-ku-band-small-form-factor-radio-design.html.

[5] Xilinx Inc. www.xilinx.com *7 Series DSP48E1 User Guide, UG479 (v1.9)* September 27, 2016 https://www.xilinx.com/support/documentation/user_guides/ug479_7Series_DSP48E1.pdf

[6] McCool, M., "Signal Processing and General-Purpose Computing on GPUs," *IEEE Signal Processing Magazine*, Vol. 24, No. 3, May 2007, http://ieeexplore.ieee.org/document/4205095/.

[7] Nivdea *GPU-accelerated Libraries for Computing* https://developer.nvidia.com/gpu-accelerated-libraries.

[8] ARM *NEON* https://developer.arm.com/technologies/neon.

[9] GNU Radio. *Welcome to GNU Radio!.* http://gnuradio.org/.

[10] Reed, J. H., *Software Radio: A Modern Approach to Radio Engineering*, Upper Saddle River, NJ: Prentice Hall PTR, 2002.

[11] Kensington, P., *RF and Baseband Techniques for Software Defined Radio*, Norwood, MA: Artech House, 2005.

[12] Grayver, E., *Implementing Software Defined Radio*, New York: Springer-Verlag, 2012.

[13] Machado, R. G. and A. M. Wyglinski, "Software-Defined Radio: Bridging the Analog to Digital Divide," *Proceedings of the IEEE*, Vol. 103, No. 3, March 2015, pp. 409–423.

[14] Mitola, J., P. Marshall, K. C. Chen, M. Mueck, and Z. Zvonar, "Software Defined Radio - 20 Years Later: Part 1 [guest editorial], *IEEE Communications Magazine*, Vol. 53, No. 9, September 2015, pp. 22–23, http://ieeexplore.ieee.org/document/7263341/?section=abstract.

[15] Mitola, J., P. Marshall, K. C. Chen, M. Mueck, and Z. Zvonar, "Software Defined Radio - 20 Years Later: Part 2 [guest editorial], *IEEE Communications Magazine*, Vol. 54, No. 1, January 2016, p. 58, http://ieeexplore.ieee.org/stamp/stamp.jsp?arnumber=7378426.

CHAPTER 2

Signals and Systems

SDR is an application-specific area of signal processing, and everyone involved in SDR development needs to not only have a solid background in signals and systems, but also RF and analog baseband processing. This chapter covers the many aspects of linear time-invariant (LTI) signals and systems, analog processing, and RF that forms the fundamental basis for the latter chapters. It is not expected that the reader will go through each topic in detail, but if any of the short topics are new, the reader should refer to the more comprehensive books on these subjects.

2.1 Time and Frequency Domains

The time and frequency domains are alternative ways of representing the same signals. The Fourier transform, named after its initial instigator, French mathematician and physicist Jean-Baptiste Joseph Fourier, is the mathematical relationship between these two representations. If a signal is modified in one domain, it will also be changed in the other domain, although usually not in the same way. For example, convolution in the time domain is equivalent to multiplication in the frequency domain. Other mathematical operations, such as addition, scaling, and shifting, also have a matching operation in the opposite domain. These relationships are called properties of the Fourier transform, which describe how a mathematical change in one domain results in a mathematical change in the other domain.

The Fourier transform is just a different way to describe a signal. For example, investigating the Gibbs phenomenon, which states if you add sine waves at specific frequency/phase/amplitude combinations you can approximate a square wave, can be expressed mathematically as (2.1), and shown in Figure 2.1.

$$x(t) = \sin(t) + \frac{\sin(3t)}{3} + \frac{\sin(5t)}{5} + \ldots = \sum_{n=1}^{n=\infty} \frac{\sin(n \times t)}{n}; \quad n = odd \tag{2.1}$$

When we look at the signal across the time axis that is perpendicular to the frequency axis, we observe the time domain. We cannot see the frequency of the sine waves easily since we are perpendicular to the frequency axis. When we transform domains and observe phenomenon across the frequency axis, which is perpendicular to the time axis, we observe the frequency or Fourier domain. We can easily make out the signal magnitude and frequency, but have lost that time aspect. Both views represents the same signal such that, they are just being observed things from different domains via transforms.

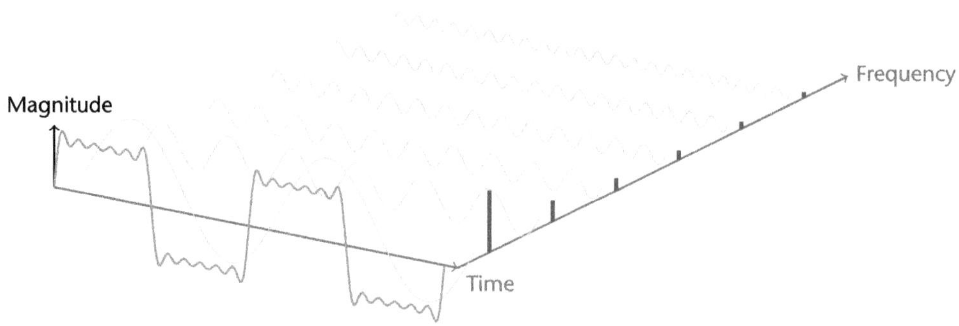

Figure 2.1 Gibbs phenomenon, looking left, the time domain, looking right, the Fourier domain.

> **Q** Investigate the Gibbs phenomenon in MATLAB by using Code 2.1 to better understand how adding sine waves with variations in frequency/phase/amplitude affect the time domain, and frequency domains

Code 2.1 Gibbs phenomenon: **gibbs.m**

```
max = 9;
fs = 1000;
for i = 1:2:max
    % dsp.SineWave(amp,freq,phase,Name,Value);
    wave = dsp.SineWave(1/i, i*2*pi, 0, ...
    'SamplesPerFrame', 5000, 'SampleRate', fs);
    y = wave();
    if i == 1
        wavesum = y;
    else
        wavesum = wavesum + y;
    end
    scope(wavesum());
    pause(.5);
    % waitforbuttonpress;
end
```

2.1.1 Fourier Transform

The Fourier transform includes four members in its family: the Fourier transform, Fourier series, discrete Fourier transform (DFT), and discrete-time Fourier transform (DTFT). The commonly referred to FFT (fast Fourier transform) and its inverse, the inverse FFT (IFFT), is a specific implementation of the DFT.

The Fourier transform of $x(t)$ is defined as [1]:

$$X(\omega) = \int_{-\infty}^{\infty} x(t)e^{-j\omega t}dt, \tag{2.2}$$

where t is the time variable in seconds across the time domain, and ω is the frequency variable in radian per seconds across frequency domain.

Applying the similar transform to $X(\omega)$ yields the inverse Fourier transform [1]:

$$x(t) = \int_{-\infty}^{\infty} X(\omega) e^{j2\pi\omega t} d\omega, \tag{2.3}$$

where we write $x(t)$ as a weighted sum of complex exponentials.

The Fourier transform pair above can be denoted as [2]:

$$x(t) \overset{\mathcal{F}}{\leftrightarrow} X(\omega), \tag{2.4}$$

where the left-hand side of the symbol $\overset{\mathcal{F}}{\leftrightarrow}$ is before Fourier transform, while the right-hand side of the symbol $\overset{\mathcal{F}}{\leftrightarrow}$ is after Fourier transform. There are several commonly used properties of Fourier transform that are useful when studying SDR Fourier domain, which have been summarized in Table 2.1 for your reference.

2.1.2 Periodic Nature of the DFT

Unlike the other three Fourier transforms, the DFT views both the time domain and the frequency domain signals as periodic (they repeat forever). This can be confusing and inconvenient since most of the signals used in many signal processing applications are not periodic. Nevertheless, if you want to use the DFT (and its fast implementation, the FFT), you must conform with the DFT's view of the world.

Figure 2.2 shows two different interpretations of the time domain signal. First, observing the upper signal, the time domain viewed as N points. This represents how signals are typically acquired by SDRs, by a buffer of N points. For instance, these 128 samples might have been acquired by sampling some analog signal at regular intervals of time. Sample 0 is distinct and separate from sample 127 because they were acquired at different times. From the way this signal was formed, there is no reason to think that the samples on the left of the signal are even related to the samples on the right.

Unfortunately, the DFT does not see things this way. As shown in the lower part of Figure 2.2, the DFT views these 128 points to be a single period of an infinitely long periodic signal. This means that the left side of the acquired signal is connected to the right side of a duplicate signal. Likewise, the right side of the acquired signal

Table 2.1 Fourier Transform Properties*

Property	*Time Signal*	*Fourier Transform Signal*
Definition	$x(t)$	$\int_{-\infty}^{\infty} x(t)e^{-j\omega t}dt$
Inversion formula	$\int_{-\infty}^{\infty} X(\omega)e^{j2\pi\omega t}d\omega$	$X(\omega)$
Linearity	$\sum_{n=1}^{N} a_n x_n(t)$	$\sum_{n=1}^{N} a_n X_n(\omega)$
Symmetry	$x(-t)$	$X(-\omega)$
Time shift	$x(t-t_0)$	$X(\omega)e^{-j\omega t_0}$
Frequency shift	$x(t)e^{j\omega_0 t}$	$X(\omega-\omega_0)$
Scaling	$x(\alpha t)$	$\frac{1}{\lvert\alpha\rvert}X(\frac{\omega}{\alpha})$
Derivative	$\frac{d^n}{dt^n}x(t)$	$(j\omega)^n X(\omega)$
Integration	$\int_{-\infty}^{\infty} x(\tau)d\tau$	$\frac{X(\omega)}{j\omega} + \pi X(0)\delta(\omega)$
Time convolution	$x(t) * h(t)$	$X(\omega)H(\omega)$
Frequency convolution	$x(t)h(t)$	$\frac{1}{2\pi}X(\omega) * H(\omega)$

* based on [2]. Suppose the time signal is $x(t)$, and its Fourier transform signal is $X(\omega)$

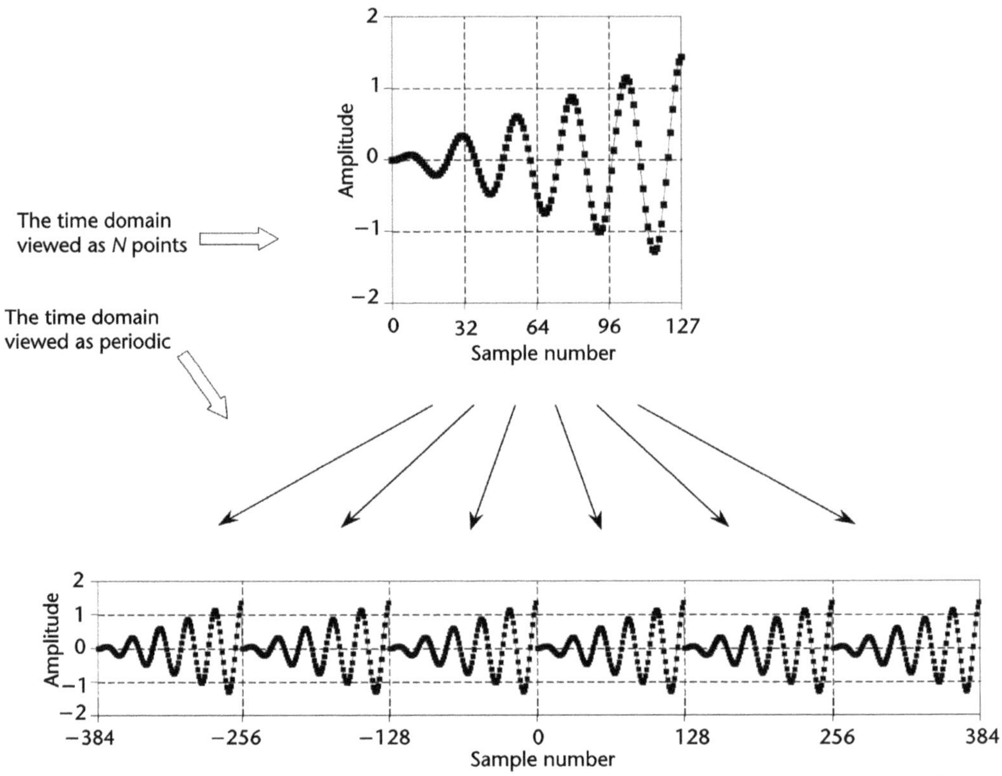

Figure 2.2 Periodicity of the DFT's time domain signal. The time domain can be viewed as N samples in length, shown in the upper figure, or as an infinitely long periodic signal, shown in the lower figures [4].

is connected to the left side of an identical period. This can also be thought of as the right side of the acquired signal wrapping around and connecting to its left side. In this view, sample 127 occurs next to sample 0, just as sample 43 occurs next to sample 44. This is referred to as being circular, and is identical to viewing the signal as being periodic. This is the reason that window [3] functions need to be preapplied to signal captures before applying an FFT function, which is multiplied by the signal and removes the discontinuities by forcing them to zero [4].

2.1.3 Fast Fourier Transform

There are several ways to calculate the DFT, such as solving simultaneous linear equations or correlation method. The FFT is another method for calculating the DFT. While it produces the same result as the other approaches, it is incredibly more efficient, often reducing the computation time by multiple orders of magnitude. While the FFT only requires a few dozen lines of code, it is one of the more complicated algorithms in signal processing, and its internal workings details are left to those that specialize in such things. You can easily use existing and proven FFT routines [4, 5] without fully understanding the internal workings as long as you understand how it is operating.

An FFT analysis using a generalized test setup shown in Figure 2.3. The spectral output of the FFT is a series of $\frac{M}{2}$ points in the frequency domain (M is the size of the FFT, the number of samples stored in the buffer memory). The spacing between

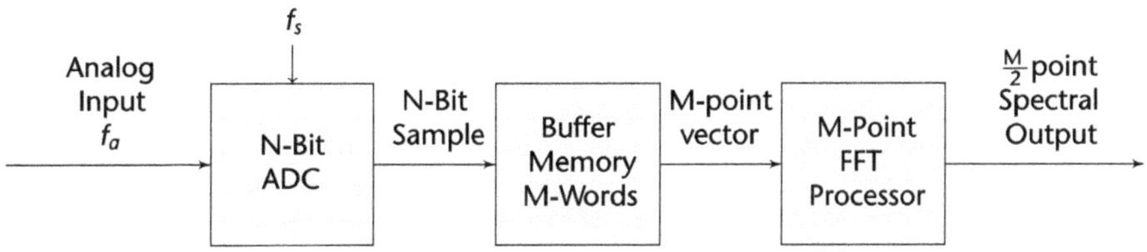

Figure 2.3 Generalized test set up for FFT analysis of ADC output [6].

the points is $\frac{f_s}{M}$, and the total frequency range covered is DC to $\frac{f_s}{2}$, where f_s is the sampling rate. The width of each frequency bin (sometimes called the resolution of the FFT) is $\frac{f_s}{M}$.

Figure 2.4 shows an FFT output for an ideal 12-bit ADC using the Analog Devices' ADIsimADC®program. Note that the theoretical noise floor of the FFT is equal to the theoretical signal-to-noise ratio (SNR) plus the FFT process gain of $10log_{10}(\frac{M}{2})$. It is important to remember the value for noise used in the SNR calculation is the noise that extends over the entire Nyquist bandwidth (DC to $\frac{f_s}{2}$), but the FFT acts as a narrowband spectrum analyzer with a bandwidth of $\frac{f_s}{M}$ that sweeps over the spectrum. This has the effect of pushing the noise down by an amount equal to the process gain—the same effect as narrowing the bandwidth of an analog spectrum analyzer.

The FFT output can be used like an analog spectrum analyzer to measure the amplitude of the various harmonics and noise components of a digitized signal. The harmonics of the input signal can be distinguished from other distortion products by their location in the frequency spectrum.

2.2 Sampling Theory

A continuous-time analog signal can be converted to a discrete-time digital signal using sampling and quantization, as shown in Figure 2.5, where a continuous analog input signal $x_a(t)$ is converted to a discrete digital output signal $x[n]$. *Sampling* is the conversion of a continuous-time signal into a discrete-time signal obtained by taking the samples of the continuous-time signal at discrete-time instants [1]. The quantization process converts the sample amplitude into a digital format. Section 2.2.1 will introduce a frequently used sampling method; namely, *uniform sampling*.

Similarly, a discrete-time signal can also be converted to a continuous-time signal using reconstruction. However, reconstruction is not always successful. Sometimes, the reconstructed signal is not the same as the original signal. Since for a given sampled signal, it can represent an infinite number of different continuous-time signals that can fit into the same quantized sample points. However, if the sampling satisfies certain criterion, the signal can be reconstructed without losing information. This criterion, called Nyquist sampling theorem, will be introduced in Sections 2.2.3 and 2.5.1.

2.2.1 Uniform Sampling

There are many ways to perform sampling of an analog signal into a digital representation. However, if we specify the sampling interval as a constant number

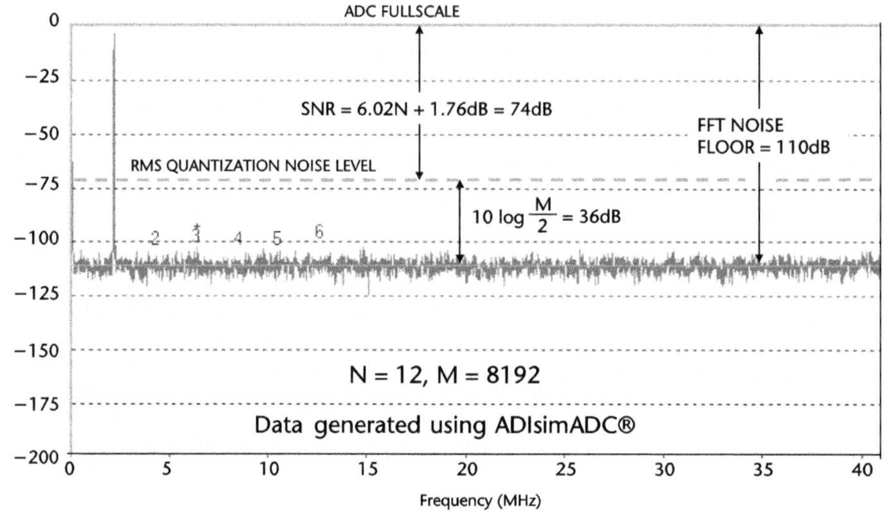

Figure 2.4 FFT output for an ideal 12-bit ADC, $f_a = 2.111$ MHz, $f_s = 82$ MSPS, average of 5 FFTs, $M = 8192$. Data generated from ADIsimADC® [6].

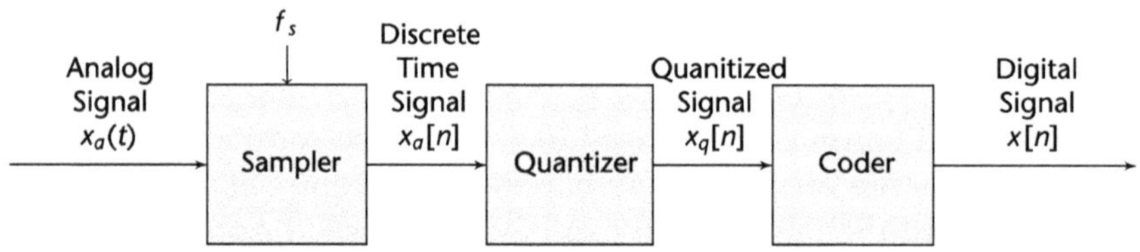

Figure 2.5 Basic parts of an analog-to-digital converter (ADC) [1]. Sampling takes place in the sampler block. $x_a(t)$ = analog continuous time signal; f_s is the digital sample rate; $x_a[n]$ is the discrete time continuous analog signal; $x_q[n]$ is the discrete time, discrete digital signal, which may come out as grey code; and $x[n]$ is the output of the coder in 2s complement form [7].

T_s, we get the most widely used sampling, called uniform sampling or periodic sampling. Using this method, we are taking samples of the continuous-time signal every T_s seconds, which can be defined as

$$x[n] = x(nT_s), \quad -\infty < n < \infty, \tag{2.5}$$

where $x(t)$ is the input continuous-time signal, $x[n]$ is the output discrete-time signal, T_s is the sampling period, and $f_s = 1/T_s$ is the sampling frequency.

An equivalent model for the uniform sampling operation is shown in Figure 2.6(a), where the continuous-time signal $x(t)$ is multiplied by an impulse train $p(t)$ to form the sampled signal $x_s(t)$, which can be defined as

$$x_s(t) = x(t)p(t), \tag{2.6}$$

where the signal $p(t)$ is referred to as the sampling function.

The sampling function is assumed to be a series of narrow pulses, which is either zero or one. Thus, $x_s(t) = x(t)$ when $p(t) = 1$, and $x_s(t) = 0$ when $p(t) = 0$. Since $p(t) = 1$ only at time instants T_s, the resulting $x_s(t) = x(nT_s) = x[n]$, which proves that this is indeed an equivalent model for the uniform sampling operation. This model will help us to obtain the frequency domain representation of uniform sampling in Section 2.2.2.

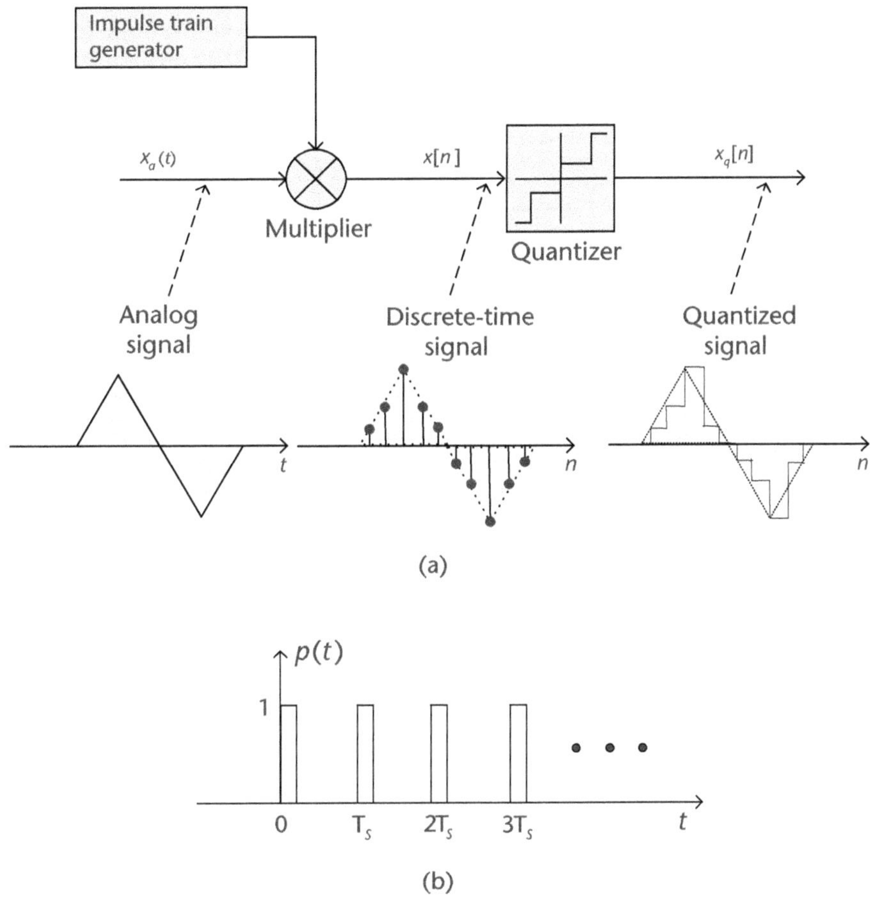

Figure 2.6 An equivalent model for the uniform sampling operation. (a) A continuous-time signal $x(t)$ is multiplied by a periodic pulse $p(t)$ to form the sampled signal $x_s(t)$, and (b) a periodic pulse $p(t)$.

2.2.2 Frequency Domain Representation of Uniform Sampling

Since it is easier to derive the Nyquist sampling theorem in frequency domain, in this section we will try to represent the uniform sampling process in frequency domain.

According to Figure 2.6(b), we can define the sampling function $p(t)$ as

$$p(t) = \sum_{k=-\infty}^{\infty} \delta(t - kT_s), \quad k = 0, 1, 2, ..., \tag{2.7}$$

where at time instants kT_s, we have $p(t) = 1$. According to [8], $p(t)$ is a Dirac comb constructed from Dirac delta functions.

Substitution of (2.7) into (2.6) gives

$$x_s(t) = x(t)p(t) = x(t) \sum_{k=-\infty}^{\infty} \delta(t - kT_s). \tag{2.8}$$

In order to understand the sampling process in frequency domain, we need to take the Fourier transform of $x_s(t)$. According to frequency-domain convolution property in Table 2.1, multiplication in time domain will lead to convolution in frequency domain. Therefore, multiplication of $x(t)$ and $p(t)$ will yield the convolution of $X(\omega)$ and $P(\omega)$:

$$X_s(\omega) = \frac{1}{2\pi} X(\omega) * P(\omega), \tag{2.9}$$

where $X(\omega)$ is the Fourier transform of $x(t)$, and $P(\omega)$ is the Fourier transform of $p(t)$.

The Fourier transform of a Dirac comb is also a Dirac comb [8], namely

$$P(\omega) = \frac{\sqrt{2\pi}}{T_s} \sum_{k=-\infty}^{\infty} \delta(\omega - k\frac{2\pi}{T_s}) = \frac{\sqrt{2\pi}}{T_s} \sum_{k=-\infty}^{\infty} \delta(\omega - k\omega_s), \tag{2.10}$$

where $\omega_s = 2\pi f_s$ is the sampling frequency.

Performing convolution with a collection of delta function pulses at the pulse location, we get

$$X_s(\omega) = \frac{1}{\sqrt{2\pi} T_s} \sum_{k=-\infty}^{\infty} X(\omega - k\omega_s). \tag{2.11}$$

Equation (2.11) tells us that the uniform sampling creates images of the Fourier transform of the input signal, and images are periodic with sampling frequency f_s.

2.2.3 Nyquist Sampling Theorem

Based on (2.11), we draw the spectrum of original signal $x(t)$ and the sampled signal $x_s(t)$ on frequency domain, as shown in Figure 2.7. We assume the bandwidth of the original signal is $[-f_h, f_h]$, as shown in Figure 2.7(a). For now, we do not pay attention to the signal amplitude, so we use A and A_s to represent them. Assuming the sampling frequency is f_s, then the sampled signal will have replicas at location kf_s. In order to reconstruct the original signal from the sampled signal, we will apply a lowpass filter on the sampled signal, trying to extract the $n = 0$ term from $X_s(f)$, as shown in Figure 2.7(b). Therefore, accomplishing reconstruction without error requires that the portion of the spectrum of $X_s(f)$ at $f = \pm f_s$ does not overlap with the portion of the spectrum at $f = 0$. In other words, this requires that $f_s - f_h > f_h$ or $f_s > 2f_h$, which leads to the Nyquist sampling theorem.

Nyquist sampling theorem applies for the *bandlimited signal*, which is a signal $x(t)$ that has no spectral components beyond a frequency B Hz [9]. That is,

$$X(\omega) = 0, \quad |\omega| > 2\pi B. \tag{2.12}$$

The Nyquist sampling theorem states that a real signal, $x(t)$, which is bandlimited to B Hz can be reconstructed without error from samples taken uniformly at a rate $R > 2B$ samples per second. This minimum sampling frequency, $F_s = 2B$ Hz, is called the Nyquist rate or the Nyquist frequency. The corresponding sampling interval, $T = \frac{1}{2B}$, is called the Nyquist interval [1]. A signal bandlimited to B Hz, which is sampled at less than the Nyquist frequency of $2B$ (i.e., which was sampled at an interval $T > \frac{1}{2B}$), is said to be undersampled.

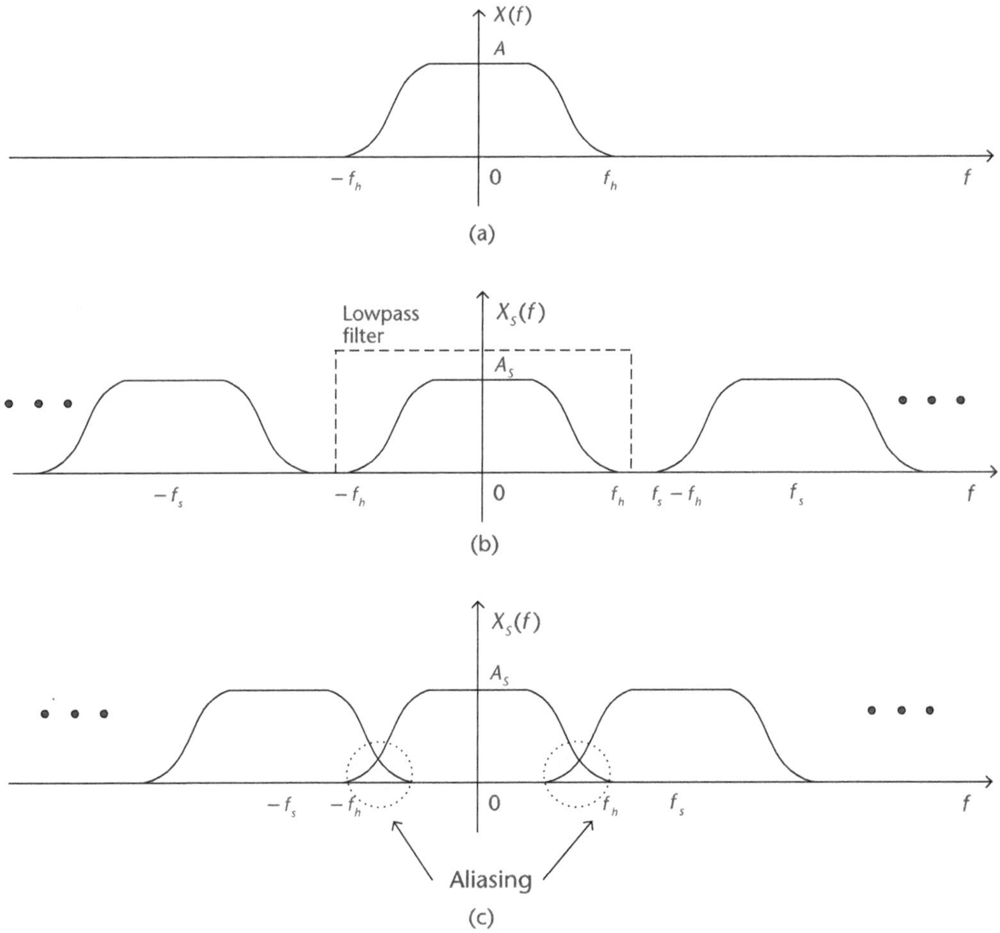

Figure 2.7 The spectrum of original signal $x(t)$ and the sampled signal $x_s(t)$ in the frequency domain. (a) The spectrum of original continuous-time signal $x(t)$, with bandwidth $-f_h$ to f_h, and amplitude A. (b) The spectrum of the digitally sampled signal $x_s(t)$, $f_s > f_h$ which satisfies Nyquist sampling theorem. (c) The spectrum of the digitally sampled signal $x_s(t)$, $f_s < f_h$ which does not satisfies Nyquist sampling theorem and has aliasing.

When a signal is undersampled, its spectrum has overlapping spectral tails, or images, where $X_s(f)$ no longer has complete information about the spectrum and it is no longer possible to recover $x(t)$ from the sampled signal. In this case, the tailing spectrum does not go to zero, but is folded back onto the apparent spectrum. This inversion of the tail is called spectral folding or aliasing, as shown in Figure 2.7(c) [10].

Hands-On MATLAB Example: Let us now explain via computer simulation how the Nyquist criteria requires that the sampling frequency be at least twice the highest frequency contained in the signal or information about the signal will be lost. Furthermore, in Section 2.5.1, the phenomena known as *aliasing* will occur and the frequency will be folded back into the first Nyquist band. In order to describe the implications of aliasing, we can investigate things in the time domain.

Consider the case of a time domain representation of a single tone sinewave sampled as shown in Figure 2.8(a). In this example, the sampling frequency (f_s) is

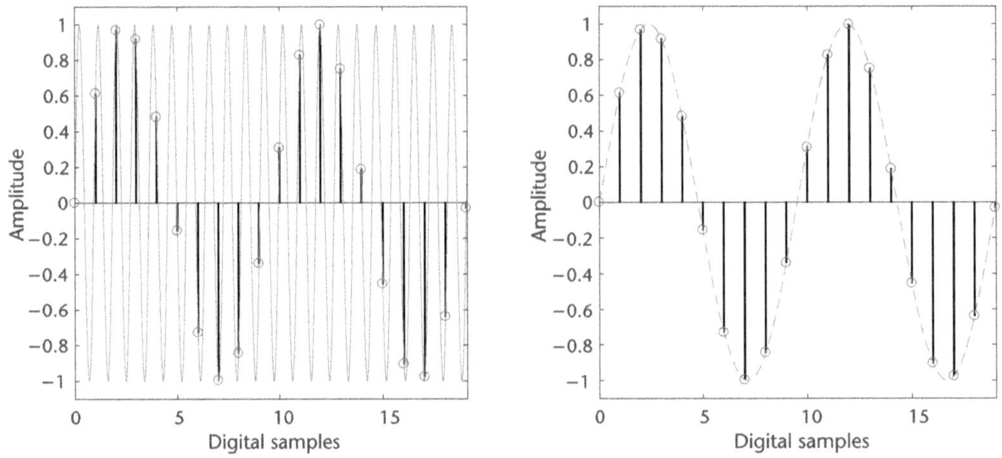

(a) (b)

Figure 2.8 Aliasing in the time domain. Digital samples are the same in both figures. (a) Analog input (F_A) solid line; digital sample data (circles) at (f_s), and (b) digital reconstruction dashed line; digital sample data (circles) at (f_s).

not at least two times the analog input frequeny (F_A), but actually slightly more than f_s. Therefore, the Nyquist criteria is violated by definition. Notice that the pattern of the actual samples produces an aliased sinewave at a lower frequency, as shown in Figure 2.8(b).

Code 2.2 can create similar figures as shown in Figure 2.8(a) and Figure 2.8(b), and may be helpful to better understand how aliasing works by manipulating `Fs` and `Fa`. Figures 2.7 and 2.25 demonstrate the same effect in the frequency domain.

Code 2.2 Time domain aliasing: **`nyquist.m`**

```
Fs = 1000;        % Sample rate (Hz)
Fa = 1105;        % Input Frequency (Hz)
% Determine Nyquist zones
zone = 1 + floor(Fa / (Fs/2));
alias = mod(Fa, Fs);
if ~mod(zone,2) % 2nd, 4th, 6th, ... Nyquist Zone
    % Its not really a negative amplitude, but it is 180 degrees out
    % of phase, which makes it harder to see on the time domain side,
    % so we cheat to make the graphs look better.
    alias = -(Fs - alias)/Fs;
else                % 3rd, 5th, 7th, ... Nyquist Zone
    alias = (alias)/Fs;
end

% Create the analog/time domain and digital sampling vectors
N = 2*1/abs(alias) + 1;          % Number of Digital samples
points = 256;                    % Analog points between digital samples
analogIndexes = 0:1/points:N-1;
samplingIndexes = 1:points:length(analogIndexes);
wave = sin(2*pi*Fa/Fs*analogIndexes);
```

2.2.4 Nyquist Zones

The Nyquist bandwidth itself is defined to be the frequency spectrum from DC to $\frac{f_s}{2}$. However, the frequency spectrum is divided into an infinite number of Nyquist zones, each having a width equal to 0.5 f_s as shown in Figure 2.9. The frequency spectrum does not just end because you are not interested in those frequencies.

This implies that some filtering ahead of the sampler (or ADC) is required to remove frequency components that are outside the Nyquist bandwidth, but whose aliased components fall inside it. The filter performance will depend on how close the out-of-band signal is to $\frac{f_s}{2}$ and the amount of attenuation required. It is important to note that with no input filtering at the input of the ideal sampler (or ADC), any frequency component (either signal or noise) that falls outside the Nyquist bandwidth in any Nyquist zone will be aliased back into the first Nyquist zone. For this reason, an analog antialiasing filter is used in almost all sampling ADC applications to remove these unwanted signals.

> **Q** How do you think the relationship changes between the measured frequency and the absolute frequency, as it goes up into the third or fourth or higher Nyquist zones as shown in Figure 2.9? See if you can confirm your hypothesis by modifying Code 2.2 to plot absolute frequency on the x-axis, and measured frequency on the y-axis.

2.2.5 Sample Rate Conversion

In real-world applications, we often would like to lower the sampling rate because it reduces storage and computation requirements. In many cases we prefer a higher sampling rate because it preserves fidelity. Sampling rate conversion is a general term for the process of changing the time interval between the adjacent elements in a sequence consisting of samples of a continuous-time function [10].

Decimation: The process of lowering the sampling rate is called *decimation*, which is achieved by ignoring all but every Dth sample. In time domain, it can be defined as

$$y[n] = x[nD], \quad D = 1, 2, 3, ..., \tag{2.13}$$

where $x[n]$ is the original signal, $y[n]$ is the decimated signal, and D is the decimation rate. According to (2.13), the sampling rates of the original signal and the decimated

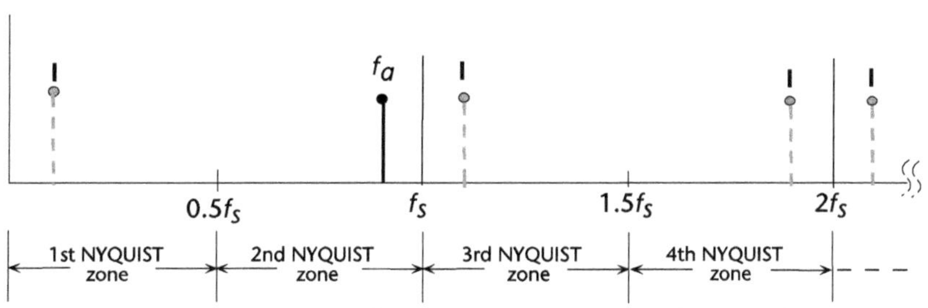

Figure 2.9 Analog signal f_a sampled at f_s has images (aliases) at $\pm kF_s \pm F_a, k = 1, 2, 3, ...$

signal can be expressed as

$$F_y = \frac{F_x}{D}, \tag{2.14}$$

where F_x is the sampling rates of the original signal, and F_y is the sampling rates of the decimated signal.

Since the frequency variables in radians, ω_x and ω_y, can be related to sampling rate, F_x and F_y, by

$$\omega_x = 2\pi F T_x = \frac{2\pi F}{F_x}, \tag{2.15}$$

and

$$\omega_y = 2\pi F T_y = \frac{2\pi F}{F_y}, \tag{2.16}$$

it follows from the distributive property that ω_x and ω_y are related by

$$\omega_y = D\omega_x, \tag{2.17}$$

which means that the frequency range of ω_x is stretched into the corresponding frequency range of ω_y by a factor of D.

In order to avoid aliasing of the decimated sequence $y[n]$, it is required that $0 \leq |\omega_y| \leq \pi$. Based on (2.17), it implies that the spectrum of the original sequence should satisfy $0 \leq |\omega_x| \leq \frac{\pi}{D}$. Therefore, in reality, decimation is usually a two-step process, consisting of a lowpass antialiasing filter and a downsampler, as shown in Figure 2.10. The lowpass antialiasing filter is used to constrain the bandwidth of the input signal to the downsampler $x[n]$ to be $0 \leq |\omega_x| \leq \frac{\pi}{D}$.

In frequency domain, the spectrum of the decimated signal, $y[n]$, can be expressed as [1]

$$Y(\omega_y) = \frac{1}{D}\sum_{k=0}^{D-1} H_D\left(\frac{\omega_y - 2\pi k}{D}\right) S\left(\frac{\omega_y - 2\pi k}{D}\right), \tag{2.18}$$

where $S(\omega)$ is the spectrum of the input signal $s[n]$, and $H_D(\omega)$ is the frequency response of the lowpass filter $h_D[n]$. With a properly designed filter $H_D(\omega)$, the aliasing is eliminated, and consequently, all but the first $k = 0$ term in (2.18) vanish [1]. Hence, (2.18) becomes

$$Y(\omega_y) = \frac{1}{D} H_D\left(\frac{\omega_y}{D}\right) S\left(\frac{\omega_y}{D}\right) = \frac{1}{D} S\left(\frac{\omega_y}{D}\right), \tag{2.19}$$

for $0 \leq |\omega_y| \leq \pi$. The spectra for the sequence $x[n]$ and $y[n]$ are illustrated in Figure 2.11, where the frequency range of the intermediate signal is $0 \leq |\omega_x| \leq \frac{\pi}{D}$, and the frequency range of the decimated signal is $0 \leq |\omega_y| \leq \pi$.

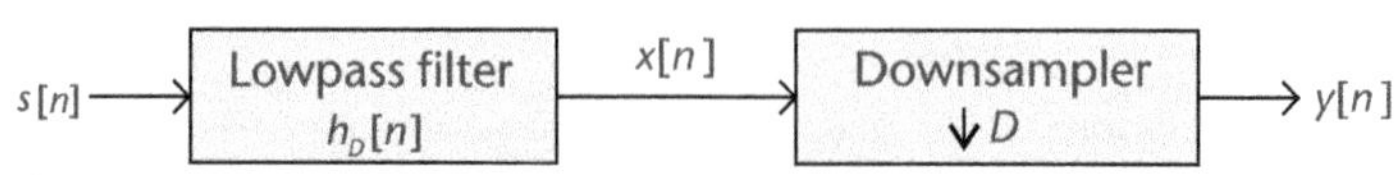

Figure 2.10 The structure of decimation, consisting of a lowpass antialiasing filter and a downsampler.

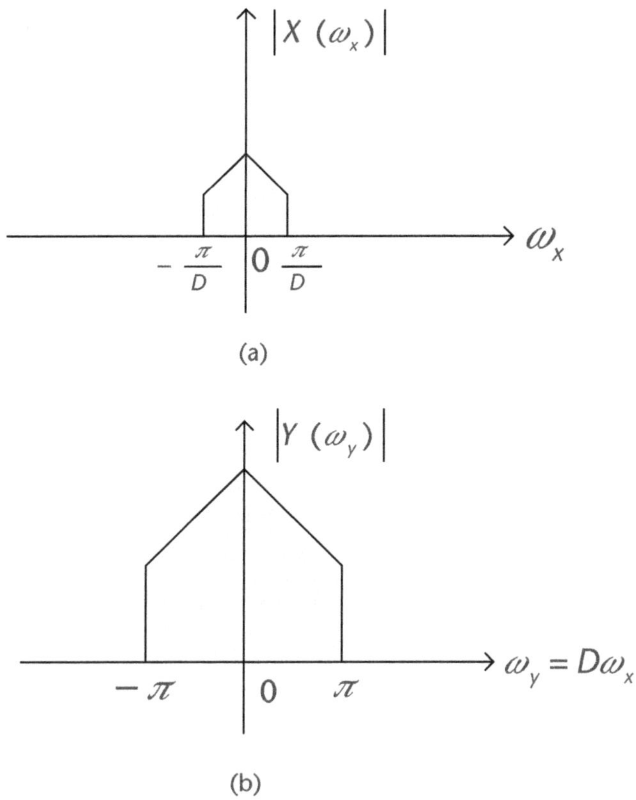

Figure 2.11 The spectra for the sequence $x[n]$ and $y[n]$, where the frequency range of ω_x is stretched into the corresponding frequency range of ω_y by a factor of D. (a) Spectrum of the intermediate sequence, and (b) spectrum of the decimated sequence.

Interpolation: The process of increasing the sampling rate is called *interpolation*, which can be accomplished by interpolating (stuffing zeros) $I-1$ new samples between successive values of signal. In time domain, it can be defined as

$$y[n] = \begin{cases} x[n/I] & n = 0, \pm I, \pm 2I, ... \\ 0 & \text{otherwise} \end{cases}, \quad I = 1, 2, 3, ..., \tag{2.20}$$

where $x[n]$ is the original signal, $y[n]$ is the interpolated signal, and I is the interpolation rate.

According to (2.20), the sampling rates of the original signal and the interpolated signal can be expressed as

$$F_y = IF_x, \tag{2.21}$$

where F_x is the sampling rates of the original signal, and F_y is the sampling rates of the interpolated signal. Since (2.15) and (2.16) also hold here, it follows that ω_x and ω_y are related by

$$\omega_y = \frac{\omega_x}{I}, \tag{2.22}$$

which means that the frequency range of ω_x is compressed into the corresponding frequency range of ω_y by a factor of I. Therefore, after the interpolation, there will be I replicas of the spectrum of $x[n]$, where each replica occupies a bandwidth of $\frac{\pi}{I}$.

Since only the frequency components of $y[n]$ in the range $0 \leq |\omega_y| \leq \frac{\pi}{I}$ are unique (i.e., all the other replicas are the same as this one), the images of $Y(\omega)$ above $\omega_y = \frac{\pi}{I}$ should be rejected by passing it through a lowpass filter with the following frequency response:

$$H_I(\omega_y) = \begin{cases} C & 0 \leq |\omega_y| \leq \frac{\pi}{I} \\ 0 & \text{otherwise} \end{cases}, \tag{2.23}$$

where C is a scale factor.

Therefore, in reality, interpolation is also a two-step process, consisting of an upsampler and a lowpass filter, as shown in Figure 2.12. The spectrum of the output signal $z[n]$ is

$$Z(\omega_z) = \begin{cases} CX(\omega_z I) & 0 \leq |\omega_z| \leq \frac{\pi}{I} \\ 0 & \text{otherwise} \end{cases}, \tag{2.24}$$

where $X(\omega)$ is the spectrum of the output signal $x[n]$.

The spectra for the sequence $x[n]$, $y[n]$ and $z[n]$ are illustrated in Figure 2.13, where the frequency range of the original signal is $0 \leq |\omega_x| \leq \pi$, and the frequency range of the decimated signal is $0 \leq |\omega_z| \leq \frac{\pi}{I}$.

Hands-On MATLAB Example: Now let us experiment with decimation and interpolation using MATLAB. The following example will manipulate two types of signals: a continuous wave (CW) signal (a sine wave) and a random signal. Furthermore, we will visualize the effects of upsampling and downsampling in the the time and frequency domains. We begin by generating these signals using MATLAB Code 2.3, and then pass the data through a lowpass filter in Code 2.4 to band-limit them. Using these band-limited versions we will observe the effects of correct and incorrect up and downsampling in the time and frequency domains.

When left unfiltered in Figure 2.14(a) and Figure 2.14(e), the sine wave signal mirrors a discrete version of a sine wave function. On the other hand in Figure 2.14(b) and Figure 2.14(f), the random binary signal consist of a string of random ones and negative ones.

Using the least-squares linear-phase FIR filter design or MATLAB's `firls` function, we are able to quickly generate FIR filter coefficients where the cut-off frequency is approximately at 0.21π.

After passing the data through Code 2.4, the resulting filtered discrete time signals (sine and random binary) represented in both the time and frequency domains are presented in Figure 2.14(c) and Figure 2.14(g). The differences between the original sine wave shown in Figure 2.14(e) and the band-limited sine wave in Figure 2.14(g) are negligible since the frequency of the sine wave is less than the low pass/band-limiting filter in Code 2.4. The differences of the random data shown in shown in Figure 2.14(b) and the band-limited in Figure 2.14(d) can been seen easily, and the reduction of bandwidth in the frequency domain show in Figure 2.14(h) is very noticeable compared to the orginal in Figure 2.14(f). In the time domain, we

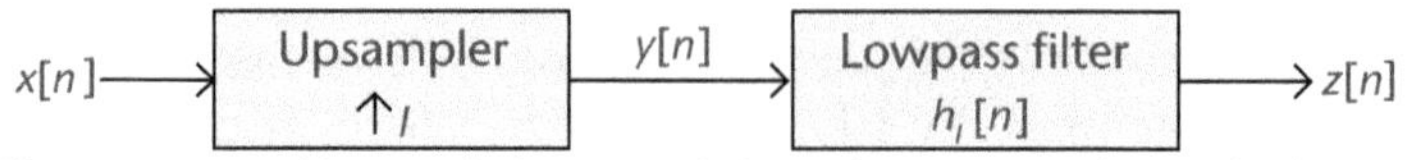

Figure 2.12 The structure of interpolation, consisting of an upsampler and a lowpass filter.

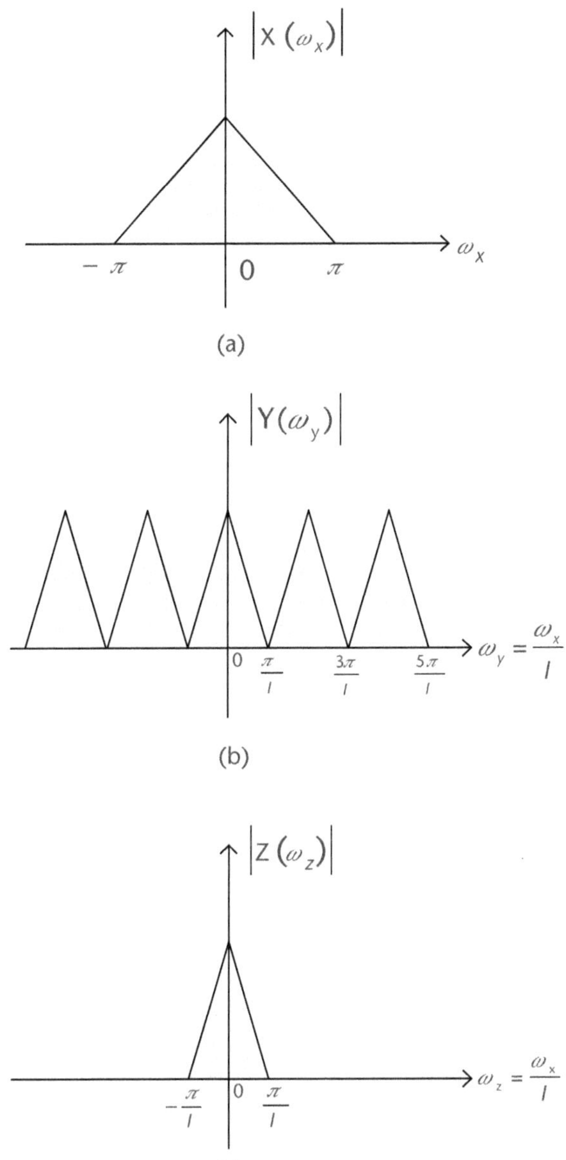

Figure 2.13 The spectra for the sequence $x[n]$, $y[n]$ and $z[n]$, where the frequency range of ω_x is compressed into the corresponding frequency range of ω_y by a factor of I. (a) Spectrum of the original sequence, (b) spectrum of the intermediate sequence, and (c) spectrum of the interpolated sequence.

Code 2.3 Create data sets: **up-down-sample.m**

```
% Create deterministic and stochastic digital data streams
n = 0:1/Fs1:100-(1/Fs1);                  % Time index vector
sin_wave = sin(5*n*2*pi);                 % Generation of sinusoidal
                                          % signal
random = 2*round(rand(1,length(n)))-1;    % Random string of +1 and
                                          % -1 values
```

Code 2.4 Create and apply lowpass filter to band-limit signal: **up-down-sample.m**

```
% Create lowpass filter and apply it to both data streams
% b = firls(n,f,a),
%     n is the FIR filter order
%     f is a vector of pairs of frequency points,
%     a is a vector containing the desired amplitude at the points in f
coeffs1 = firls(taps,[0 0.2 0.22 1],[1 1 0 0]);    % FIR filter
                                                    % coefficients
sin_bwlimited = filter(coeffs1,1,sin_wave);
random_bwlimited = filter(coeffs1,1,random);
```

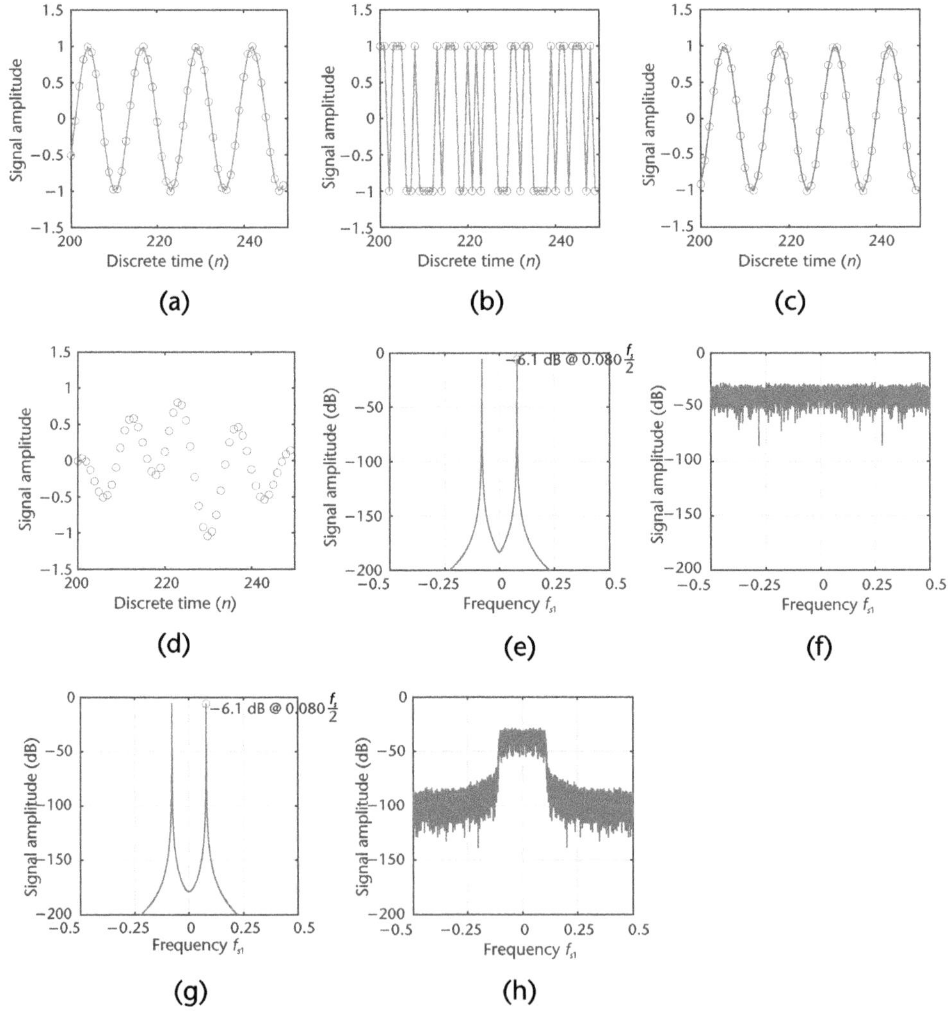

Figure 2.14 Sine and random data, created by Code 2.3, then bandlimited by Code 2.4. (a) Original sine wave: time domain, (b) original random data: time domain, (c) band-limited sine wave: time domain, (d) band-limited random data: time domain, (e) original sine wave: Fourier domain, (f) original random data: Fourier domain, (g) band-limited sine wave: Fourier domain, and (h) band-limited random data: Fourier domain.

examine the signal starting at a sample large enough to ignore the initial artifacts of the lowpass filter, which is why we do not look at things when $t = 0$. In the frequency domain, the lowpass filtering effectively limits the bandwidth of these signals, which will make them more clearly visible in the subsequent exercises in this section.

With the filtered sine wave and filtered random binary signals, we now want to explore the impact of upsampling these signals by observing the resulting outcomes in the frequency domain. Using Code 2.5, we can use the function `upsample` to take the filtered signals and upsample them by a factor of N (in this case, `N=5`). It should be noted that all `upsample` simply inserts `N-1` zeros between each original input sample. This function is quite different than the `interp` interpolation function, which combines upsampling and filtering operations.

The impact of upsampling can be clearly observed from the before-and-after frequency responses of the signals. Specifically, we expect that the frequency responses should be compressed by the upsampling factor of N, and also contain N periodic replicas across the original frequency band. Referring to Figures 2.15(a) and 2.15(b), we can observe this phenomena after upsampling our filtered sine wave and random binary signals, which are upsampled by a factor of `N=5` from Code 2.5.

Notice the difference between Figure 2.14(g) and Figure 2.15(e) for the filtered sine wave signal, or between Figure 2.14(h) and Figure 2.15(f) for the filtered random data signal. In both cases, we can readily see that the spectra of these signals have been compressed by the upsampling factor, and that we now have periodic replicas across frequency. It is also obvious that the amplitude as changed, as the average signal amplitude has been effected by this insertion of zeros. Although the literature may discuss this effect as *compression*, it is interesting to note that the actual signal has not changed in frequency. However, it appears in a different location with respect to the $-\frac{f_s}{2}$ to $\frac{f_s}{2}$ scale. Therefore, it is important to remember that f_s in both figures are not the same (differs by a factor of 5), which is a slight abuse of notation.

Now that we see how upsampling can compress the frequency responses of signals and make periodic replicas across the spectrum, let us now explore how downsampling these signals can either result in frequency expansion without any aliasing or with substantial amounts of aliasing. Recall that the downsampling process involves the periodic removal of $M - 1$ samples, which results in a frequency response that expands by a factor of M. This frequency expansion can be problematic when the frequency spectra begins to overlap with its periodic replicas

Code 2.5 Upsample the signals: **`up-down-sample.m`**

```
% y = upsample(x,n)
%     increases the sampling rate of x by inserting (n  1) zeros
%     between samples.
N = 5;
sin_up = upsample(sin_bwlimited,N);
random_up = upsample(random_bwlimited,N);
```

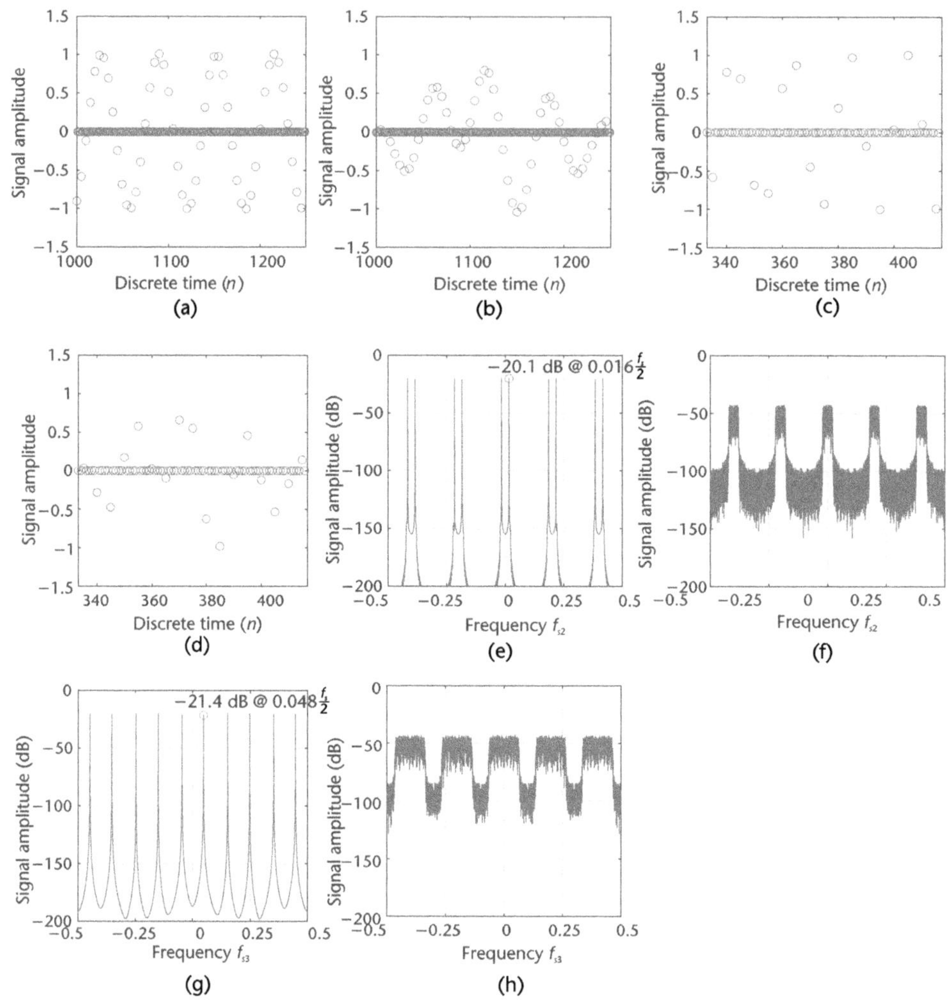

Figure 2.15 Upsampled data: Code 2.5, then processed by a downsampler: Code 2.6. (a) Upsampled, band-limited, sine wave: time domain, (b) band-limited random data: Fourier domain, (c) incorrect, upsampled, then downsampled, band-limited, sine wave: Fourier domain, (d) incorrect, upsampled, then downsampled, band-limited, random data: Fourier domain, (e) upsampled, band-limited, sine wave: Fourier domain, (f) band-limited random data: Fourier domain, (g) incorrect, upsampled, then downsampled, band-limited, sine wave: Fourier domain, and (h) incorrect, upsampled, and then downsampled, band-limited random data: Fourier domain.

centered at every multiple of 2π (sampled signals have spectra that repeat every 2π). This overlapping results in the aliasing of the signal, thus distorting it and making it very difficult to recover at the receiver. Code 2.6 will downsample the band-limited signals and Code 2.5 will upsample the created signals by a factor of `M=3`. Note that the MATLAB function `downsample` is different than the MATLAB function `decimate`, where the latter combines the downsampling and filtering operations to perform signal decimation.

In Code 2.6, which produces Figure 2.15(h) and Figure 2.15(g), we observed the incorrect case in which downsampling without filtering out one of the periodic replicas caused aliasing. Next, in Code 2.7 we will perform the necessary filtering to remove the periodic replicas before downsampling. In this code we apply an amplitude correct of the upsampling rate to compensate for these operations.

Code 2.6 Downsample the signals: **up-down-sample.m**

```
% Attempt to downsampling by M without filtering
% This is incorrect, but is instructive to show what artifacts occur
M = 3;
sin_up_down = downsample(sin_up,M);
random_up_down = downsample(random_up,M);
```

Code 2.7 Lowpass filter, then downsample the data: **up-down-sample.m**

```
% Lowpass filtering of baseband periodic replica followed by
% downsampling
% (correct approach)
coeffs2 = firls(taps,[0 0.15 0.25 1],[N N 0 0]); % FIR filter
                                                 % coefficients
sin_up_filtered = filter(coeffs2,1,sin_up);
sin_up_filtered_down = downsample(sin_up_filtered,M);
random_up_filtered = filter(coeffs2,1,random_up);
random_up_filtered_down = downsample(random_up_filtered,M);
```

When the signal is upsampled, the periodic replicas generated by this process span across the entire $-\pi$ to π radians (or $-\frac{f_s}{2}$ to $\frac{f_s}{2}$) of spectra. Without adequate filtering of these replicas before downsampling, these periodic replicas will begin to expand into other periodic replicas and result in aliasing. This phenomena is illustrated in Figure 2.15(d) and Figure 2.15(h), where we downsample the upsampled filtered sine wave and random binary signals previously described. For the downsampling of the upsampled filtered sine wave signal, we observe aliasing in Figure 2.15(g) with the spectra from other integers of Nyquist bands. When performed properly, we should only have one replica that is expanded by the downsampling factor, as observed in Figure 2.16(g) and Figure 2.17(b)

Since it is difficult to observe the spectra expansion of the sine wave signal because it is narrow, let us also observe the frequency responses of the random binary signal shown in Figures 2.15(h) and 2.17(b). In these figures, it is clearly evident that aliasing is occurring since the replicas from the upsampling process were not filtered out. On the other hand, when the upsampling replicas are filtered, we observe a clean, unaliased, frequency response of the downsampled signal shown in Figure 2.17(b).

We can do a final comparison of the signals in the time domain and see that the shape of the time domain signal is nearly exactly the same in amplitude and absolute time (seconds). It just has been sample rate converted from f_s to $\frac{5}{3}f_s$ adding more samples to the same signal.

2.3 Signal Representation

Understanding how a signal is represented can greatly enhance one's ability to analyze and design digital communication systems. We need multiple convenient numeric mathematical frameworks to represent actual RF, baseband, and noise

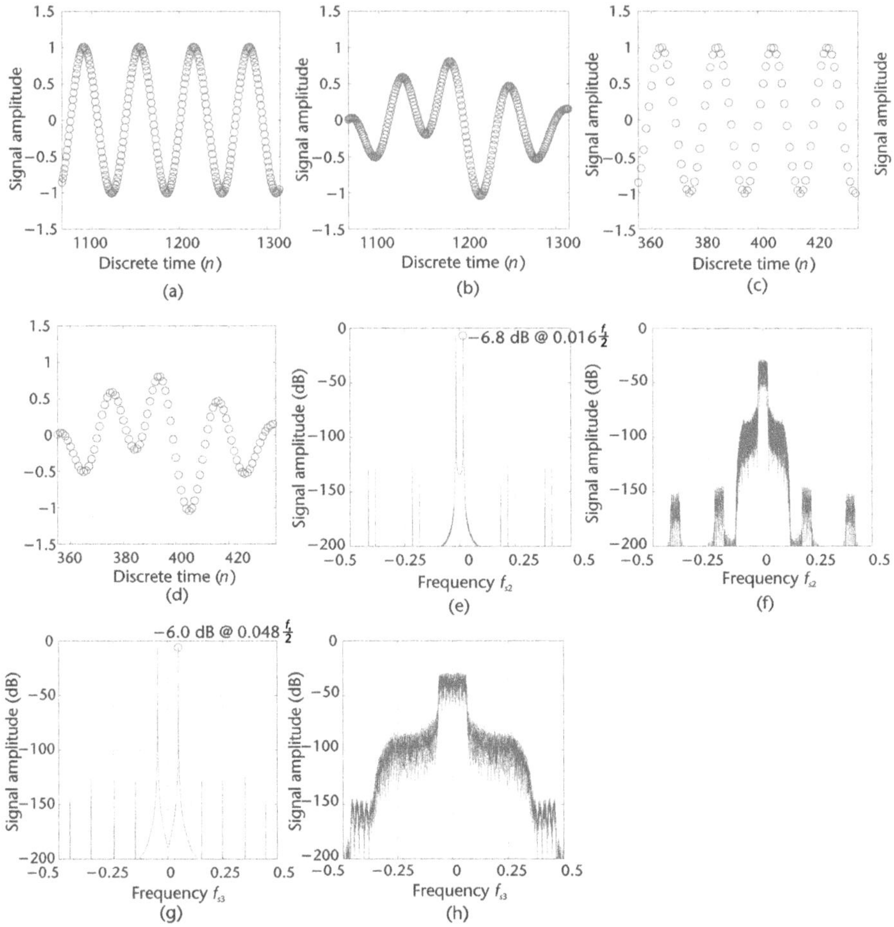

Figure 2.16 Upsampled by $N = 5$, then filtered band-limited waveforms produced by Code 2.7. You can still see the replicas of the signals, but the filter has suppressed it to very low levels. The filtering has corrected the amplitude loss from the upsampling. (a) Upsampled, then filtered band-limited, sine wave: Fourier domain, (b) upsampled, filtered, then downsampled, band-limited sine wave: Fourier domain, (c) upsampled, then filtered band-limited, sine wave: Fourier domain, (d) upsampled, filtered, then downsampled, band-limited sine wave: Fourier domain, (e) upsampled, then filtered band-limited, random data: Fourier domain, (f) upsampled, then filtered band-limited, random data: Fourier domain, (g) upsampled, then filtered band-limited, random Data: Fourier domain, and (h) upsampled, filtered, then band-limited random data: Fourier domain.

signals. We usually have two: envelope/phase and in-phase/quadrature, and both can be expressed in the time and Fourier domains.

2.3.1 Frequency Conversion

To understand how we can move signals from baseband to RF and from RF to baseband, let us look more closely at modulators and demodulators. For example, in Figure 2.18 we see a very classical quadrature modulator. The ADL5375 accepts two differential baseband inputs and a single-ended LO, which generates a single-ended output. The LO interface generates two internal LO signals in quadrature (90° out of phase) and these signals are used to drive the mixers, which simply multiply the LO signals with the input.

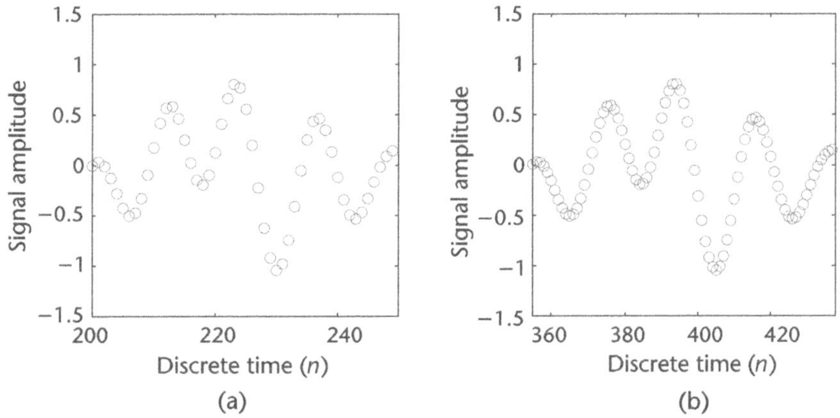

Figure 2.17 Upsampled by $N = 5$, then filtered, and then downsampled by $M = 3$, band-limited random data. A phase shift (time shift) occurs from the filtering. (a) Original band-limited random data time domain at f_s. (b) Upsampled, filtered, and the band-limited random data: time domain at $\frac{5}{3}f_s$.

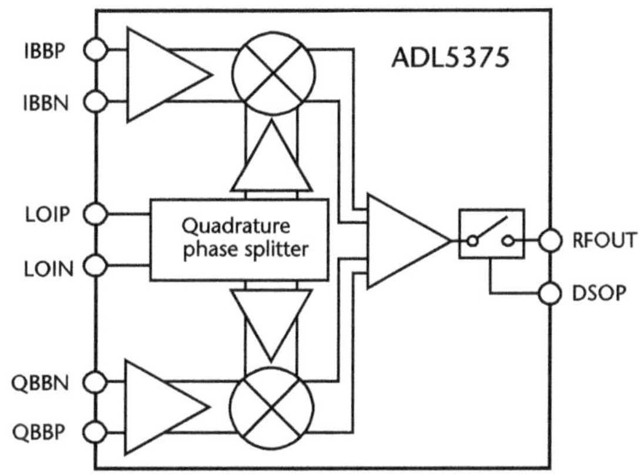

Figure 2.18 ADL5375 broadband quadrature modular with range from 400 MHz to 6 GHz.

Mathematically, this mixing process will take the two inputs IBB and QBB, which we will denote by $I(t)$ and $Q(t)$, multiply them by our LO at frequency ω_c, and add the resulting signal to form our transmitted signal $r(t)$. The LO used to multiply $Q(t)$ is phase shifted by 90° degree to make it orthogonal with the multiplication of the $I(t)$ signal. Consequently, this yields in the following equation:

$$r(t) = I(t)cos(\omega_c t) - Q(t)sin(\omega_c t). \tag{2.25}$$

The LO frequency is denote as ω_c since it will be typically called the carrier frequency, which exploits the phase relationship between the in-phase $(I(t)cos(\omega_c t))$ and quadrature $(Q(t)sin(\omega_c t))$ components. Therefore, the transmitted signal will contain both components but will appear as a single sinusoid.

At the receiver, we will translate or down mix $r(t)$ back into our in-phase and quadrature baseband signals through a similar process but in reverse. By applying the same LO with a second phase-shifted component to $r(t)$ with a lowpass filter, we arrive at

$$I_r(t) = LPF\{r(t)cos(\omega_c t)\} = LPF\{(I(t)cos(\omega_c t) - Q(t)sin(\omega_c t))cos(\omega_c t)\} = \frac{I(t)}{2}, \tag{2.26}$$

$$Q_r(t) = LPF\{r(t)sin(\omega_c t)\} = LPF\{(-I(t)cos(\omega_c t) + Q(t)sin(\omega_c t))sin(\omega_c t)\} = \frac{Q(t)}{2}. \tag{2.27}$$

In practice, there will be some phase difference between the transmitter LO and receiver LO, which can cause rotation to $r(t)$. However, the phase relation between $I(t)$ and $Q(t)$ will alway be maintained.

2.3.2 Imaginary Signals

Discussing signals as *quadrature* or *complex* signal is taking advantage of the mathematical constructs that Euler and others have created to make analysis easier. We try to refer to signals as in-phase (I) and quadrature (Q) since that is actually pedantically correct. As described in Section 2.3.1, the in-phase (I) refers to the signal that is in the same phase as the local oscillator, and the quadrature (Q) refers to the part of the signal that is in phase with the LO shifted by 90°.

It is convenient to describe this as I being *real* and Q being *imaginary* since it enables many mathematical techniques but at the end of the day is just a construct. A prime example of this convience is frequency translation, which is performed by the mixer. We start from the Euler relation of

$$e^{jx} = \cos(x) + j\sin(x), \tag{2.28}$$

where we can define x as target frequency plus time. Taking the conventions from Section 2.3.1 but redefining $I(t)$ and $Q(t)$ as real and imaginary, we arrive at

$$y(t) = I(t) + jQ(t). \tag{2.29}$$

Now, if we assume $y(t)$ is a CW tone at frequency ω_a for illustration purposes, $y(t)$ becomes

$$y(t) = cos(\omega_a t) + jsin(\omega_a t). \tag{2.30}$$

Now applying (2.28) we can frequency shift $y(t)$ by the desired frequency ω_c:

$$\begin{aligned} y(t)e^{j\omega_c t} &= \big(I(t)cos(\omega_c t) - Q(t)sin(\omega_c t)\big) + j\big(Q(t)cos(\omega_c t) + I(t)sin(\omega_c t)\big) \\ &= cos((\omega_c + \omega_a)t) + jsin((\omega_c + \omega_a)t). \end{aligned} \tag{2.31}$$

Now our resulting signal will exist at frequency $\omega_a + \omega_c$ through a simple application of Euler's identity.

Let us motivate the usefulness of a complex number representation further from the perspective of hardware. If we consider the mixer itself as in Figure 2.18, the IQ mixer will transmit signals with arbitrary phase and amplitude (within power constraints). This is an example of how we encode information into the data we transmit through differences in phase and amplitude. This is actually accomplished through the relation of our in-phase and quadrature signal, which can be used to create a single sinusoid with arbitrary phase and amplitude. Mathematically, we can produce a sinusoid with a specific envelope and phase ($\mathbf{A}$,ϕ) with two orthogonal components sine and cosine. This relationship is written as

$$\mathbf{A}\sin(\omega t + \phi) = (\mathbf{Acos}\phi)\sin(\omega t) + (\mathbf{Asin}\phi)\cos(\omega t). \tag{2.32}$$

Therefore, by just modifying the amplitude of our sine and cosine components over time, we can create the desired waveform from a fixed frequency and phase LO. Alternatively, we can consider others coordinate systems to visualize complex values. Expanding these complex numbers (rectangular coordinates) can be translated in order to be represent a magnitude and angle (polar) or even plotted as a vector.

The in-phase and quadrature sine waves plotted in Figure 2.19 show how things look via a phasor plot as time increases, with the vector indicating magnitude rotating around the axis. One can clearly see the phase shift between I and Q. In the time domain, you can also see the differences between in-phase and magnitude, although phase differences can be a little more subtle to notice. In a Cartesian plane, the signal appears as a rotating circle over time. The phasor plot will always rotate counterclockwise with time, and the Cartesian plot can rotate in either direction depending on if the phase difference between I and Q is positive or negative. While the time domain plot shows things moving as time changes, the phasor plot and Cartesian plane are snapshots in time (at $t = 0$ on the time domain plot). Finally, the frequency domain plot provides the spectrum of the phasor but only communicates magnitude and loses phase information. This happens because we are only plotting the real component of the spectrum. However, comparing the two waveforms in Figures 2.19 and 2.20, changes in the I and Q components (amplitude and phase relationship), will effect the other domains as well.

2.4 Signal Metrics and Visualization

Before an engineer can decide whether a project is done, he or she needs to conduct some form of verification. However, communications systems are complex to evaluate given their integrated nature and depth of transmit and receive chains. As referenced in Section 1.4, communication systems may have a variety of metrics beyond size, weight, power, and cost (SWaP-C). Performance metrics such as bit error rate (BER), data throughput, and distance are also only top-level system

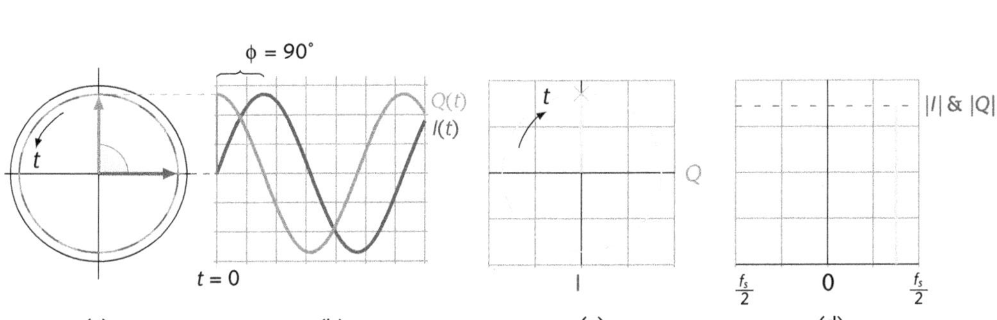

Figure 2.19 Same continuous wave signal, plotted in multiple domains. (a) Phasor *rad*(t), (b) time $x(t) \rightarrow$, (c) Cartesian $(I, Q)(t)$, and (d) frequency $X(\omega)$.

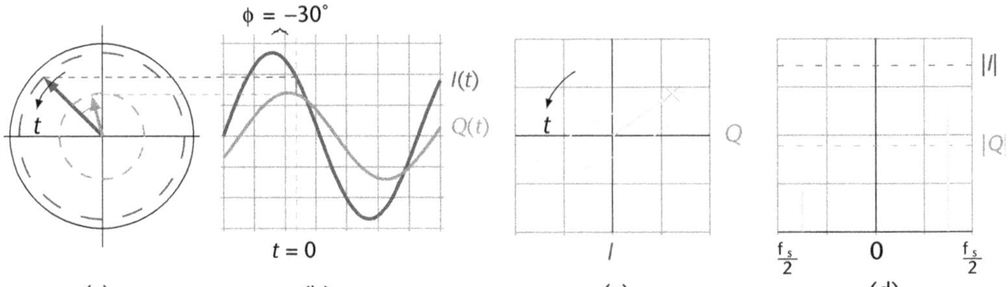

Figure 2.20 Continuous wave signal differences in magnitude and phase cause shifts in various domains. (a) Phasor $rad(t)$, (b) time $x(t) \rightarrow$, (c) Cartesian $(I, Q)(t)$, and (d) frequency $X(\omega)$.

specifications. Just as we have system-level specifications for the entire system, we have specifications and measurement techniques for other subsystems and SDR building blocks. In this way, we know we are not over- or underdesigning specific components. However, trade-offs should always be considered at the system level since upstream modifications can effect downstreaming components or implementations.

System-level specifications are met by ensuring each block in the system will allow those specifications to be met. Making a world-class communications system requires world-class hardware and world-class algorithmic design and implementation. That is the issue with many aspects of engineering—quantitatively determining when something is complete or functional and that it can be connected to the rest of the system. It is never more true than in communications systems that a system is only as good as the weakest link. If you have bolted everything together, and a system-level specification like bit error rate is not meeting your top-level specifications, unless you understand how to measure each part of the communications system, from the RF to the SDR to the algorithmic design, you will be lost and unable to determine what to do next.

Depending on what you are looking at, there are numerous techniques and tools to measure almost everything. Studying communications is not for those who do not want to be rigorous.

2.4.1 SINAD, ENOB, SNR, THD, THD + N, and SFDR

Six popular specifications for quantifying analog dynamic performance are found in Table 2.2 [6]; namely, list out by using and understanding these measurements will help you analyze your designs and make sure you are designing something to be the most robust. Although most device and system manufacturers have adopted the same definitions for these specifications, some exceptions still exist. Due to their importance in comparing devices and systems, it is important not only to understand exactly what is being specified, but the relationships between the specifications.

- *Spurious free dynamic range* (SFDR) is the ratio of the root mean squared (RMS) value of the signal to the rms value of the worst spurious signal regardless of where it falls in the frequency spectrum. The worst spur may or may not be a harmonic of the original signal. This is normally measured over the bandwidth of interest, which is assumed to be the Nyquist bandwidth

Table 2.2 Six Popular Specifications

Property	*Definition*	*MATLAB Function*
SFDR	Spurious free dynamic range	`sfdr`
SINAD	Signal-to-noise-and-distortion ratio	`sinad`
ENOB	Effective number of bits	
SNR	Signal-to-noise ratio	`snr`
THD	Total harmonic distortion	`thd`
THD + N	Total harmonic distortion plus noise	

unless otherwise stated; DC to $f_s/2$ (for baseband), and $-f_s/2$ to $f_s/2$ for complex (RF) converters, which are found on devices like the Pluto SDR. SFDR is an important specification in communications systems because it represents the smallest value of signal that can be distinguished from a large interfering signal (blocker). SFDR is generally plotted as a function of signal amplitude and may be expressed relative to the signal amplitude (dBc) or the ADC full-scale (dBFS) as shown in Figure 2.21. MATLAB's `sfdr` function provides results in terms of dBc. For a signal near full-scale, the peak spectral spur is generally determined by one of the first few harmonics of the fundamental. However, as the signal falls several dB below full-scale, other spurs generally occur that are not direct harmonics of the input signal. This is due to the differential nonlinearity of the systems transfer functions normally dominates at smaller signals. Therefore, SFDR considers all sources of distortion regardless of their origin, and is a useful tool in evaluating various communication systems.

- *Total harmonic distortion* (THD) is the ratio of the rms value of the fundamental signal to the mean value of the root-sum-square of its harmonics (generally, only the first five harmonics are significant). THD of an ADC is also generally specified with the input signal close to full-scale, although it can be specified at any level.
- *Total harmonic distortion plus noise* (THD + N) is the ratio of the rms value of the fundamental signal to the mean value of the root-sum-square of its harmonics plus all noise components (excluding DC). The bandwidth over which the noise is measured must be specified. In the case of an FFT, the bandwidth is DC to $\frac{f_s}{2}$. (If the bandwidth of the measurement is DC to $\frac{f_s}{2}$ (the Nyquist bandwidth), THD + N is equal to SINAD).
- *Signal-to-noise-and-distortion* (SINAD, or S/(N + D) is the ratio of the rms signal amplitude to the mean value of the root-sum-square (rss) of all other spectral components, including harmonics, but excluding DC. SINAD is a good indication of the overall dynamic performance of an analog system because it includes all components that make up noise and distortion. SINAD is often characterized for various input amplitudes and frequencies. For a given input frequency and amplitude, SINAD is equal to THD + N, provided the bandwidth for the noise measurement is the same for both (the Nyquist bandwidth)
- *Signal-to-noise ratio* (SNR, or sometimes called SNR-without-harmonics) is calculated from the FFT data the same as SINAD, except that the signal harmonics are excluded from the calculation, leaving only the noise terms. In practice, it is only necessary to exclude the first five harmonics, since they

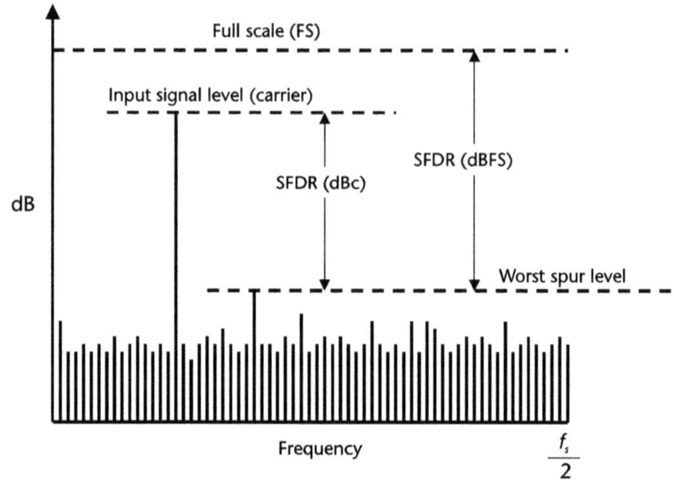

Figure 2.21 Spurious free dynamic range (SFDR) for BW DC to $f_s/2$.

dominate. The SNR plot will degrade at high input frequencies, but generally not as rapidly as SINAD because of the exclusion of the harmonic terms.

2.4.2 Eye Diagram

Although its obvious to state, time domain plots are used to observe changes of an electrical signal over time. Any number of phenomena such as amplitude, frequency, rise time, time interval, distortion, noise floor, and others can be empirically determined, and how these characteristics change over time. In telecommunication, an *eye diagram*, also known as an *eye pattern*, is an time domain display in which a digital data signal from a receiver is repetitively sampled and applied to the vertical input, while the data rate is used to trigger the horizontal sweep [9]. It is called an eye diagram because the pattern looks like a series of eyes between a pair of rails.

Several system performance measures can be derived by analyzing the display, especially the extent of the intersymbol-interference (ISI). As the eye closes, the ISI increases; as the eye opens, the ISI decreases. Furthermore, if the signals are too long, too short, poorly synchronized with the system clock, too high, too low, too noisy, or too slow to change, or have too much undershoot or overshoot, this can be observed from the eye diagram. For example, Figure 2.22 shows a typical eye pattern for the noisy quadrature phase-shift keying (QPSK) signal.

Since the eye diagram conveys and measures many different types of critical data, this can help quantify how well an algorithm or system is working. The two key measurements are the *vertical opening*, which is the distance between BER threshold points, and the *eye height*, which is the minimum distance between eye levels. Larger vertical and horizontal openings in the eye are always better.

Hands-On MATLAB Example: To provide some hands-on experience with eye diagrams, let us use the MATLAB function `eyediagram`, which is a very handy way of visually analyzing a transmission regarding the amount of noise and intersymbol interference present in the signal. Using the pulse shaped signals, we should be able to observe any distortion present within the transmission.

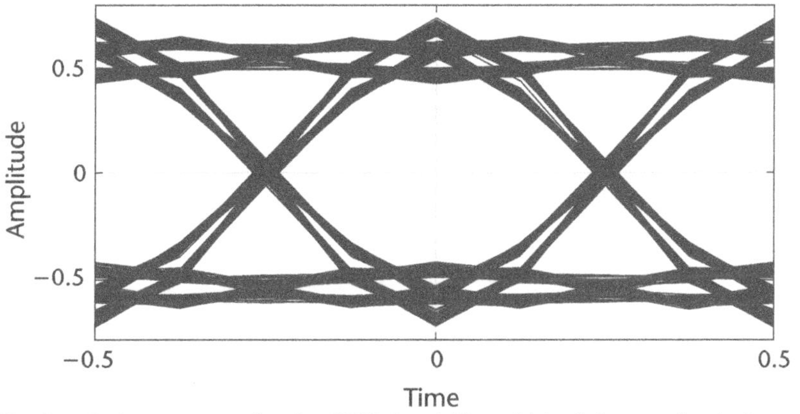

Figure 2.22 A typical eye pattern for the BPSK signal. The width of the opening indicates the time over which sampling for detection might be performed. The optimum sampling time corresponds to the maxmum eye opening, yielding the greatest protection against noise.

From Figure 2.23, we can see that the pulse shaped transmissions do not have any distortion present (we know this in advance since we have intentionally omitted any sort of noise and distortion from the transmission in the first place). When we have eye diagrams such as those shown in Figures 2.23(a) and 2.23(b), we refer to these situations as the eye being open. The prime indicator of having some sort of distortion present within the transmission is when the aperture at time instant 0 is not at its maximum.

Let us explore the scenarios when distortion is present within the transmission and how this translates into an eye diagram. Suppose with take the `y_impulse1` and `y_impulse2` output signals from Code 2.8 and introduce some noise to it. In Code 2.9, we introduced some Gaussian noise using the function `randn`.

We can clearly see in Figure 2.24 the impact of the additional noise on the transmitted signals via the eye diagram. In both Figures 2.24(a) and 2.24(b), it is observed that the eye walls of the diagram are no longer smooth when compared with Figured 2.23(a) and 2.23(b). Although the eye is still open in both cases, we only introduced as small amount of noise into the transmission; the impact of a large amount of noise introduced into the transmission could potential close the eye, meaning the desired sampling instant at time 0 could potentially be corrupted and translate into bit errors. Later on in this book, we will explore how other forms of distortion will affect the aperture of the eye diagram.

2.5 Receive Techniques for SDR

The study of modern communications maintains a great duality when considering both the analog and digital domains. Both domains are manipulated efficiently and with great speed. However, analog signals maintain a perspective of infinite precision but will always contain some degree of randomness due to their very nature. Digital signals, on the other hand, are exact and precisely defined, but are limited by the boundaries of computational complexity and their fundamental foundations. Digital communications must effectively manage both of these world

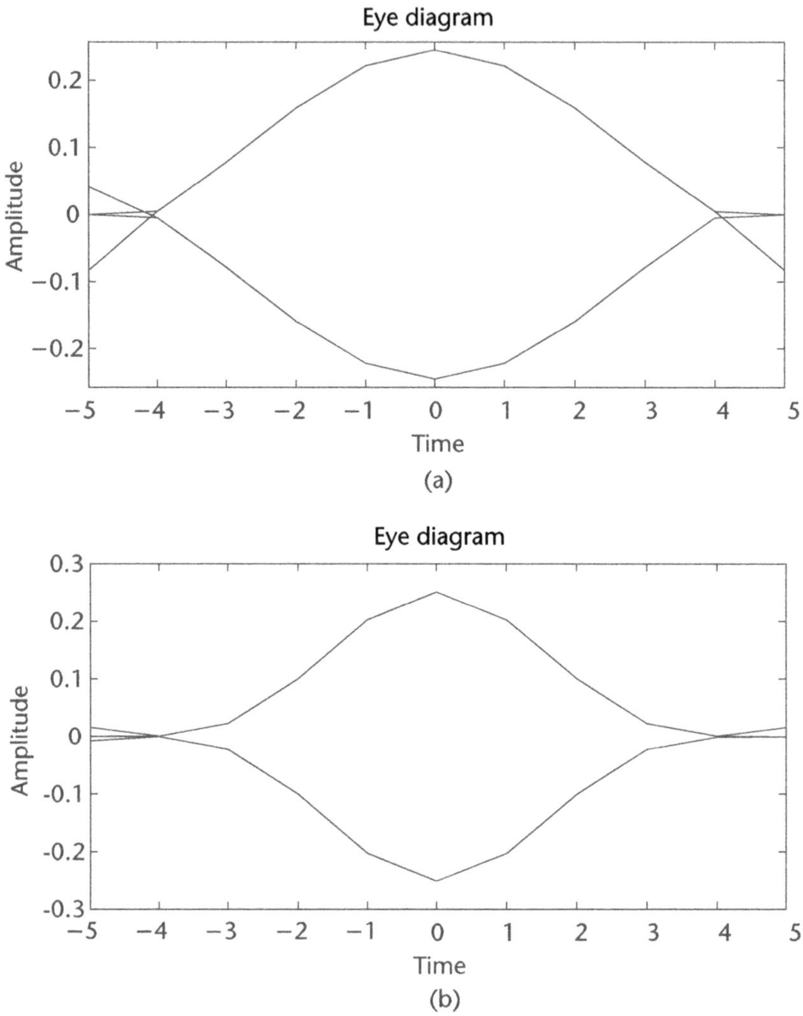

Figure 2.23 Eye diagrams of signals filtered by a system possessing a rectangular frequency response and a triangular frequency response. (a) Rectangular frequency response pulse filtering, and (b) triangular frequency response pulse filtering.

to design robust links between points. Therefore, both domains are highly dependent on one another.

Even in today's world of abundant and cost-effective digital signal processing (DSP) devices, an analog signal is processed, amplified, filtered, and only then converted into binary form by an ADC. The output of the ADC is just a binary representation of the analog signal and is processed on a number of computational units from FPGAs to general purpose CPUs. After processing, the information obtained from the digitized signal, it may be converted back into analog form using a digital-to-analog converter (DAC). Signals physically recovered by a SDR start out as a time-varying electric field, which induces a current in the receiving antenna and resulting in a detectable voltage at the receiver. Transmission by the SDR, on the other hand, are time-varying voltages being applied to an antenna, which causes movement of electrons as an outwardly radiating electric field.

Code 2.8 Eye diagram example: **eye_example.m**

```
% Create impulse train of period L and length len with random +/- one
% values
L = 10;
impulse_num = 100; % Total number of impulses in impulse train
len = L*impulse_num;
temp1 = [2*round(rand(impulse_num,1))-1 zeros(impulse_num,L-1)];
x_impulse = reshape(temp1.',[1,L*impulse_num]);
% Create two transmit filter pulse shapes of order L
% Approximate rectangular frequency response --> approximate
% sinc(x) impulse response
txfilt1 = firls(L,[0 0.24 0.25 1],[4 4 0 0]);
% Approximate triangular frequency response --> approximate
% sinc(x)^2 impulse response
txfilt2 = firls(L,[0 0.5 0.52 1],[4 0 0 0]);
% Pulse shape impulse train
y_impulse1 = filter(txfilt1,1,x_impulse);
y_impulse2 = filter(txfilt2,1,x_impulse);
% eyediagram(x,n,period,offset)
% creates an eye diagram for the signal x, plotting n samples in each
% trace horizontal axis range between -period/2 and period/2.
eyediagram(y_impulse1,L,L,floor(L/2));
eyediagram(y_impulse2,L,L,floor(L/2));
```

Code 2.9 Eye diagram example: **eye_example.m**

```
eyediagram((y_impulse1+0.1*randn(1,length(y_impulse1))),L,L,floor(L/2));
eyediagram((y_impulse2+0.1*randn(1,length(y_impulse2))),L,L,floor(L/2));
```

2.5.1 Nyquist Zones

In Section 2.2.3, we considered the case of baseband sampling (i.e., all the signals of interest lie within the first Nyquist zone). Figure 2.25 shows such a case, where the band of sampled signals is limited to the first Nyquist zone and images of the original band of frequencies appear in each of the other Nyquist zones. Consider the case shown in Figure 2.25 B, where the sampled signal band lies entirely within the second Nyquist zone. The process of sampling a signal outside the first Nyquist zone is often referred to as undersampling, or harmonic sampling. Note that the image, which falls in the first Nyquist zone, contains all the information in the original signal with the exception of its original location.

Figure 2.25 shows the sampled signal restricted to the third Nyquist zone. Note that the image that falls into the first Nyquist zone has no frequency reversal. In fact, the sampled signal frequencies may lie in any unique Nyquist zone, and the image falling into the first Nyquist zone is still an accurate representation. At this point we can clearly restate the Nyquist criteria:

A signal must be sampled at a rate equal to or greater than twice its bandwidth in order to preserve all the signal information.

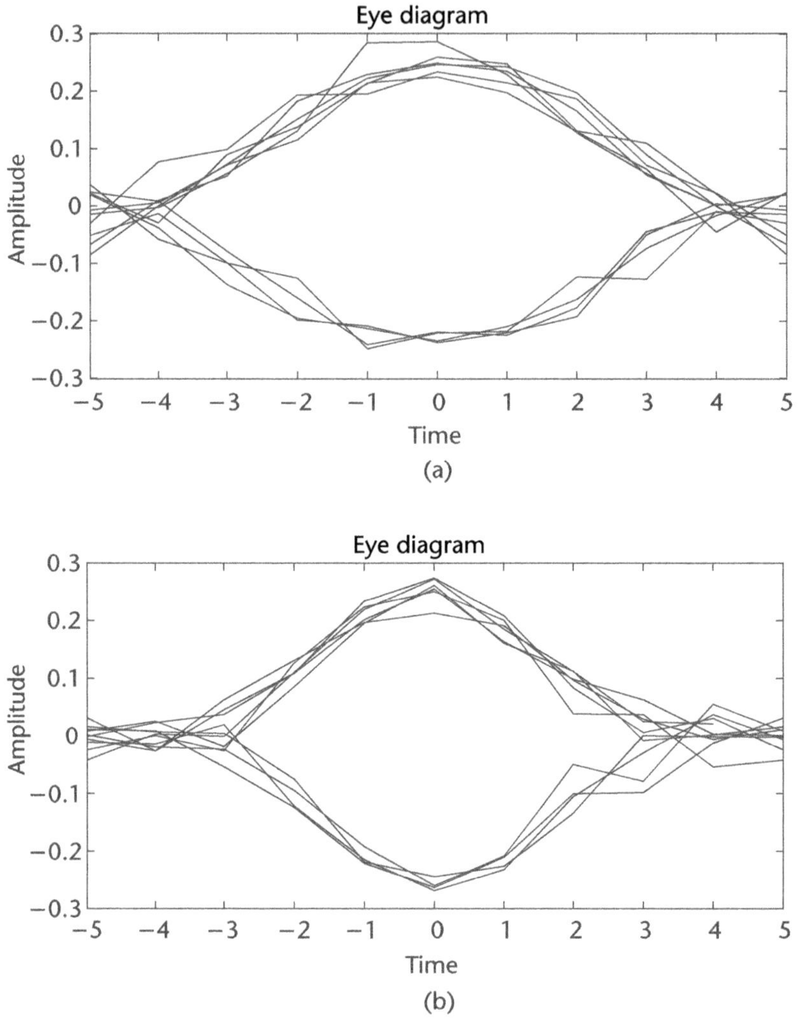

Figure 2.24 Eye diagrams of signals filtered by a system possessing a rectangular frequency response and a triangular frequency response with additive white Gaussian noise present. (a) Rectangular frequency response pulse filtering, and (b) triangular frequency response pulse filtering.

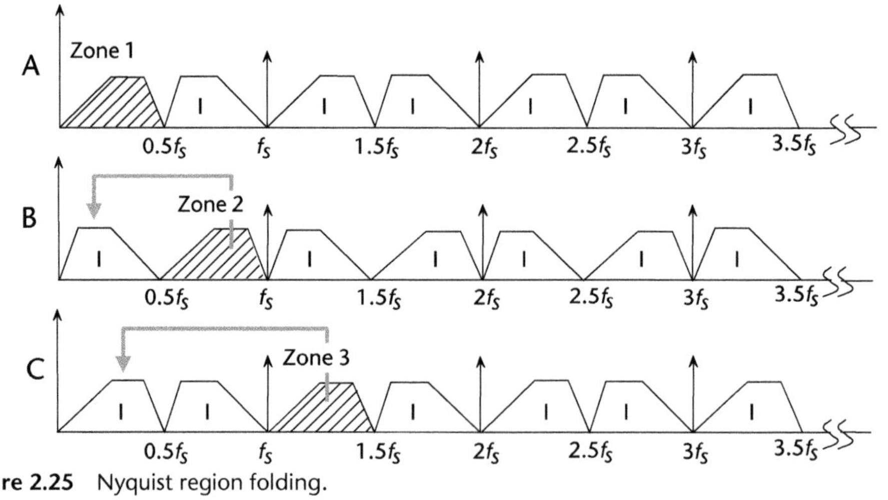

Figure 2.25 Nyquist region folding.

Notice that there is no mention of the absolute location of the band of sampled signals within the frequency spectrum relative to the sampling frequency. The only constraint is that the band of sampled signals be restricted to a single Nyquist zone (i.e., the signals must not overlap any multiple of $\frac{f_s}{2}$). In fact, this is the primary function of the antialiasing filter. Sampling signals above the first Nyquist zone has become popular in communications because the process is equivalent to analog demodulation. It is becoming common practice to sample IF signals directly and then use digital techniques to process the signal, thereby eliminating the need for an IF demodulator and filters. However, as the IF frequencies become higher, the dynamic performance requirements (bandwidth, linearity, distortion, etc.) on the ADC become more critical as performance must be adequate at the second or third Nyquist zone, rather than only baseband. This presents a problem for many ADCs designed to process signals in the first Nyquist zone. Therefore, an ADC suitable for undersampling applications must maintain dynamic performance into the higher-order Nyquist zones. This is specifically important in devices like the Pluto SDR, which includes *DC correction* to remove local oscillator leakage. This DC correction can be through of a highpass filter, which is set close to DC (25 kHz). For those modulation schemes, which do not strictly transmit information at DC, such as QPSK and quadrature amplitude modulation (QAM), this does not matter. For those modulation schemes that pass information at DC, this can be very difficult to work around without using an undersampling technique as described above. Capturing the data above DC, and then digitally moving it down can improve performance.

2.5.2 Fixed Point Quantization

The only errors (DC or AC) associated with an ideal N-bit data converter are those related to the sampling and quantization processes. The maximum error an ideal converter makes when digitizing a signal is $\pm\frac{1}{2}$ LSB, directly in between two digital values. This is obvious from the transfer function of an ideal N-bit ADC, which is

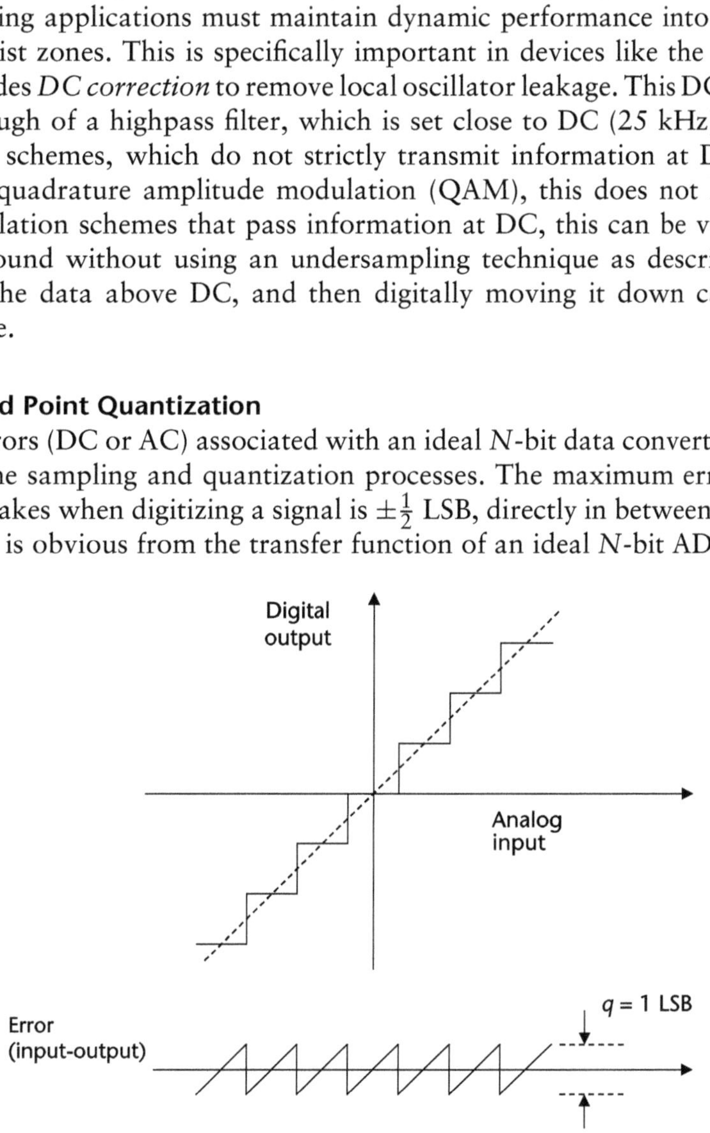

Figure 2.26 Ideal N-bit ADC quantization noise.

shown in Figure 2.26. The quantization error for any AC signal, which spans more than a few LSBs, can be approximated by an uncorrelated sawtooth waveform having a peak-to-peak amplitude of q, the weight of an LSB. Although this analysis is not precise it is accurate enough for most applications.

The quantization error as a function of time is shown in Figure 2.27. Again, a simple sawtooth waveform provides a sufficiently accurate model for analysis. The equation of the sawtooth error is given by

$$e(t) = st, \quad \frac{-q}{2s} < t < \frac{q}{2s}, \tag{2.33}$$

where s is the slope of the quantized noise. The mean-square value of $e(t)$ can be written:

$$\bar{e}^2(t) = \frac{q}{s}\int_{\frac{q}{2s}}^{\frac{-q}{2s}} (st)^2 dt = \frac{q^2}{12}. \tag{2.34}$$

The square root of (2.34), the root mean squared (RMS) noise quantization error, is approximately Gaussian and spread more or less uniformly over the Nyquist bandwidth of DC to $\frac{f_s}{2}$.

The theoretical SNR can now be calculated assuming a full-scale input sine wave $v(t)$

$$v(t) = \frac{q2^N}{2} sin(\omega t), \tag{2.35}$$

by first calculating the RMS value of the input signal defined as

$$\sqrt{\overline{v}(t)^2} = \frac{q2^N}{2\sqrt{2}}. \tag{2.36}$$

Therefore, the RMS signal-to-noise ratio for an ideal N-bit converter is

$$SNR = 20\,log_{10}\left(\frac{RMS\ of\ full\ scale\ input}{RMS\ of\ quatization\ noise}\right) = 20\,log_{10}\left(\frac{\frac{q2^N}{2\sqrt{2}}}{\frac{q}{\sqrt{12}}}\right). \tag{2.37}$$

After some simplification of (2.37) we arrive at our SNR in dB:

$$SNR = 20\,log_{10}\left(\sqrt{\frac{3}{2}}2^N\right) = 6.02N + 1.76. \tag{2.38}$$

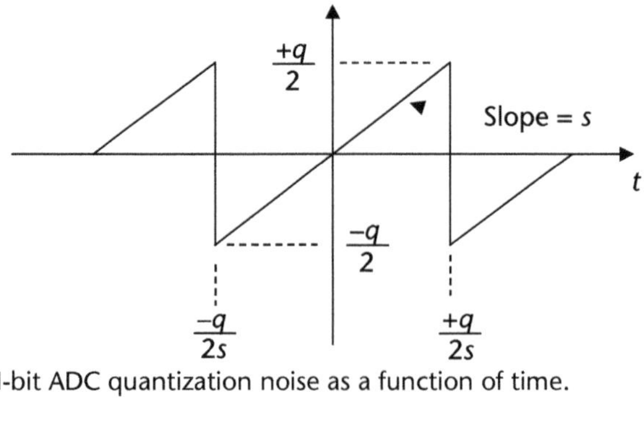

Figure 2.27 Ideal N-bit ADC quantization noise as a function of time.

Again, this is for over DC to $\frac{f_s}{2}$ bandwidth. In many applications the actual signal of interest occupies a smaller bandwidth (BW). For example, if digital filtering is used to filter out noise components outside BW, then a correction factor (called *process gain*) must be included in the equation to account for the resulting increase in SNR. The process of sampling a signal at a rate, which is greater than twice its bandwidth, is often referred to as oversampling. In fact oversampling in conjunction with quantization noise shaping and digital filtering is a key concept in sigma-delta converters, which will be discussed in Section 2.5.4.

Hands-On MATLAB Example: In Section 2.5.2 we made the following assumptions [11]:

- The sequence of error samples $e(t)$ is a sample sequence of a stationary random process;
- The error sequence is uncorrelated with the sequence of exact samples, $v(t)$;
- The random variables of the error process are uncorrelated; that is, the error is a white-noise process;
- The probability of the errpr process is uniform over the range of quantization error.

The underlying assumption here is that the quantization noise is uncorrelated to the input signal. We will see in certain common trivial examples that is not true. Under certain conditions where the sampling clock and the signal are harmonically related, the quantization noise becomes correlated and the energy is concentrated at the harmonics of the signal. In a practical ADC application, the quantization error generally appears as random noise because of the random nature of the wideband input signal and the additional fact that there is a usually a small amount of system noise that acts as a dither signal to further randomize the quantization error spectrum. It is important to understand the above point because single-tone sinewave FFT testing of ADCs is one of the universally accepted methods of performance evaluation. In order to accurately measure the harmonic distortion of an ADC, steps must be taken to ensure that the test setup truly measures the ADC distortion, not the artifacts due to quantization noise correlation.

We will use variations on the MATLAB code from Code 2.10 to explore the SFDR. We will expand on this concept with regards to the precision allowed for the computations. We begin with a double-precision floating point number

Code 2.10 SFDR test: **sfdr_test.m**

```
deltat = 1e-8;
fs = 1/deltat;
t = 0:deltat:1e-5-deltat;
fundamental = 3959297;
x = 10e-3*sin(2*pi*fundamental*t);
r = sfdr(x,fs)
sfdr(x,fs);
```

representation (MATLAB default). To keep the signal uncorrelated, we choose a tone at 3,959,297 Hz (a prime number).

This provides the results in Figure 3.28, where a SFDR measurement of 288.96 dBc is obtained, which is a very impressive measurement. However, a real radio such as Pluto SDR can not accept double-precision floating-point numbers, and thus fixed-point (12-bit) representation must be used.

The fixed-point format Pluto SDR and many other devices use is a signed format. The MSB bit is for sign and 11 remaining bits are for the magnitude. In our MATLAB script, we use 2^{11} as the magnitude to maximize the dynamic range of our signal before transmission. Therefore, we multiply to our integer value, round, and then scale down to ± 1 to normalize the amplitude.

This provides an underwhemling 45.97 dBc for the SFDR, as shown in Figure 2.29, which is not even 8-bits of performance. This is because in the example we scaled our signal to 10^{-3}, and the dynamic range of a fixed-point number, does not scale equally with a double-precision floating-point number.

To resolve the dynamic range issue, we will remove the 10^{-3} scaling and use 12-bit full scale. This results in Figure 2.30(a) with a respectable 86.82 dBc.

To improve this result even more, we can take advantage of a concept known as the FFT processing gain. We simply increase the number of samples to $10,000$ by using the following code in code 2.15, which simply changes the length of `t`. This provides the results in Figure 2.30(b), with an SFDR of 93.98 dBc. This is accomplished by simply increasing the number of samples, which increases the number of FFT bins, which in turn decreases the energy accumulated in each bin.

The example code in Code 2.10, Code 2.11, Code 2.12 and Code 2.15 utilized an uncorrelated F_A of $3,959,297$ Hz. What happens when it is correlated? If we simply round F_A to 4 MHz in the example (see Codes 2.13 and 2.14), we can see the results in Figure 2.31(a), which yields an SFDR of 82.58 dBc, a loss of 11.4 dB from our previous result in Figure 2.30(b).

To regain this loss in Figure 2.31(b), we can use a technique known as dithering. This moves the energy accumulated in the harmonics and pushes it out into the rest of the noise floor. The noise floor is higher but the worse case spur is lower, which is the figure of merit when calculating SFDR. This results in an SFDR of 91.68 dBc, only a 2.3-dB difference from our uncorrelated results in Figure 2.30(b).

Code 2.11 SFDR test: **sfdr_test.m**

```
bits=2^11;
x = round(10e-3*bits*sin(2*pi*fundamental*t))/bits;
sfdr(x,fs)
```

Code 2.12 SFDR test: **sfdr_test.m**

```
bits=2^11;
x = round(bits*sin(2*pi*fundamental*t))/bits;
sfdr(x,fs)
```

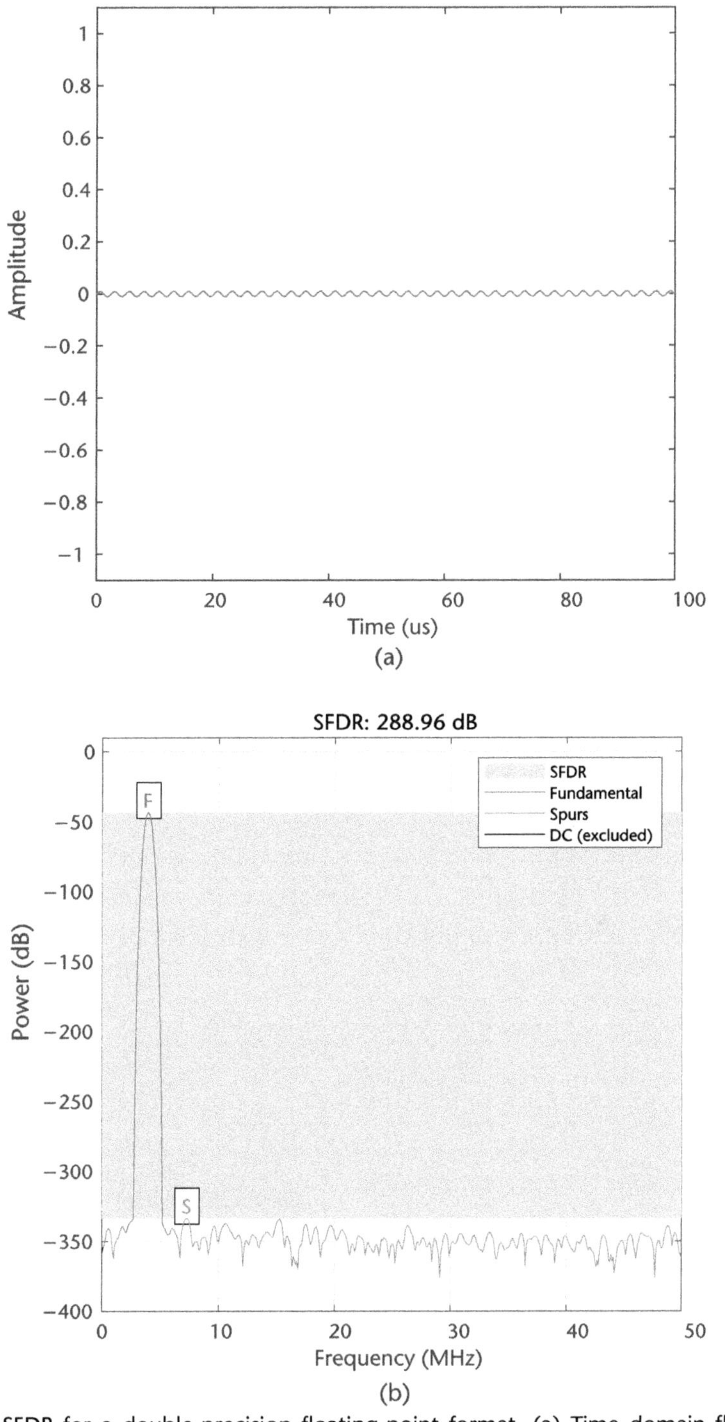

Figure 2.28 SFDR for a double-precision floating-point format. (a) Time domain floating-point representation, and (b) SFDR of double-precision floating-point, 1k points, F_A of 3,959,297 Hz.

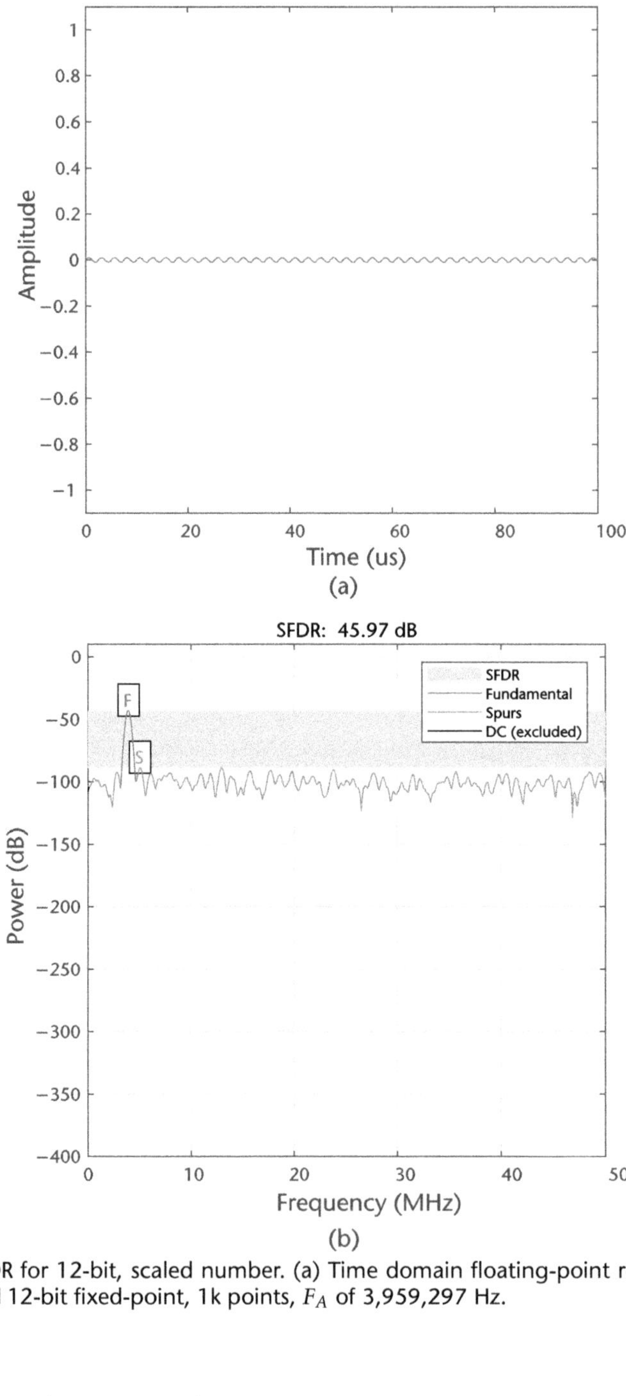

Figure 2.29 SFDR for 12-bit, scaled number. (a) Time domain floating-point representation, and (b) SFDR of scaled 12-bit fixed-point, 1k points, F_A of 3,959,297 Hz.

Rather than rounding up or down in a repeating pattern, we randomly round up or down by applying a ± 0.5 offset to the vector before we round. This is a very simple dither algorithm, but more complex implementations will exist in hardware.

The effects of finite bit length and data correlation should be understood before sending data to the hardware. The hardware will only make effects worse, not better.

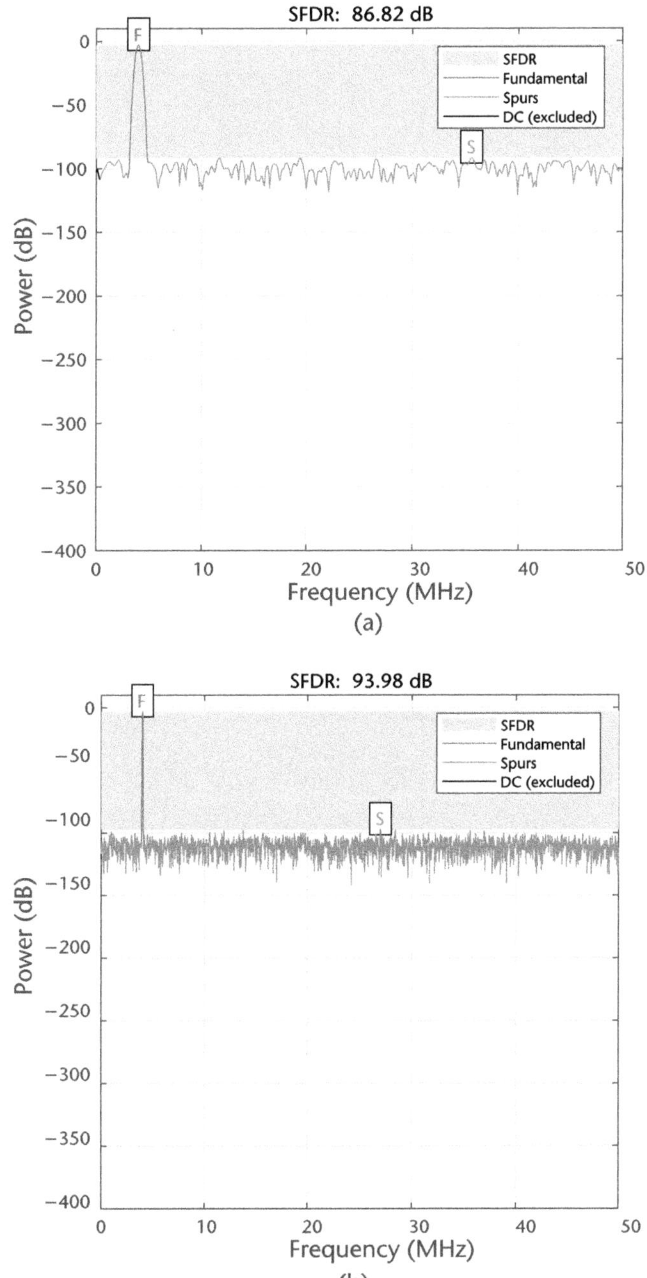

(b)

Figure 2.30 SFDR for 12-bit, scaled number. (a) SFDR of fullscale 12-bit fixed point, 1k points, F_A of 3,959,297 Hz, and (b) SFDR of full scale 12-bit fixed point, 10k points, F_A of 3,959,297 Hz.

2.5.3 Design Trade-offs for Number of Bits, Cost, Power, and So Forth

The most important aspect to remember about both receive chains (I/Q) is the effect of quantization from the ADC itself. That is, an N-bit word represents one of 2^N possible states, and therefore an N-bit ADC (with a fixed reference) an have only 2^N possible digital outputs. The resolution of data converters may be expressed in several different ways: the weight of the least significant bit (LSB), parts per million of full-scale (ppm FS), and millivolts (mV). Different devices, even from the same

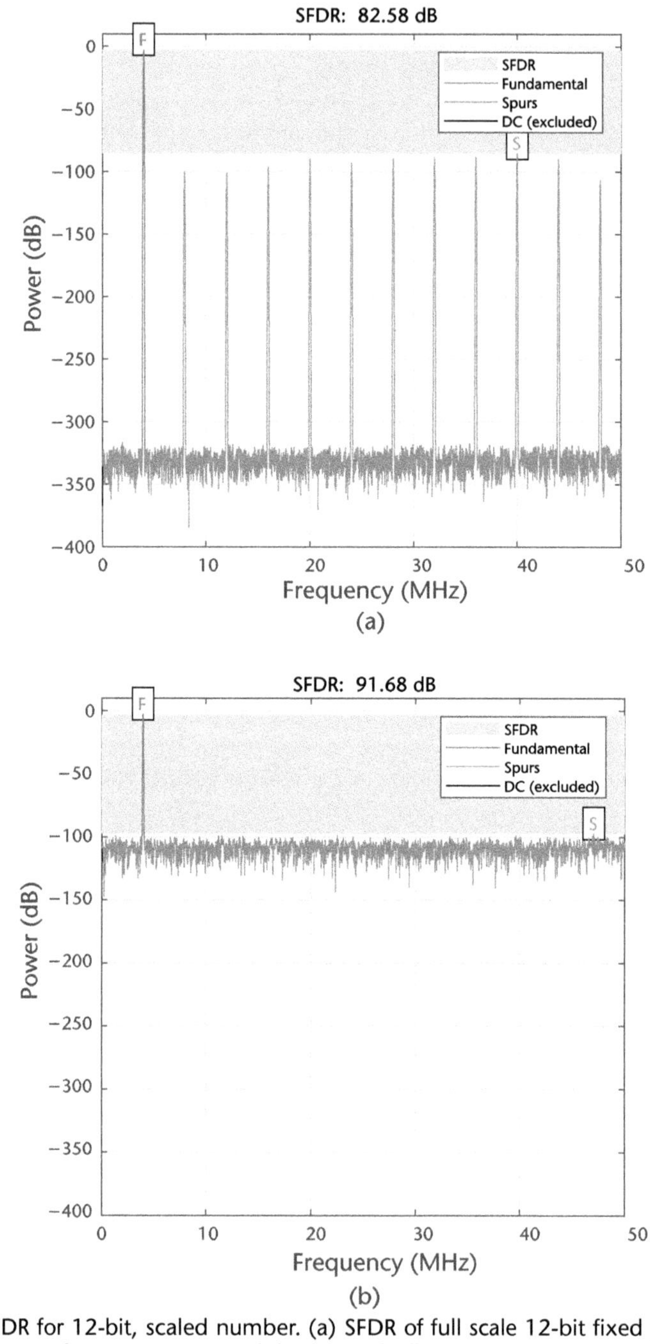

Figure 2.31 SFDR for 12-bit, scaled number. (a) SFDR of full scale 12-bit fixed point, 10k points, without dithering F_A of 4,000,000 Hz, and (b) SFDR of full scale 12-bit fixed point, 10k points, with dithering F_A of 4,000,000 Hz.

manufacturer, will be specified differently. Therefore, converter users must learn to translate between the different types of specifications if they are to compare devices successfully.

The size of the least significant bit for various resolutions for a 10 watt (20 V peak-to-peak) input is shown in Table 2.3.

Code 2.13 SFDR test: **sfdr_test.m**

```
t = 0:deltat:1e-4-deltat;
x = round(bits*sin(2*pi*fundamental*t))/bits;
r = sfdr(x,fs)
sfdr(x,fs);
```

Code 2.14 SFDR test: **sfdr_test.m**

```
fundamental=4000000;
x = round(bits*sin(2*pi*fundamental*t))/bits;
r = sfdr(x,fs)
sfdr(x,fs);
```

Code 2.15 SFDR test: **sfdr_test.m**

```
ran = rand(1,length(t)) - 0.5;
x = round(bits*sin(2*pi*fundamental*t) + ran)/bits;
sfdr(x,fs);
```

Table 2.3 Quantization: The Size of a Least Significant Bit

Resolution (N)	2^N	*Voltage (20 Vpp)*[1]	*PPM FS*	*%FS*	*dBFS*
2-bit	4	5.00 V	250,000	25	−12
4-bit	16	1.25 V	62,500	6.25	−24
6-bit	64	313 mV	15,625	1.56	−36
8-bit	256	78.1 mV	3,906	.391	−48
10-bit	1,024	19.5 mV	977	.097	−60
12-bit	4,096	4.88 mV	244	.024	−72
14-bit	16,384	1.22 mV	61.0	.0061	−84
16-bit	65,536	305 μV	15.2	.0015	−96
18-bit	262,144	76.2 μV	3.81	.00038	−108
20-bit	1,048,576	19.0 μV	.953	.000095	−120
22-bit	4,194,304	4.77 μV	.238	.000024	−132
24-bit	16,777,216	1.19 μV	.0596	.0000060	−144
26-bit	67,108,864	298 nV [1]	.0149	.0000015	−156

[1] 600 nV is the Johnson (thermal) noise in a 10-kHz BW of a 2.2 $k\Omega$ resistor at 25°C.

While a 24-bit converter with −144 dB of performance may sound like a good idea, it is not practical from a power or speed perspective. Althrough many 24-bit ADCs exist, they are not wideband such as the AD7177, which is a state-of-the-art 32-bit converter that is limited to 5 SPS to 10 kSPS output data rate and is not suitable for SDR but is a great solution for things like temperature and pressure measurement, chromatography, or weigh scales [12]. On the other hand, higher-speed ADCs do exist, such as the AD9208, which is a dual-channel 14-Bit 3GSPS ADC, providing an SFDR of 70 dBFS with 9-GHz analog input full-power bandwidth (the sixth Nyquist band). However, the AD9208 power draw is over 3W, which exceeds the entire power consumption of the Pluto SDR while streaming data over USB [13]. Nonetheless, a system based on the AD9208 could provide up to 12 Gbytes/second, which is more data than could be processed by software and would require custom signal processing hardware inside a FPGA. Devices like the AD9208 are used in 60-GHz bands RF bands, where single channels have bandwidths of over 2 GHz. These high-speed and wideband links use the same concepts and techniques described in the remaining text, but possess higher data rates, are more power-hungry, and are much more expensive. Nevertheless, learning the basic techniques of digital communications can be performed in with cost-effective devices such as the Pluto SDR in 20 MHz, or in many cases with much less of bandwidth. With regard to the ADC, in order to enable 12-bit devices to be used for a radio applications such as Pluto SDR, the signal chain for the AD9361 includes programmable analog gain as shown in Figure 2.32. This allows the input to the ADC to be driven to full scale as much as possible, which as we learned in Section 2.4.1 provides the best possible performance.

In a brief recap from operational amplifier theory, two types of gain are associated with amplifiers: signal gain and noise gain. We want to increase the signal but at the same time keep the noise as as low as possible. This is accomplished by increasing the signal in the analog domain before digitizing it with the ADC, as shown in Figure 2.32.

2.5.4 Sigma-Delta Analog-Digital Converters

Sigma-delta (Σ-Δ) analog-digital converters (ADCs) have been known for over 50 years, but only recently has the technology (high-density digital VLSI) existed to manufacture them as inexpensive monolithic integrated circuits. They are now used in many applications where a low-cost, medium-bandwidth, low-power, high-resolution ADC is required. There have been innumerable descriptions of the architecture and theory of Σ-Δ ADCs, but most commence with a deep description of the math, starting at the integrals and go on from there. Since this is not an ADC textbook, we will try to refrain from the mathematical development and explore things based on the previous topics covered in this chapter.

There is nothing particularly difficult to understand about Σ-Δ ADCs. The Σ-Δ ADC contains very simple analog electronics (a comparator, voltage reference, a switch, and one or more integrators and analog summing circuits), and digital computational circuitry. This circuitry consists of a filter, which is generally, but not invariably, a lowpass filter. It is not necessary to know precisely how the filter works to appreciate what it does. To understand how a Σ-Δ ADC works, familiarity with the concepts of oversampling, quantization noise shaping, digital filtering, and decimation is required, all topics covered earlier in this chapter.

Let us consider the technique of oversampling with an analysis in the frequency domain. Where a DC conversion has a quantization error of up to $\frac{1}{2}$ LSB, a sampled data system has quantization noise. A perfect classical N-bit sampling ADC has an RMS quantization noise of $\frac{q}{\sqrt{12}}$ uniformly distributed within the Nyquist band of DC to $\frac{f_s}{2}$, where q is the value of an LSB, as shown in Figure 2.33(a). Therefore, its SNR with a full-scale sine wave input will be $(6.02N + 1.76)$ dB. If the ADC is less than perfect and its noise is greater than its theoretical minimum quantization noise, then its effective resolution will be less than N-bits. Its actual resolution, often known as its effective number of bits (ENOB), will be defined by

$$ENOB = \frac{SNR - 1.76dB}{6.02dB} \tag{2.39}$$

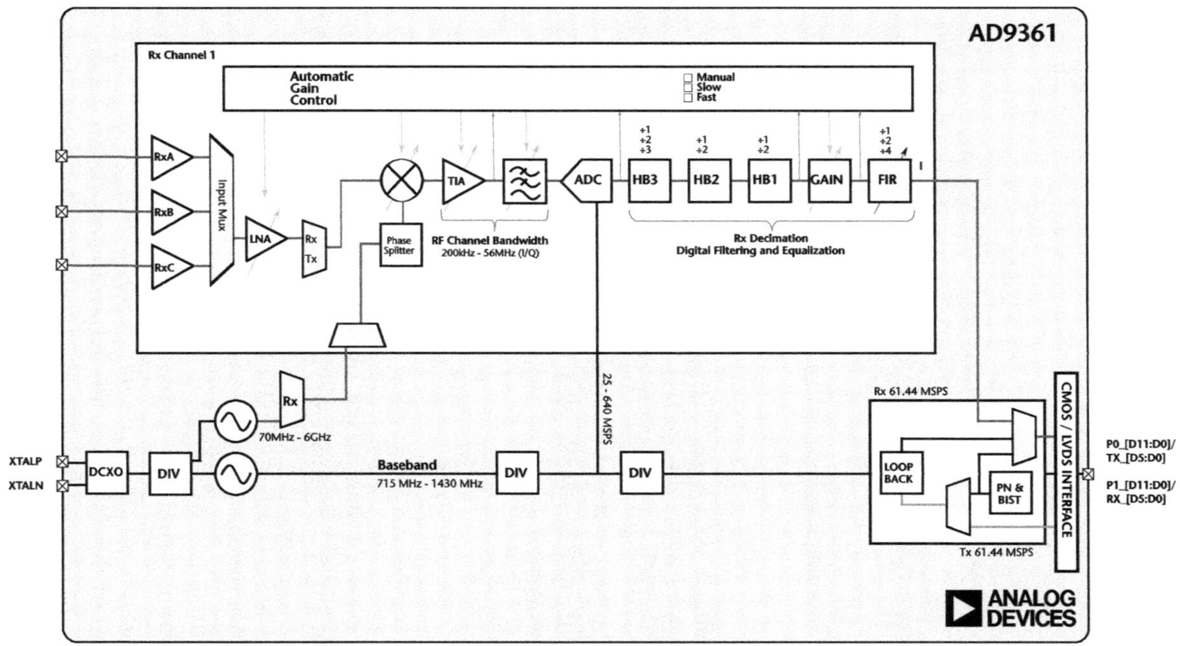

Figure 2.32 Simplified AD9361 receive block diagram.

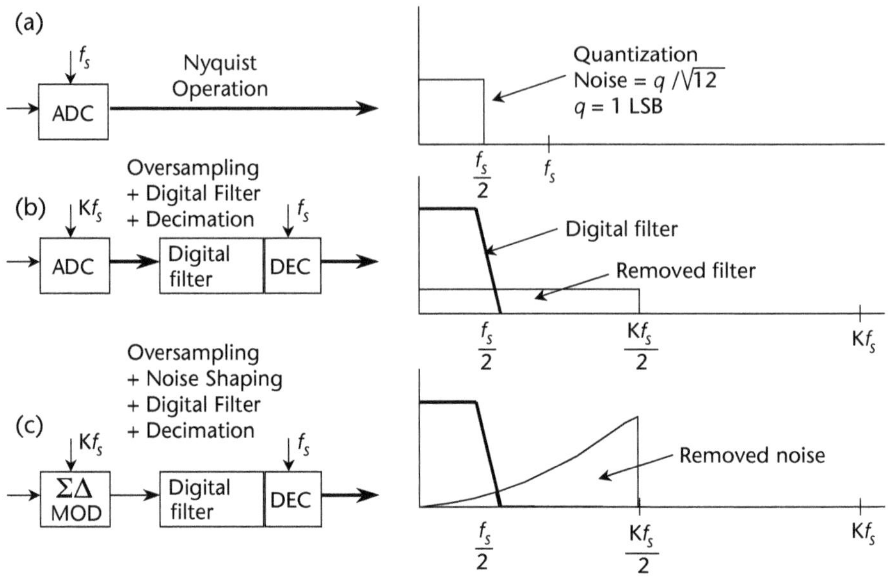

Figure 2.33 Oversampling, digital filtering, noise shaping, and decimation in a Σ-ΔADC.

Practically, ENOB is calculated from measuring signal-to-noise-and-distortion (SINAD, or S/(N + D)), which is the ratio of the RMS signal amplitude to the mean value of the root-sum-square (RSS) of all other spectral components, including harmonics but excluding DC, and correcting for a nonfull-scale input signal [6]. We can modify (2.39) to take into account the full-scale amplitude A_{FS} and the true input amplitude A_{IN} as

$$ENOB = \frac{SINAD - 1.76dB + 20log_{10}\frac{A_{FS}}{A_{IN}}}{6.02dB}. \tag{2.40}$$

If we choose a much higher sampling rate, Kf_s (see Figure 2.33[b]), the RMS quantization noise remains $\frac{q}{\sqrt{12}}$ but the noise is now distributed over a wider bandwidth DC to $\frac{Kf_s}{2}$. If we then apply a digital lowpass filter (LPF) to the output, we can remove much of the quantization noise but do not affect the wanted signal, resulting in an improved ENOB. Therefore, we can accomplished a high-resolution A/D conversion with a low-resolution ADC. The factor K is generally referred to as the oversampling ratio. It should be noted at this point that oversampling has an added benefit in that it relaxes the requirements on the analog antialiasing filter.

Since the bandwidth is reduced by the digital output filter, the output data rate may be lower than the original sampling rate (Kf_s) and still satisfy the Nyquist criterion. This may be achieved by passing every M^{th} result to the output and discarding the remainder. The process is known as decimation by a factor of M. Decimation does not cause any loss of information (see Figure 2.33[b]) as long as the decimation does not violate the Nyquist criterion. For a given input frequency, higher-order analog filters offer more attenuation. The same is true of Σ-Δ modulators, provided certain precautions are taken. By using more than one integration and summing stage in the Σ-Δ modulator, we can achieve higher orders of quantization noise shaping and even better ENOB for a given oversampling ratio as is shown in Figure 2.34.

The actual Σ-Δ ADC found in the AD9363 used in the Pluto SDR is a fourth order, as shown in Figure 2.35 and described in the *Analog Devices Transceiver Support* Simulink model. As can be seen, reality is always a little more complicated than theory or first-order approximations.

2.6 Digital Signal Processing Techniques for SDR

DSP is a field always on the edge of mathematical complexity, computational performance, and growing mobility, influencing communications, medical imagining, radar, entertainment, and even scientific exploration. However, all these fields rely on the concept of translating analog information into digital representations and by some mechanisms processing that data. To do so, engineers and scientists rely on common tools and languages including C and Verilog, all of which enable manipulation of digital information in an efficient and procedural way. Regardless of the language, many important DSP software issues are specific to hardware, such as truncation error, bit patterns, and computational speed and efficiency of processors [14]. For now, we will mostly ignore those issues and focus on the algorithmic issues of signal processing and discuss the commonly used algorithms.

2.6.1 Discrete Convolution

Convolution is a mathematical tool of combining two signals to form a third signal, and forms the foundation for all DSP. Using the strategy of impulse decomposition, systems are described by a signal called the *impulse response*. Convolution is important because it relates the three signals of interest: the input signal, the output signal, and the impulse response.

Figure 2.36 presents the notation of convolution as applied to linear systems. A discrete sampled input signal, $x[n]$, enters a linear system with an impulse

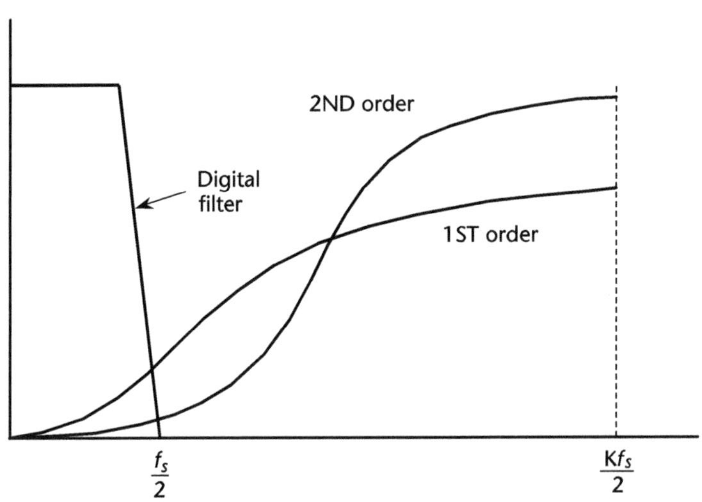

Figure 2.34 Σ-Δ modulators shape quantization noise.

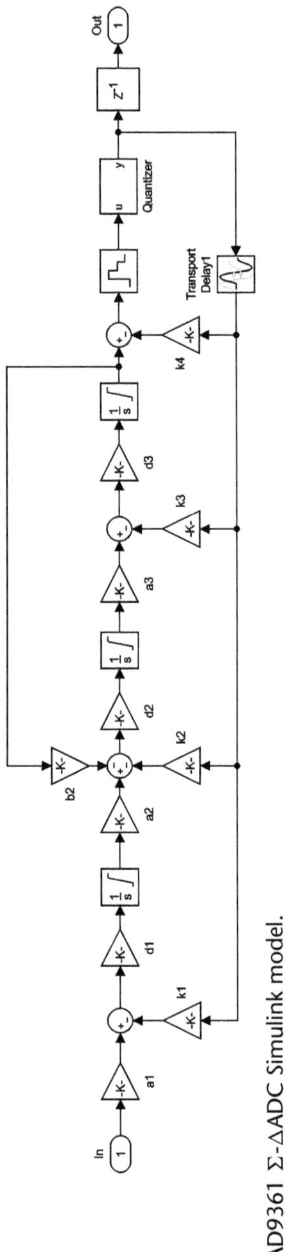

Figure 2.35 Σ-ΔADC Simulink model.

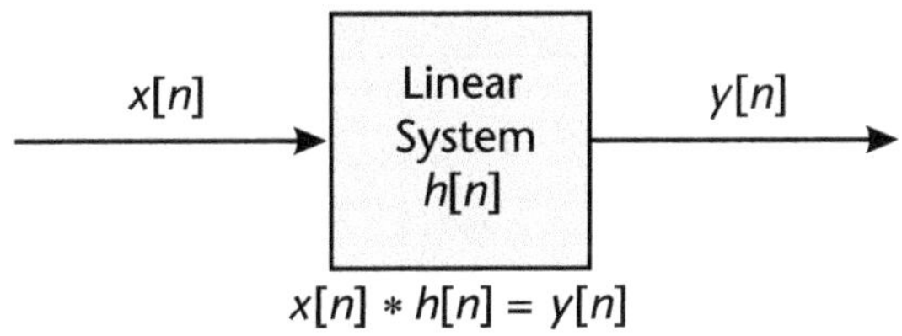

Figure 2.36 How convolution is used in DSP. The output signal from a linear system is equal to the input signal convolved with the system's impulse response.

response, $h[n]$ resulting in an output signal, $y[n]$. Expressed in words, the input signal convolved with the impulse response is equal to the output signal. As denoted in (2.41), convolution is represented by the $*$ operator. It is unfortunate that most programming languages, such as MATLAB, use the star to indicate multiplication and use special functions like MATLAB's `conv` function to indicate convolution. A star in a computer program means multiplication, while a star here notes convolution.

Fundamentally, the mathematics of convolution consists of several multiplications and additions. If $x[n]$ is an N point signal running from data sample 0 to $N-1$, and $h[n]$ is an M point signal running from 0 to $M-1$, the convolution of the two $y[n] = x[n] * h[n]$, is an $N + M - 1$ point signal running from 0 to $N + M - 2$, given by

$$y[i] = \sum_{j=0}^{M-1} h[j] \times x[i-j] = h[i] * x[i], \tag{2.41}$$

This equation is called the *convolution sum*. It allows each point in the output signal to be calculated independently of all other points in the output signal. The index, i, determines which sample in the output signal is being calculated. The use should not be confused by the n in $y[n] = x[n] * h[n]$, which is merely a placeholder to indicate that some variable is the index into the array. An implementation of the convolution sum of two vectors in MATLAB is shown in Code 2.16.

As used in signal processing, convolution can be understood in two separate ways. The first looks at convolution from the viewpoint of the input signal. This involves analyzing how each sample in the input signal contributes to many points in the output signal. The second way looks at convolution from the viewpoint of the output signal. This examines how each sample in the output signal has received information from many points in the input signal. Keep in mind that these two perspectives are different ways of thinking about the same mathematical operation. The first viewpoint is important because it provides a conceptual understanding of how convolution pertains to signal processing. The second viewpoint describes the mathematics of convolution. This typifies one of the most difficult tasks you will encounter in the signal processing field, making your conceptual understanding fit with the jumble of mathematics used to communicate the ideas.

Figure 2.37 shows convolution being used for lowpass and highpass filtering, which we will cover in more detail in Section 2.6.4. The example input signal is the sum of two components: three cycles of a sine wave (representing a high frequency),

Code 2.16 Convolution: `my_convolution.m`

```
% Receive two vectors and return a vector resultant of
% convolution operation
function conv = simple_conv(f, g)
    % Transform the vectors f and g in new vectors with the same length
    F = [f,zeros(1,length(g))];
    G = [g,zeros(1,length(f))];

    % FOR Loop to put the result of convolution between F and G vectors
    % in a new vector C. According to the convolution operation
    % characteristics, the length of a resultant vector of convolution
    % operation between two vector is the sum of vectors length minus 1
    for i=1:length(g)+length(f)-1
        % Create a new vector C
        C(i) = 0;
        % FOR Loop to walk through the vector F ang G
        for j=1:length(f)
           if(i-j+1>0)
               C(i) = C(i) + F(j) * G(i-j+1);
           end
        end
    end
    out = C;
end
```

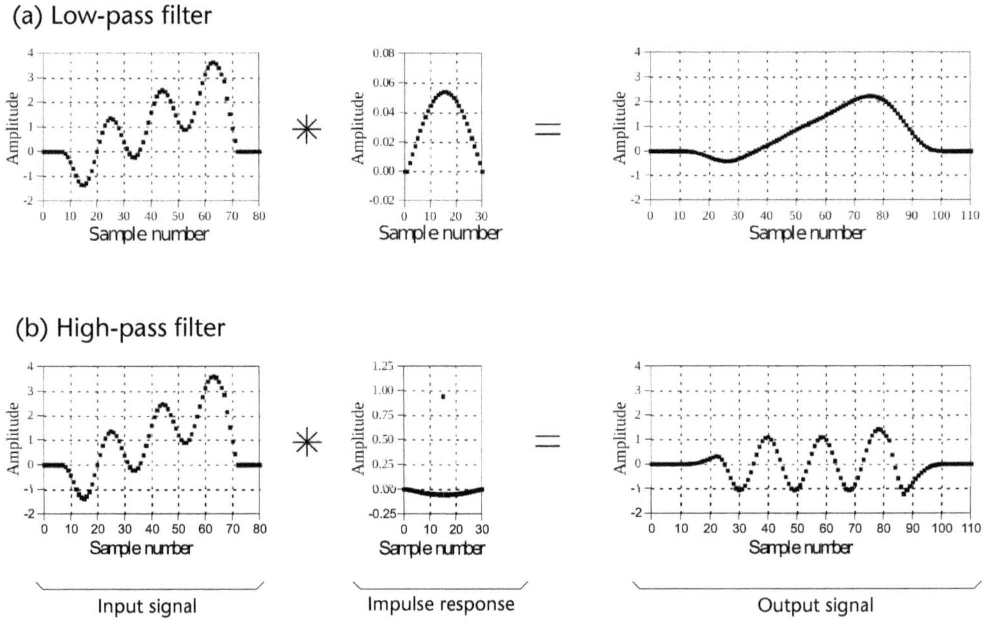

Figure 2.37 Examples of (a) lowpass and (b) highpass filtering using convolution. In this example, the input signal is a few cycles of a sine wave plus a slowly rising ramp. These two components are separated by using properly selected impulse responses.

plus a slowly rising ramp (composed of low frequencies). In (a), the impulse response for the lowpass filter is a smooth arch, resulting in only the slowly changing ramp waveform being passed to the output. Similarly, the highpass filter, (b), allows only the more rapidly changing sinusoid to pass.

2.6.2 Correlation

Cross-correlation and *autocorrelation* are two important concepts in SDR. Cross-correlation is a measure of similarity of two series as a function of the displacement of one relative to the other.

The concept of correlation can best be presented with an example. Figure 2.38 shows the key elements of a radar system. A specially designed antenna transmits a short burst of radio wave energy in a selected direction. If the propagating wave strikes an object, such as the helicopter in this illustration, a small fraction of the energy is reflected back toward a radio receiver located near the transmitter. The transmitted pulse is a specific shape that we have selected, such as the triangle shown in this example. The received signal will consist of two parts: (1) a shifted and scaled version of the transmitted pulse, and (2) random noise, resulting from interfering radio waves, thermal noise in the electronics, and so forth. Since radio signals travel at a known rate, the speed of light, the shift between the transmitted and received

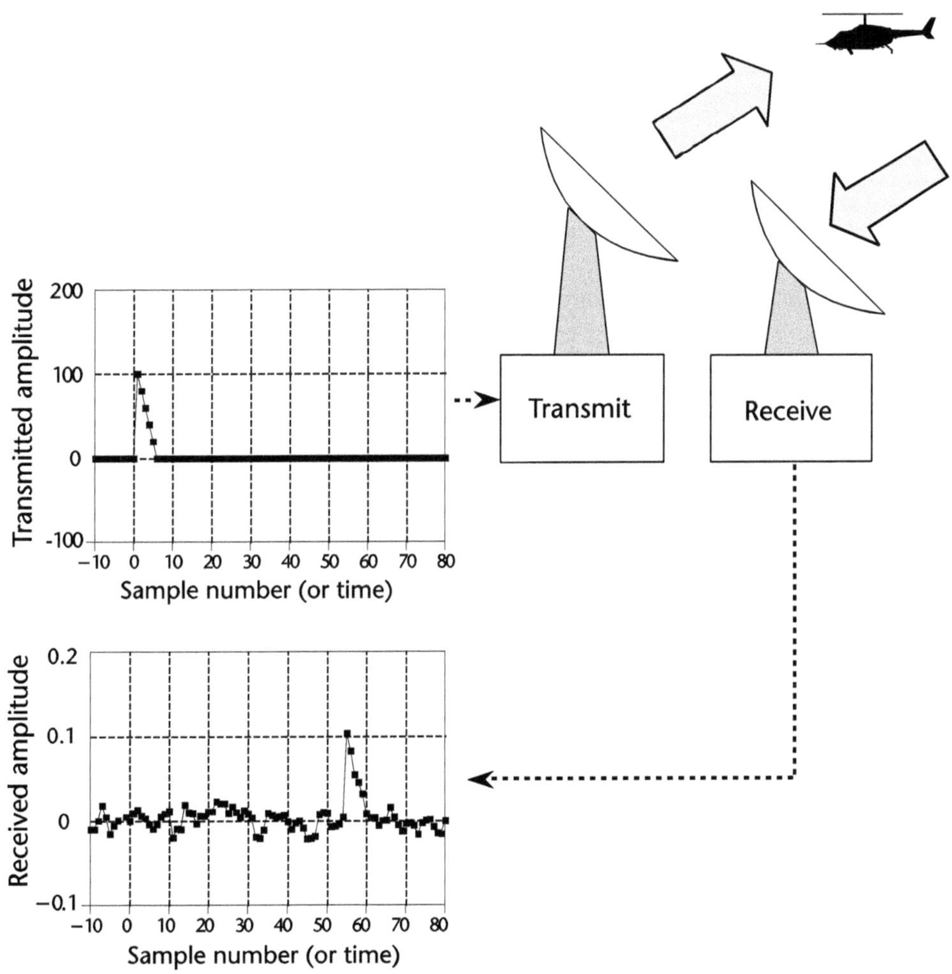

Figure 2.38 Key elements of a radar system. Like other echo location systems, radar transmits a short pulse of energy that is reflected by objects being examined. This makes the received waveform a shifted version of the transmitted waveform, plus random noise. Detection of a known waveform in a noisy signal is the fundamental problem in echo location. The answer to this problem is *correlation*.

pulse is a direct measure of the distance to the object being detected given a signal of some known shape. The challenge is determine the best way which signal occurs in another signal. Correlation is the solution to this problem.

Correlation is a mathematical operation that is very similar to convolution. Just as with convolution, correlation uses two signals to produce a third signal. This third signal is called the cross-correlation of the two input signals. If a signal is correlated with itself, the resulting signal is instead called the autocorrelation. The amplitude of each sample in the cross-correlation signal is a measure of how much the received signal resembles the target signal at that location. This means that a peak will occur in the cross-correlation signal for every target signal that is present in the received signal. In other words, the value of the cross-correlation is maximized when the target signal is aligned with the same features in the received signal.

One of the optimal techniques for detecting a known waveform in random noise is correlation. That is, the peak is higher above the noise using correlation than can be produced by any other linear system. (To be perfectly correct, it is only optimal for random white noise.) Using correlation to detect a known waveform is frequently called *matched filtering*. More on this in Section 4.7.1.

For discrete functions f and g, the cross-correlation is defined as

$$(f \star g)[n] = \sum_{j=-\infty}^{\infty} f^*[m] \times g[m+n], \tag{2.42}$$

where h^* denotes the complex conjugate of h. The cross-correlation of functions f(t) and g(t) is equivalent to the convolution of f*(â˘’t) and g(t). That is

$$(f \star g)[n] = f^*(-t) * g(t) \tag{2.43}$$

Do not let the mathematical similarity between convolution and correlation fool you; they represent very different signal processing concepts. Convolution is the relationship between a system's input signal, output signal, and impulse response. Correlation is a way to detect a known waveform in a noisy background. The similar mathematics is a convenient coincidence that allows for algorithmic optimizations.

2.6.3 Z-Transform

In Section 2.1.1, we have introduced the Fourier transform, which deals with continuous-time signals on frequency domain. Since we are focusing on digital filter design in this section, where discrete-time signals are involved, we need to introduce a new type of transform; namely, the z-transform.

The z-transform of a discrete-time signal $x[n]$ is defined as the power series:

$$X(z) = \sum_{n=-\infty}^{\infty} x[n]z^{-n}, \tag{2.44}$$

where z is a complex variable [1].

The z-transform is used to analyze discrete-time systems. Its continuous-time counterpart is the Laplace transform, defined as following:

$$X(s) = \int_{-\infty}^{\infty} x(t)e^{-st}dt, \tag{2.45}$$

where t is the time variable in seconds across the time domain and $s = \sigma + j\omega$ is a complex variable. When evaluated along the $j\omega$ axis (i.e., $\sigma = 0$), the Laplace transform reduces to the Fourier transform defined in (2.2). Thus, the Laplace transform generalizes the Fourier transform from the real line (the frequency axis $j\omega$) to the entire complex plane.

According to Section 2.2.2, we know that if a continuous-time signal $x(t)$ is uniformly sampled, its sampling signal $x_s(t)$ can be expressed as

$$x_s(t) = \sum_{n=-\infty}^{\infty} x(nT)\delta(t - nT), \tag{2.46}$$

where T is the sampling interval. If we take the Laplace transform of both sides, we will get

$$X_s(s) = \int_{-\infty}^{\infty} x_s(t)e^{-st}dt = \int_{-\infty}^{\infty} \left[\sum_{n=-\infty}^{\infty} x(nT)\delta(t - nT)\right] e^{-st}dt. \tag{2.47}$$

Since integration and summation are both linear operators, we can exchange their order. Then, based on the sampling property of the delta function, we can further get

$$X_s(s) = \sum_{n=-\infty}^{\infty} x(nT)\left[\int_{-\infty}^{\infty} \delta(t - nT)e^{-st}dt\right] = \sum_{n=-\infty}^{\infty} x(nT)e^{-snT}. \tag{2.48}$$

Let $z = e^{sT}$, or $s = \frac{1}{T}\ln z$, then (2.48) becomes

$$X(z) = \sum_{n=-\infty}^{\infty} x(nT)z^{-n}. \tag{2.49}$$

Since T is the sampling interval, $x(nT) = x[n]$. The equation above can be further written as

$$X(z) = \sum_{n=-\infty}^{\infty} x[n]z^{-n}, \tag{2.50}$$

which is exactly the definition of z-transform in (2.44). Therefore, the z-transform and the Laplace transform can be connected by

$$z = e^{sT}, \tag{2.51}$$

or

$$s = \frac{1}{T}\ln(z). \tag{2.52}$$

According to (2.44), we know that z-transform is the series of z^{-1}. Actually, the z-transform has no meaning unless the series converge. Given a limitary-amplitude sequence $x[n]$, the set of all the z values that makes its z-transform converge is called region of convergence (ROC). Based on the theory of series, these z-values must satisfy

$$\sum_{n=-\infty}^{\infty} \left| x[n] z^{-n} \right| < \infty. \tag{2.53}$$

The frequently used z-transform pairs and their region of convergence are listed in Table 2.4.

When discussing a linear time-invariant system, the z-transform of its system impulse response can be expressed as the ratio of two polynomials:

$$H(z) = \frac{b_m z^m + b_{m-1} z^{m-1} + \cdots + b_1 z + b_0}{a_n z^n + a_{n-1} z^{n-1} + \cdots + a_1 z + a_0} = \frac{B(z)}{A(z)}, \tag{2.54}$$

where the roots of $A(z) = 0$ are called the *poles* of the system and the roots of $B(z) = 0$ are called the *zeros* of the system. It is possible that the system can have multiple poles and zeros.

If we factorize the numerator $B(z)$ and denominator $A(z)$, (2.54) can be written as:

$$H(z) = C \frac{(z - z_1)(z - z_2) \cdots (z - z_m)}{(z - p_1)(z - p_2) \cdots (z - p_n)}, \tag{2.55}$$

where C is a constant, $\{p_k\}$ are all the poles, and $\{z_k\}$ are all the zeros. It will help us draw the pole-zero plot of $H(z)$.

Suppose we have a linear time-invariant system whose system impulse response is defined as

$$h[n] = n^2 a^n u[n]. \tag{2.56}$$

According to Table 2.4, its z-transform is as follows:

$$H(z) = a \frac{z(z + a)}{(z - a)^3}. \tag{2.57}$$

Comparing (2.57) with (2.55), we can easily get that this system has two zeros, $z_1 = 0$ and $z_2 = -a$, and three poles, $p_1 = p_2 = p_3 = a$. Therefore, its pole-zero plot is shown in Figure 2.39.

Table 2.4 Z-Transform Table: Selected Pairs[1]

$x[n]$	$X(z)$	*Region of Convergence*
$\delta[n]$	1	all z
$a^n u[n]$	$\frac{z}{z-a}$	$\lvert z\rvert > \lvert a\rvert$
$n a^n u[n]$	$\frac{az}{(z-a)^2}$	$\lvert z\rvert > \lvert a\rvert > 0$
$n^2 a^n u[n]$	$\frac{az(z+a)}{(z-a)^3}$	$\lvert z\rvert > a > 0$
$\left(\frac{1}{a^n} + \frac{1}{b^n}\right) u[n]$	$\frac{az}{az-1} + \frac{bz}{bz-1}$	$\lvert z\rvert > \max(\frac{1}{\lvert a\rvert}, \frac{1}{\lvert b\rvert})$
$a^n u[n] \sin(\omega_0 n)$	$\frac{az \sin \omega_0}{z^2 - 2az \cos \omega_0 + a^2}$	$\lvert z\rvert > a > 0$
$a^n u[n] \cos(\omega_0 n)$	$\frac{z(z - a \cos \omega_0)}{z^2 - 2az \cos \omega_0 + a^2}$	$\lvert z\rvert > a > 0$
$e^{an} u[n]$	$\frac{z}{z-e^a}$	$\lvert z\rvert > e^{-a}$
$e^{-an} u[n] \sin(\omega_0 n)$	$\frac{ze^a \sin \omega_0}{z^2 e^{2a} - 2ze^a \cos \omega_0 + 1}$	$\lvert z\rvert > e^{-a}$
$e^{-an} u[n] \cos(\omega_0 n)$	$\frac{ze^a (ze^a - \cos \omega_0)}{z^2 e^{2a} - 2ze^a \cos \omega_0 + 1}$	$\lvert z\rvert > e^{-a}$

[1] From [15]

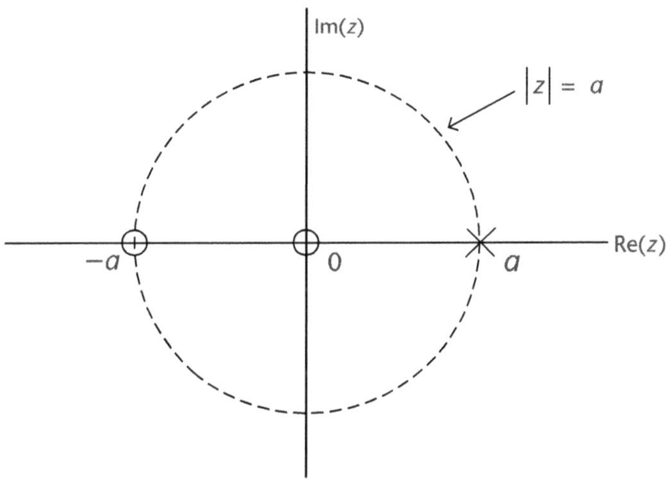

Figure 2.39 Pole-zero plot of the system defined in (2.57). Poles are denoted using crossings, and zeros are denoted using circles. The region of convergence of this system is the region outside the circle $z = |a|$.

There are several properties of the z-transform that are useful when studying signals and systems in the z-transform domain. Since these properties are very similar to those of the Fourier transform introduced in Section 2.1.1, we list them in Table 2.5 without further discussion.

2.6.4 Digital Filtering

When doing signal processing, we usually need to get rid of the noise and extract the useful signal. This process is called filtering, and the device employed is called filter, which discriminates, according to some attribute of the objects applied at its input, what passes through. A typical filter is a frequency-selective circuit. If noise and useful signal possess different frequency distributions and are present together at input of the filter, then by applying this circuit, the noise will be attenuated or even eliminated while useful signal is retained.

Filters can be classified from different aspects. For example, according to its frequency response, filter can be classified as lowpass, highpass, bandpass and bandstop. According to the signal it deals with, a filter can be classified as a analog filter or a digital filter [1]. Specifically, an analog filter deals with continuous-time signals, while a digital filter deals with discrete-time signals. This section will focus on digital filters. The ideal magnitude response characteristics of these types of filters are shown in Figure 2.40. According to Figure 2.40, the magnitude response characteristics of an ideal filter can be generalized as follows: In pass-band, the magnitude is a constant, while in stop-band, the magnitude falls to zero. However, in reality, this type of ideal filter cannot be achieved, so a practical filter is actually the optimum approximation of an ideal filter.

In order to perform digital filtering, input and output of a digital system must both be discrete-time series. If the input series is $x[n]$, the impulse response of the filter is $h[n]$, then the output series $y[n]$ will be

$$y[n] = x[n] * h[n]. \tag{2.58}$$

Table 2.5 Z-Transform Properties[1]

Property	*Time Signal*	*Z-Transform Signal*
Linearity	$\sum_{m=1}^{N} a_m x_m(t)$	$\sum_{m=1}^{N} a_m X_m(z)$
Symmetry	$x[-n]$	$X(z^{-1})$
Shifting	$x[n-m]$	$z^{-m}X(z)$
Scaling	$a^n x[n]$	$X\left(\frac{z}{a}\right)$
Derivative	$nx[n]$	$-z\frac{dX(z)}{dz}$
Integration	$\sum_{m=-\infty}^{n} x[m]$	$\frac{z}{z-1}X(z)$
Time convolution	$x[n] * h[n]$	$X(z)H(z)$
Frequency convolution	$x[n]h[n]$	$\frac{1}{2\pi j}\int X(v)H\left(\frac{z}{v}\right)\frac{dv}{v}$

[1] Based on [15]. Suppose the time signal is $x[n]$, and its z-transform signal is $X(z)$.

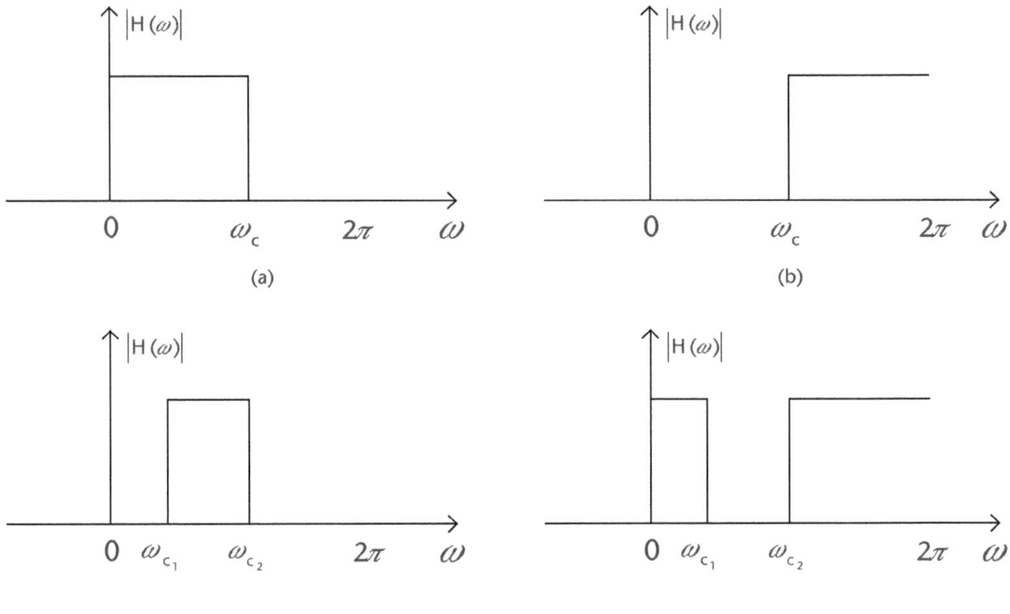

Figure 2.40 Ideal magnitude response characteristics of four types of filters on the frequency range $[0, 2\pi]$. (a) Lowpass filter, (b) highpass filter, (c) bandpass filter, where $(\omega_{c_1}, \omega_{c_2})$ is passband, and (d) bandstop filter, where $(\omega_{c_1}, \omega_{c_2})$ is stopband.

According to the time convolution property in Table 2.5, on the frequency domain, (2.58) is equivalent to

$$Y(z) = X(z)H(z), \tag{2.59}$$

where $X(z)$ and $Y(z)$ are the z-transforms of the input and output series, $x[n]$ and $y[n]$, and $H(z)$ is the z-transform of $h[n]$.

Since ideal brick wall filters are not achievable in practice, we limit our attention to the class of linear time-invariant systems specified by the difference equation [1]:

$$y[n] = -\sum_{k=1}^{N} a_k y[n-k] + \sum_{k=0}^{M} b_k x[n-k], \tag{2.60}$$

where $y[n]$ is the current filter output, the $y[n-k]$ are previous filter outputs, and the $x[n-k]$ are current or previous filter inputs. This system has the following frequency response:

$$H(z) = \frac{\sum_{k=0}^{M} b_k e^{-z}}{1 + \sum_{k=1}^{N} a_k e^{-z}}, \tag{2.61}$$

where the $\{a_k\}$ are the filter's feedback coefficients corresponding to the poles of the filter, and the $\{b_k\}$ are the filter's feed-forward coefficients corresponding to the zeros of the filter, and N is the filter's order.

The basic digital filter design problem is to approximate any of the ideal frequency response characteristics with a system that has the frequency response (2.61), by properly selecting the coefficients $\{a_k\}$ and $\{b_k\}$ [1].

There are two basic types of digital filters, finite impulse response (FIR) and infinite impulse response (IIR) filters. When excited by an unit sample $\delta[n]$, the impulse response $h[n]$ of a system may last a finite duration, as shown in Figure 2.41(a), or forever even before the input is applied, as shown in Figure 2.41(b). In the former case, the system is finite impulse response, and in the latter case, the system is infinite impulse response.

An FIR filter of length M with input $x[n]$ and output $y[n]$ can be described by the difference equation [1]:

$$y[n] = b_0 x[n] + b_1 x[n-1] + \cdots + b_{M-1} x[n-M+1] = \sum_{k=0}^{M-1} b_k x[n-k], \tag{2.62}$$

where the filter has no feedback coefficients $\{a_k\}$, so $H(z)$ has only zeros.

IIR filter has been defined in (2.60), which has one or more nonzero feedback coefficients $\{a_k\}$. Therefore, once the filter has been excited with an impulse, there is always an output.

2.6.4.1 Case Study: Cascaded Integrator-Comb Filters

Cascaded integrator-comb filters (CIC filters) play an important role in the SDR hardware. They were invented by Eugene B. Hogenauer and are a class of FIR filters used in multirate signal processing. The CIC filter finds applications in interpolation and decimation. However, unlike most FIR filters, it has a decimator or interpolator built into the architecture [16].

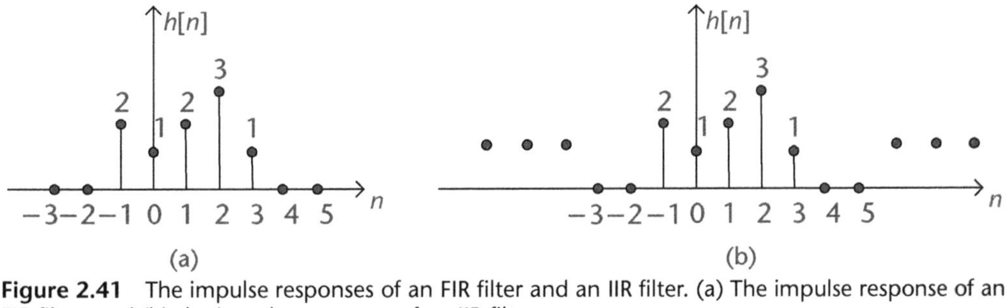

Figure 2.41 The impulse responses of an FIR filter and an IIR filter. (a) The impulse response of an FIR filter, and (b) the impulse response of an IIR filter.

A CIC filter consists of one or more integrator and comb filter pairs. In the case of a decimating CIC, the input signal is fed through one or more cascaded integrators, and then a downsampler, followed by one or more comb sections, whose number equals to the number of integrators. An interpolating CIC is simply the reverse of this architecture, with the downsampler replaced with a upsampler, as shown in Figure 2.42.

To illustrate a simple form of a comb filter, consider a moving average FIR filter described by the difference equation:

$$y[n] = \frac{1}{M+1}\sum_{k=0}^{M} x[n-k]. \tag{2.63}$$

The system function of this FIR filter is

$$H(z) = \frac{1}{M+1}\sum_{k=0}^{M} z^{-k} = \frac{1}{M+1}\frac{[1-z^{-(M+1)}]}{(1-z^{-1})}. \tag{2.64}$$

Suppose we replace z by z^L, where L is a positive integer; then the resulting comb filter has the system function:

$$H_L(z) = \frac{1}{M+1}\frac{[1-z^{-L(M+1)}]}{(1-z^{-L})}, \tag{2.65}$$

where L is decimation or interpolation ratio, M is number of samples per stage, usually 1 or 2.

This filter has zeros on the unit circle at

$$z_k = e^{j2\pi k/L(M+1)}, \tag{2.66}$$

for all integer value of k except $k = 0, L, 2L, ..., ML$, as shown in Figure 2.43.

The common form of the CIC filter usually consists of several stages, then the system function for the composite CIC filter is

$$H(z) = H_L(z)^N = \left(\frac{1}{M+1}\frac{1-z^{-L(M+1)}}{1-z^{-L}}\right)^N, \tag{2.67}$$

where N is number of stages in the composite filter.

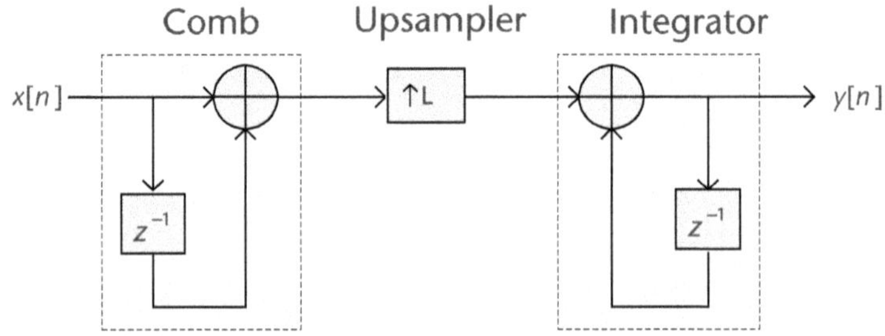

Figure 2.42 The structure of an interpolating cascaded integrator-comb filter [17], with input signal $x[n]$ and output signal $y[n]$. This filter consists of a comb and integrator filter pair, and an upsampler with interpolation ratio L.

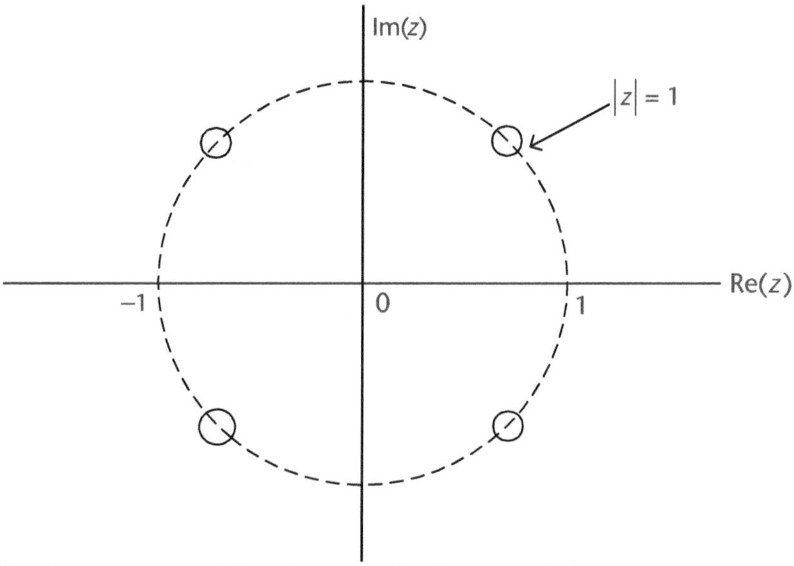

Figure 2.43 The zeros of a CIC filter defined in (2.65), where all the zeros are on the unit circle.

Characteristics of CIC filters include linear phase response and utilizing only delay and addition and subtraction. In other words, it requires no multiplication operations, thus making it suitable for hardware implementation.

2.6.4.2 Case Study: FIR Halfband Filter

FIR *halfband filters* are also widely used in multirate signal processing applications when interpolating/decimating. Halfband filters are implemented efficiently in polyphase form because approximately half of its coefficients are equal to zero. Halfband filters have two important characteristics, the passband and stopband ripples must be the same, and the passband-edge and stopband-edge frequencies are equidistant from the halfband frequency $\frac{f_s}{4}$.

For example the Pluto SDR has multiple programmable halfband filters in the receive and transmit chains. The RX HB3/DEC3 provides the choice between two different fixed-coefficient decimating filters, decimating by a factor of 2, 3, or 1 (bypassed). The input to the filter (the output of the ADC) is 2^4, or 16 values.

When the RX HB3 filter is used, the decimation factor is set to 2, and the following coefficients are used : [1, 4, 6, 4, 1]. Note that the full- scale range for the RX HB3 filter is 16 (2^4). When the RX DEC3 filter is used, the decimation factor is set to 3. and the following coefficients: [55, 83, 0, -393, -580, 0, 1914, 4041, 5120, 4041, 1914, 0, -580, -393, 0, 83, 55]. The full-scale range for the RX DEC3 filter is 16384 (2^{14}).

2.7 Transmit Techniques for SDR

In Section 2.2, it was described how an analog signal is converted to a digital signal using an ADC, as illustrated in Figure 2.5. Although these digital signals, which consist of 0 and 1, can be processed with various digital signal processing techniques, these digital signals cannot be directly used for transmission. These signals must be first conditioned then converted back into an analog signal DAC.

It can be seen from this that the extra decimation will allow bit growth and extra fidelity to gain in the system.

In Figure 2.44(a) a continuous analog sine wave is shown, which has been sampled at sample rate f_s. These samples goes through a quantizer, described in Section 2.2, which provides the output as shown in Figure 2.44(b). In this example,

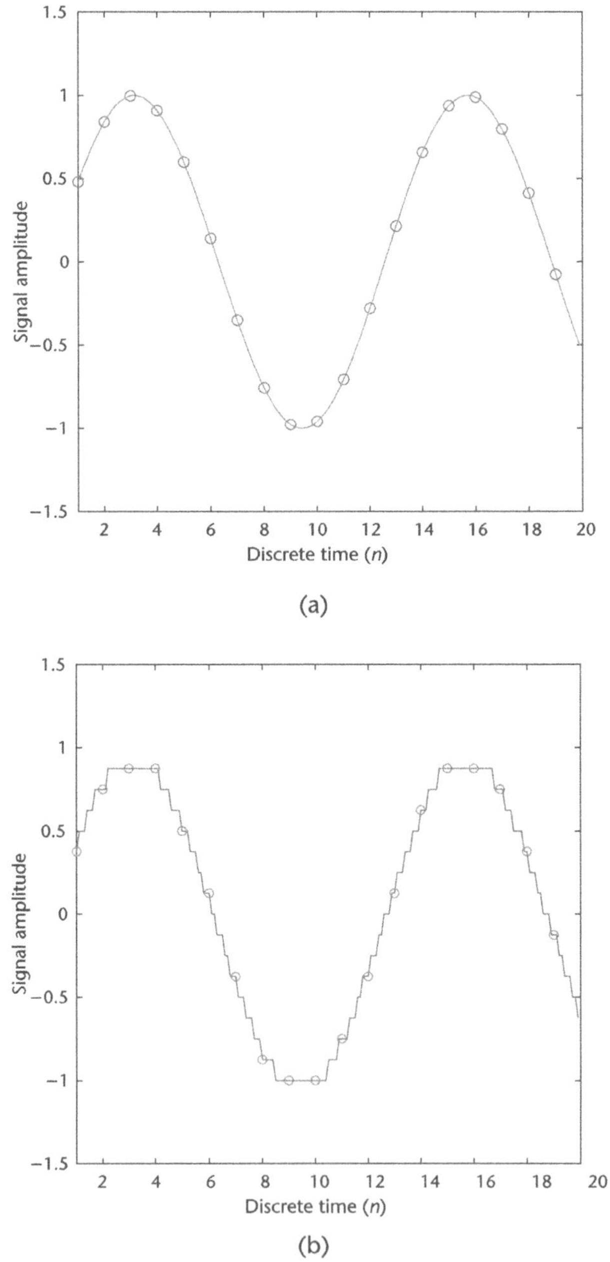

Figure 2.44 Time domain. (a) Continuous analog sine wave: time domain, and (b) quanitized analog sine wave: time domain.

a 4-bit converter is used, which only provides 16 states (to make things easier to visualize). It is natural that many people want to connect the samples with smooth lines. However, the true waveform does have multiple discrete steps between the samples, as shown in Figure 2.44(b). Those values between samples are ignored and are not passed out of the ADC.

What comes out the ADC and is processed by the digital signal processing algorithms and eventually the digital to analog converter have no need for waveforms to be smooth. Therefore, they experience what is shown in Figure 2.45(a). Signals are not continuous and are not smooth. This is adequate as long as the topics described in Section 2.2.3 are understood and do not violate the Nyquist sampling theorem. The only time this is a problem is when we want to actually convert the sampled data back into the continuous analog domain. The bandwidth of the continuous analog domain is infinite and does not stop at $\frac{f_s}{2}$, as shown in Figure 2.45(b). When the time domain data is consider beyond the digital limits, aliases of signals are still visible.

This is why we traditionally have two filters in the transmit portion of a SDR, a digital transmit or *pulse-shaping* filter, which changes each symbol in the digital message into a digital pulse stream for the DAC. This is followed by an *analog reconstruction filter*, which removes aliasing caused by the DAC [9]. This process is shown in Figure 2.46.

2.7.1 Analog Reconstruction Filters

In Section 2.5.4, it was discussed how oversampling can ease the requirements on the antialiasing filter and how a sigma-delta ADC has this inherent advantage. In a DAC-based system, the concept of interpolation can be used in a similar manner with the analog reconstruction filter. This is common in digital audio CD players, where the basic update rate of the data from the CD is about 44.1 kSPS. As described in Section 2.2.5, zeros are inserted into the data, which is passed through a digital filter thereby increasing the effective sample update rate to four times, eight times, or sixteen times the original rate. The high oversampling rate moves the image frequencies higher, allowing a less complex filter with a wider transition band.

The same concept can be applied to a high-speed DAC. Suppose that a traditional DAC is driven at an input word rate of 30 MSPS, as shown in Figure 2.47 A; the DAC output frequency f_s is 10 MHz. The image frequency component at 20 MHz must be attenuated by the analog reconstruction filter, and the transition band of the filter is 10 to 20 MHz. The image frequency must be attenuated by 60 dB. Therefore, the filter must cover a passband of 10 MHz with 60-dB stopband attenuation over the transition band lying between 10 and 20 MHz (one octave). An analog Butterworth filter design gives 6 dB attenuation per octave for each pole. Therefore, a minimum of 10 poles are required to provide the desired attenuation. The necessary filter becomes even more complex as the transition band becomes narrower.

Next, let us assume that we increase the DAC update rate to 60 MSPS and interpolate the original data samples by 2, resulting in a data stream at 60 MSPS. The response of the DAC relative to the two-times oversampling frequency is shown in Figure 2.47 B. The analog reconstruction filter transition zone is now 10 to

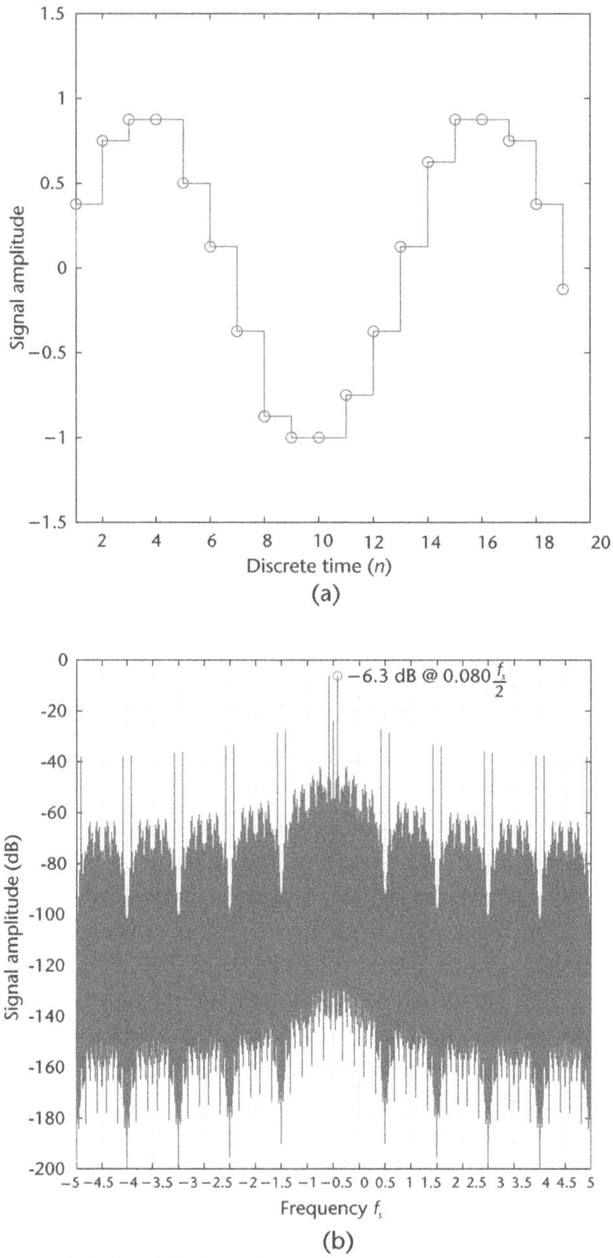

Figure 2.45 (a) Upsampled, band-limited, sine wave: time domain, and (b) band-limited random data: fourier domain.

50 MHz (the first image occurs at $2f_c - f_o = 60 - 10 = 50$ MHz). This transition zone is now larger than two octaves, implying that a five- or six-pole Butterworth filter is sufficient, which is much easier to design and build.

2.7.2 DACs

In theory, the simplest method for digital-to-analog conversion is to pull the samples from memory and convert them into an impulse train. Similar to the sampling

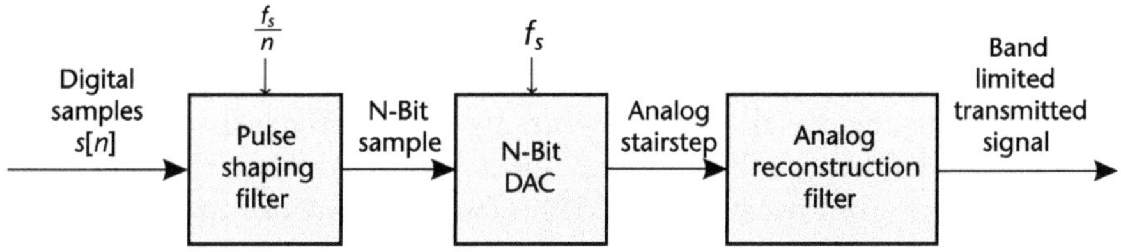

Figure 2.46 On the transmitter side, the DAC converts the digital symbols into an analog signal for transmission.

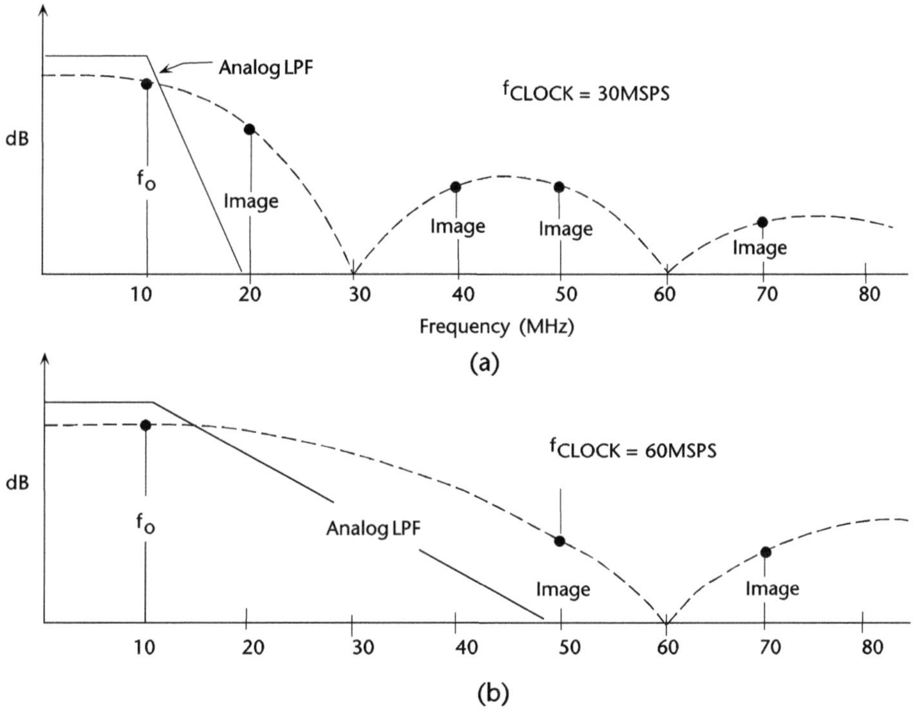

Figure 2.47 Analog reconstruction filter requirements for $f_o = 10$ MHz, with $f_s = 30$ MSPS, and $f_s = 60$ MSPS [18].

function in Section 2.2.2, the impulse modulator can be defined as

$$p(t) = \sum_{k=-\infty}^{\infty} \delta(t - kT), \tag{2.68}$$

where $p(t) = 1$ for $t = kT$, and $p(t) = 0$ for all the other time instants. Therefore, the analog pulse train after the impulse modulator is

$$s_a(t) = s[n]p(t) = \sum_{k=-\infty}^{\infty} s(kT)\delta(t - kT) = \begin{cases} s(kT) & t = kT \\ 0 & t \neq kT \end{cases}, \tag{2.69}$$

where each digital symbol $s[n]$ initiates an analog pulse that is scaled by the value of the symbol. The original analog signal can be reconstructed by passing this impulse train through a lowpass filter, with the cutoff frequency equal to one-half of the sampling rate.

Although this method is mathematically pure, it is difficult to generate the required infinitively narrow impulse pulses even in modern in electronics. To get around this, nearly all DACs operate by holding the last value until another sample is received. This is called a *zeroth-order hold*, which is the DAC equivalent of the sample-and-hold used during ADC. The zeroth-order hold produces the staircase appearance in the time domain, as shown in Figure 2.44(b).

In the frequency domain, the zeroth-order hold results in the spectrum of the impulse train being multiplied by sinc function, given by the equation

$$H(f) = \left| \frac{\sin(\pi f / f_s)}{\pi f / f_s} \right|. \tag{2.70}$$

If you already have a background in this material, the zeroth-order hold can be understood as the convolution of the impulse train with a rectangular pulse, having a width equal to the sampling period. This results in the frequency domain being multiplied by the Fourier transform of the rectangular pulse, that is, the sinc function. The dashed line in Figure 2.47 shows the sinc function of the 30 MHz and 60 MHz DACs. It can be seen from Figure 2.47 that the sinc function attenuates signals in the passband. However, something in the system needs to compensate for this effect by the reciprocal of the zeroth-order hold's effect, $\frac{1}{sinc(x)}$, or simply accept this nonideality. Many ignore this problem, but it is very trivial to do with the Pluto SDR. Allowing a user to easily transmit with a very flat passband.

2.7.3 Digital Pulse-Shaping Filters

To understand why we need to include digital pulse-shaping filters in all radio designs, a short example will be shown. Phase-shift keying (PSK) is a simple but common digital modulation scheme that transmits data by changing the phase of a reference signal. This is shown in Figures 2.48(a) and 2.48(d), where the time and frequency domain for the carrier is shown. The frequency is very narrow and should be easy to transmit.

The alternating bitstream of ones and zeros shown in Figures 2.48(b) and 2.48(e) causes the problem. Examining this, we observe that the square wave has infinite frequency information, which is something very difficult to transmit in a practical consideration.

When we multiply the carrier in Figure 2.48(a) with the bitstream in Figure 2.48(b) to attempt to transmit this data out the antenna, it results in Figures 2.48(c) and 2.48(f), which is a signal with infinite bandwidth in the continuous time analog domain. Not only will we have difficulty transmitting things, our nearest neighbors in adjacent RF channels will not like either.

Going back to our mathematical model of impulses, (2.70) indicates that without a digital pulse-shaping filter, these pulses $s_a(t)$ will be transmitted through the channel right away. In theory, given infinite bandwidth on the receiver side, we should be able to get the same pulse train although with some delay. However, in reality we actually cannot recover such a signal due to finite bandwidth and interference between adjacent symbols.

There are two situations when adjacent symbols may interfere with each other: when the pulse shape is wider than a single symbol interval T, and when there is a

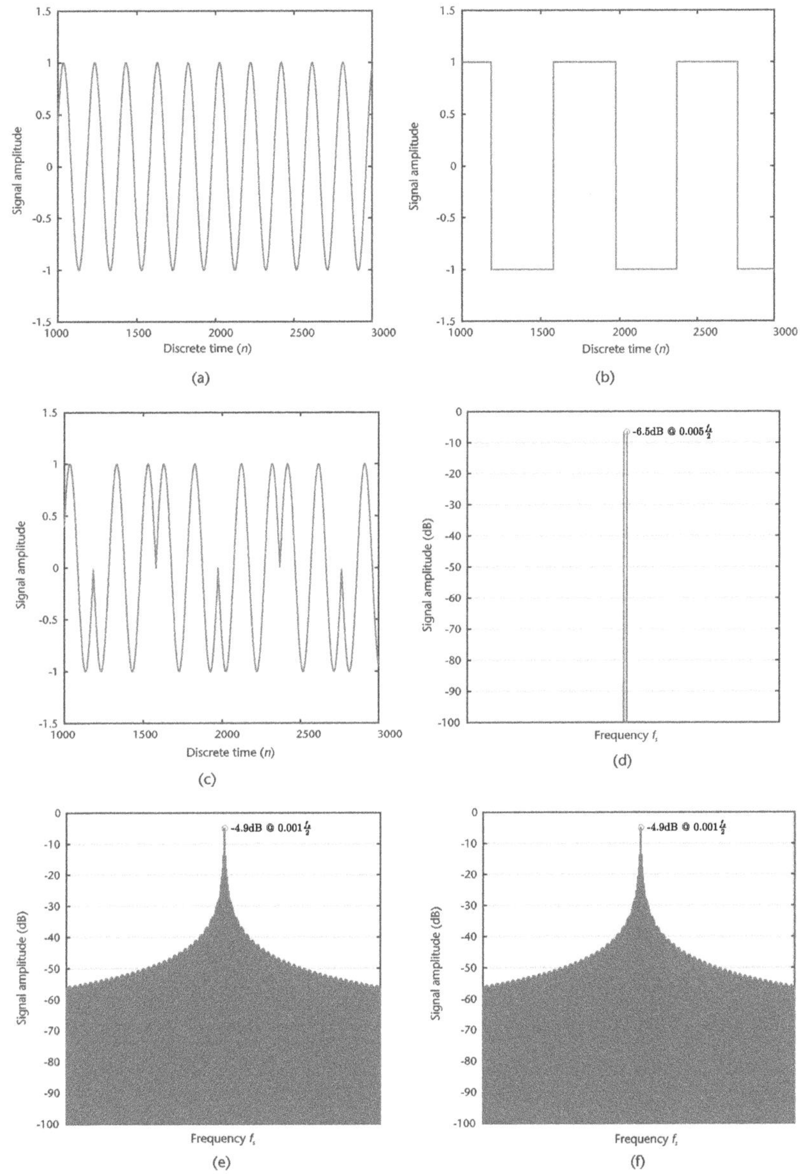

Figure 2.48 Phase-shift keyed modulated signal, carrier, data, and resulting modulated waveform, time and fourier domain. (a) Carrier: time domain, (b) PSK data: time domain, (c) PSK modulation: time domain, (d) carrier: Fourier domain, (e) PSK data: Fourier domain, and (f) PSK modulation: Fourier domain.

nonunity channel that smears nearby pulses, causing them to overlap. Both of these situations are called *intersymbol interference* (ISI) [9].

In order to solve these problems, pulse-shaping filters are introduced to bandlimit the transmit waveform.

2.7.4 Nyquist Pulse-Shaping Theory

In a communication system, there are normally two pulse-shaping filters, one on the transmitter side, and the other on the receiver side, as shown in Figure 2.49, where

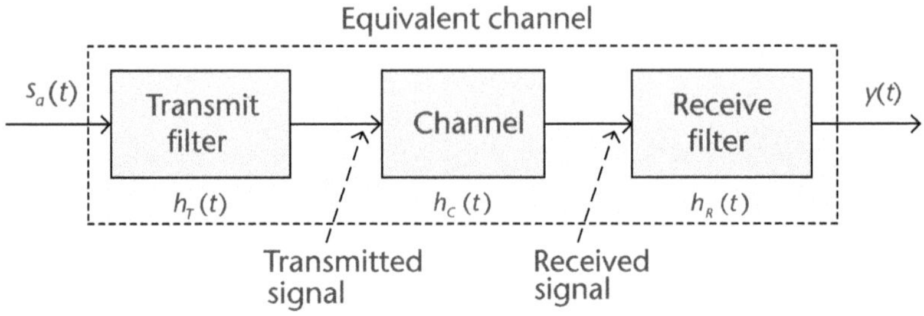

Figure 2.49 The equivalent channel of a communication system, which consists of the transmit filter, the channel, and the receive filter.

we use $h_T(t)$ and $h_R(t)$ to denote the impulse response of the transmit filter and receive filter. For simplicity, when considering the Nyquist pulse-shaping theory, we usually use the concept of equivalent channel, which not only includes the channel itself, but also the two pulse-shaping filters. Therefore, the impulse response of the equivalent channel is

$$h(t) = h_T(t) * h_C(t) * h_R(t). \tag{2.71}$$

Now, let us consider under which condition we can assure that there is no intersymbol interference between symbols. The input to the equivalent channel, $s_a(t)$, has been defined in (2.70). As mentioned before, each analog pulse is scaled by the value of the symbol, so we can express $s_a(t)$ in another way:

$$s_a(t) = \sum a_k \delta(t - kT), \tag{2.72}$$

where a_k is the value of the kth symbol. It yields the output to the equivalent channel, $y(t)$, which is

$$y(t) = s_a(t) * h(t) = \sum a_k[\delta(t - kT) * h(t)] = \sum a_k h(t - kT). \tag{2.73}$$

Therefore, given a specific time instant, for example, $t = mT$, where m is a constant, the input $s_a(t)$ is

$$s_a(mT) = \sum a_k \delta(mT - kT) = a_m. \tag{2.74}$$

Consequently, the output $y(t)$ becomes

$$y(mT) = \sum a_k h(mT - kT) = a_0 h(mT) + a_1 h(mT - T) + \ldots + a_m h(mT - mT). \tag{2.75}$$

Since we do not want the interference from the other symbols, we would like the output to contain only the a_m term, which is

$$y(mT) = a_m h(mT - mT). \tag{2.76}$$

Moreover, it means at a time instant $t = mT$, we need to have

$$h(mt - kT) = \begin{cases} C & k = m \\ 0 & k \neq m \end{cases}, \tag{2.77}$$

where C is some nonzero constant.

If we generalize (2.77) to any time instant t, we can get the Nyquist pulse-shaping theory as below. The condition that one pulse does not interfere with other pulses at subsequent T-spaced sample instants is formalized by saying that $h(t)$ is a *Nyquist pulse* if it satisfies

$$h(t) = h(kT) = \begin{cases} C & k = 0 \\ 0 & k \neq 0 \end{cases}, \tag{2.78}$$

for all integers k.

2.7.5 Two Nyquist Pulses

In this section, we will introduce two important Nyquist pulses; namely, *sinc pulse* and *raised cosine pulse*. When considering (2.78), the most straightforward pulse we can think of is a rectangular pulse with time width less than T, or any pulse shape that is less than T wide. However, the bandwidth of the rectangular pulse (and other narrow pulse shapes) may be too wide. Narrow pulse shapes do not utilize the spectrum efficiently, but wide pulse shapes may cause ISI, so what is needed is a signal that is wide in time (and narrow in frequency) that also fulfills the Nyquist condition in (2.78) [9].

In mathematics, the sinc function is defined as

$$\text{sinc}(x) = \frac{\sin(\pi x)}{\pi x}, \tag{2.79}$$

and is shown in Figure 2.50, when variable x takes an integer value k, the value of the sinc function will be

$$\text{sinc}(k) = \begin{cases} 1 & k = 0 \\ 0 & k \neq 0 \end{cases}. \tag{2.80}$$

In other words, zero crossings of the normalized sinc function occur at nonzero integer values.

Another important property of sinc function is that sinc pulse is the inverse Fourier transform of a rectangular signal, as shown in Figure 2.51(a). Suppose the

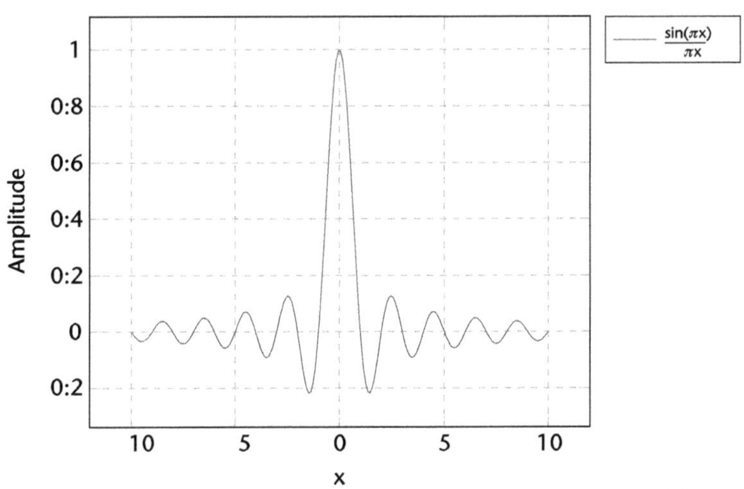

Figure 2.50 The plot of sinc function as defined in (2.79). The x-axis is x, and the y-axis is sinc(x).

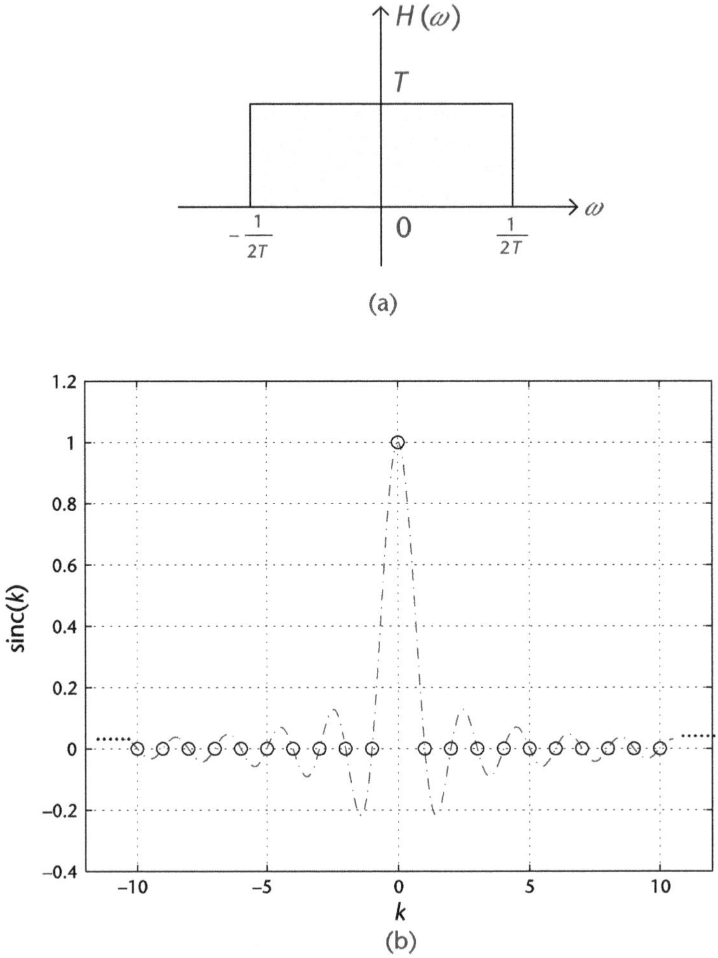

Figure 2.51 The sinc pulse on time domain and its Fourier transform, rectangular pulse, on frequency domain. (a) The rectangular pulse on frequency domain, defined in (2.81), and (b) the sinc pulse defined in (2.84). The x-axis is k, where $k = \frac{t}{T}$, and the y-axis is sinc(k).

rectangular signal is defined as [19]:

$$H(\omega) = \begin{cases} T & |\omega| < \frac{1}{2T} \\ 0 & \text{otherwise} \end{cases}. \tag{2.81}$$

Taking the inverse Fourier transform of the rectangular signal will yield the sinc signal as

$$h(t) = \text{sinc}\left(\frac{t}{T}\right). \tag{2.82}$$

Change the variable $t = kT$ in (2.82) yields

$$h(t) = h(kT) = \text{sinc}\left(\frac{kT}{T}\right) = \text{sinc}(k). \tag{2.83}$$

Since k is an integer here, according to (2.80), we can continue writing (2.83) as

$$h(t) = h(kT) = \text{sinc}\left(\frac{kT}{T}\right) = \text{sinc}(k) = \begin{cases} 1 & k = 0 \\ 0 & k \neq 0 \end{cases}. \tag{2.84}$$

Comparing (2.84) with (2.78), we can easily find that if we make $t = kT$, the sinc pulse exactly satisfies Nyquist pulse-shaping theory in Section 2.7.4. In other words, by choosing sampling time at kT (sampling frequency equals $\frac{1}{T}$), our sampling instants are located at the equally spaced zero crossings, as shown in Figure 2.51(b), so there will be no intersymbol interference.

Recall the Nyquist sampling theorem in Section 2.2.3 states that a real signal, $x(t)$, which is bandlimited to B Hz can be reconstructed without error using a minimum sampling frequency of $F_s = 2B$ Hz. In this case, the minimum sampling frequency is $F_s = \frac{1}{T}$ Hz. Therefore, the corresponding minimum bandwidth is

$$B = \frac{F_s}{2} = \frac{1}{2T}, \tag{2.85}$$

which is exactly the bandwidth of the rectangular signal defined in (2.81). Based on the discussion above, this choice of sinc pulse $h(t)$ yields the minimum bandwidth $B = B_{\text{min}} = \frac{1}{2T}$, so it is called the Nyquist-I Pulse [20].

Sinc pulses are a very attractive option because they are wide in time and narrow in frequency, which means they have the advantages of both spectrum efficiency and no ISI. However, sinc pulses are not practical since they have ISI sensitivity due to timing errors. For instance, for large t, the sinc pulse defined in (2.82) has the following approximation:

$$h(t) \sim \frac{1}{t}, \tag{2.86}$$

so it is obvious that timing error can cause large ISI. We must also note that sinc pulse are infinite in time, making them unrealizable.

Consequentially, we need to introduce Nyquist-II pulses, which have a larger bandwidth $B > B_{\text{min}}$, but with less ISI sensitivity. Since this type of pulse is more practical, it is much more widely used in practice.

The raised cosine pulse is one of the most important type of Nyquist-II pulses, which has the frequency transfer function defined as

$$H_{\text{RC}}(f) = \begin{cases} T & 0 \leq |f| \leq \frac{1-\beta}{2T} \\ \frac{T}{2}\left(1 + \cos\left(\frac{\pi T}{\beta}(|f| - \frac{1-\beta}{2T})\right)\right) & \frac{1-\beta}{2T} \leq |f| \leq \frac{1+\beta}{2T} \\ 0 & |f| \geq \frac{1+\beta}{2T} \end{cases}, \tag{2.87}$$

where β is the rolloff factor, which takes value from 0 to 1, and $\frac{\beta}{2T}$ is the excess bandwidth.

The spectrum of raised cosine pulse is shown in Figure 2.52. In general, it has the bandwidth $B \geq 1/(2T)$. When $\beta = 0$, the bandwidth $B = 1/(2T)$, and there is no excess bandwidth, which is actually a rectangular pulse. On the other end, when $\beta = 1$, it reaches the maximum bandwidth $B = 1/T$.

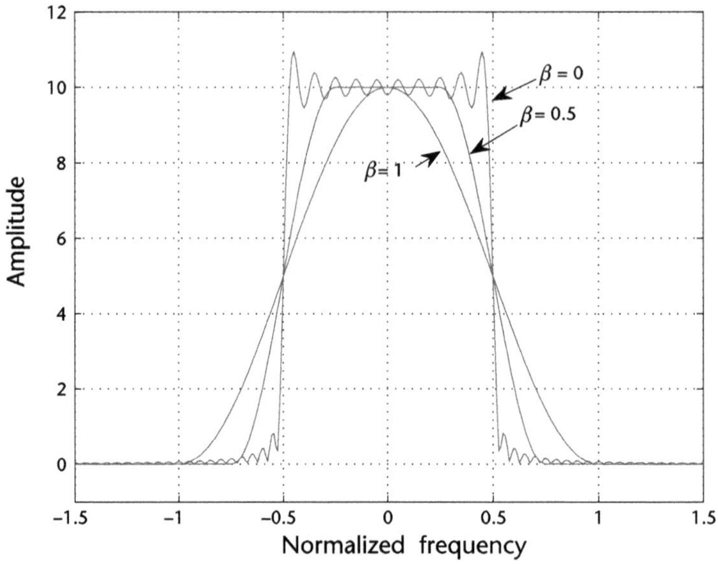

Figure 2.52 Spectrum of a raised cosine pulse defined in (2.87), which varies by the rolloff factor β. The x-axis is the normalized frequency f_0. The actual frequency can be obtained by f_0/T.

By taking the inverse Fourier transform of $H_{RC}(f)$, we can obtain the impulse response of raised cosine pulse, defined as

$$h_{RC}(t) = \frac{\cos\left(\pi\frac{\beta t}{T}\right)}{1-\left(2\frac{\beta t}{T}\right)^2}\text{sinc}\left(\frac{\pi t}{T}\right). \tag{2.88}$$

Nyquist-II pulses do not have an ISI sensitivity because its peak distortion, the tail of $h_{RC}(t)$, converges quickly, which can be expressed as

$$D_p = \sum_{n=-\infty}^{\infty} |h_{RC}(\epsilon' + (n-k))| \sim \frac{1}{n^3}. \tag{2.89}$$

Therefore, when timing error occurs, the distortion will not accumulate to infinity in a similar fashion related to Nyquist-I pulses [20].

Actually, in many practical communications systems, *root raised cosine filters* are usually used [21]. If we consider the communication channel as an ideal channel and we assume that the transmit filter and the receive filter are identical, we can use root raised cosine filters for both of them, and their net response must equal to $H_{RC}(f)$ defined in (2.87). Since the impulse response of the equivalent channel can be expressed as

$$h(t) = h_T(t) * h_C(t) * h_R(t), \tag{2.90}$$

where $h_C(t)$ is the impulse response of the communication channel, and $h_T(t)$ and $h_R(t)$ are the impulse responses of the transmit filter and the receive filter, it means on frequency domain, we have

$$H_{RC}(f) = H_T(f)H_R(f). \tag{2.91}$$

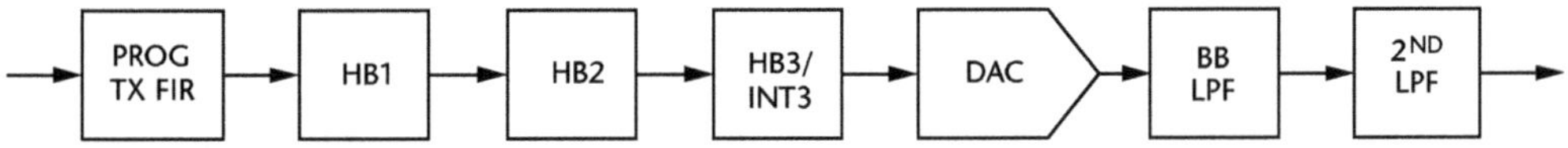

Figure 2.53 Internal AD9361 Tx signal path.

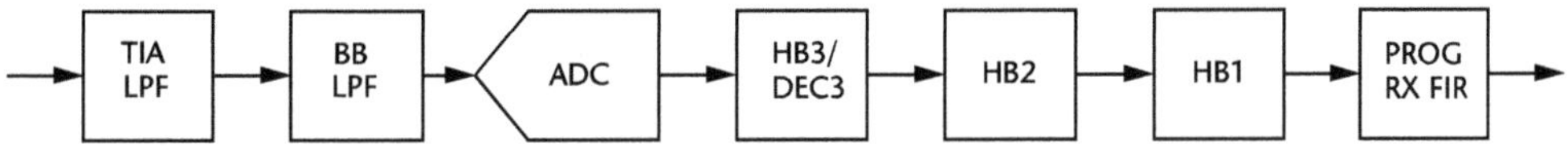

Figure 2.54 Internal AD9361 Rx signal path.

Therefore, the frequency response of root raised cosine filter must satisfy

$$|H_T(f)| = |H_R(f)| = \sqrt{|H_{RC}(f)|}. \tag{2.92}$$

2.8 Chapter Summary

SDR experimentation requires both strong computer skills and extensive knowledge of digital signal processing. This chapter lists some useful topics including sampling, pulse shaping, and filtering. The purpose of this chapter is to help the readers to get prepared for the experimentation chapters, especially for the Pluto SDR hardware. For example, on the Pluto SDR transmit path (see Figure 2.53), there is a programmable 128-tap FIR filter, interpolating halfband filters, a DAC, followed by two lowpass analog reconstruction filters (LPF). In order to understand how these work together and properly configure things, you need to understand sampling theory.

On the Pluto SDR receive path (see Figure 2.54), the signal flows through an antialiasing filter, the ADC, through digital decimating half band filters, and eventually a 128-tap programmable FIR, where the filtering knowledge is useful. This 128-tap programmable FIR can be used to compensate for the loss in the passband because of the antialiasing filter.

In addition, when you build a real communication system, you will need additional transmit and receive filters, which requires expertise in pulse shaping.

References

[1] Proakis, J. G., and Dimitris G. Manolakis, *Digital Signal Processing: Principles, Algorithms, and Applications*, Third Edition, Upper Saddle River, NJ: Prentice Hall, 1996.

[2] Elali, T. S., and M. A. Karim, *Continuous Signals and Systems with MATLAB*, Boca Raton, FL: CRC Press, 2001.

[3] Harris, F. J., "On the Use of Windows for Harmonic Analysis with the Discrete Fourier Transform," *Proceedings of the IEEE,* Vol. 66, No. 1, January 1978.

[4] Smith, S. W., *The Scientist and Engineers Guide to Digital Signal Processing*, Second Edition, 1999, http://www.analog.com/en/education/education-library/scientist_engineers_guide.html.

[5] Frigo, M., "A Fast Fourier Transform Compiler," Proceedings of the 1999 ACM SIGPLAN Conference on Programming Language Design and Implementation (PLDI '99), Atlanta, GA, May 1999, http://www.fftw.org/.

[6] Kester, W., Understand SINAD, ENOB, SNR, THD, THD + N, and SFDR so You Don't Get Lost inthe Noise Floor, Analog Devices, 2008, http://www.analog.com/MT-003.

[7] Kester, W., Mixed Signal and DSP Design Techniques, Analog Devices, 2002, http://www.analog.com/en/education/education-library/mixed_signal_dsp_design_book.html.

[8] James, J. F., *A Student's Guide to Fourier Transforms: With Applications in Physics and Engineering*, Third Edition, Cambridge, UK: Cambridge University Press, 2011.

[9] Johnson, C. R., Jr., and W. A. Sethares, *Telecommunications Breakdown: Concepts of Communication Transmitted via Software-Defined Radio*. Prentice Hall, 2004.

[10] Cavicchi, T. J., Digital Signal Processing. John Wiley and Sons, 2000.

[11] Oppenheim, A. V., and R. W., Schafer, *Digital Signal Processing*, Prentice-Hall, 1975.

[12] Analog Devices AD7177 Product Page, 2016 http://www.analog.com/AD7177-2.

[13] Analog Devices AD9208 Product Page, 2017 http://www.analog.com/AD9208.

[14] Smith, S., *The Scientist and Engineer's Guide to Digital Signal Processing*, San Diego: California TechnicalPublishing, 1997.

[15] Jeffrey, A., and D. Zwillinger, *Table of Integrals, Series, and Products*, Seventh edition, San Diego: Academic Press, 2007.

[16] Hogenauer, E. B., "An Economical Class of Digital Filters for Decimation and Interpolation," *IEEE Transactions on Acoustics, Speech and Signal Processing*, Vol. 29, No. 2, 1981, pp. 155–162, 1981.

[17] Lyons, R. G., Streamlining Digital Signal Processing: A Tricks of the Trade Guidebook, Piscataway, NJ: Wiley-IEEE Press, 2007.

[18] Kester, W., *The Data Conversion Handbook*, 2005, http://www.analog.com/en/education/education-library/data-conversion-handbook.html.

[19] Johnson, C. R., W. A. Sethares, and A. G. Klein, *Software Receiver Design: Build Your Own Digital Communications System in Five Easy Steps*, Cambridge, UK: Cambridge University Press, 2011.

[20] Wyglinski, A. M., M. Nekovee, and Y. T. Hou, *Cognitive Radio Communications and Networks: Principles and Practice*, Burlington, MA: Academic Press, 2009.

[21] Barry, J., E. A. Lee, and D. G. Messerschmitt, Digital Communication, Third Edition, Kluwer AcademicPress, 2004.

CHAPTER 3

Probability in Communications

The previous chapter provided us with a range of insights, tools, and algorithms for the modeling and processing of signals and systems from a purely deterministic perspective. However, there are numerous elements within the communication system environment that possess random characteristics, such as the digital values of the binary data stream from an information source, the noise introduced to a transmission when transversing a communication channel, and the sources of interference resulting from multiple users operating within the same environment. Whereas Chapter 2 presented a collection of useful tools for the understanding and modeling of deterministic signals and systems, in this chapter we will introduce a range of techniques and approaches that can be used to model and study probabilistic signals and systems. Starting with an introduction to the concept of both the continuous and *discrete random variable*, this chapter will proceed with an explanation of time-varying random phenomena, called *random processes*, followed by the modeling of various random noise channels.

3.1 Modeling Discrete Random Events in Communication Systems

A discrete random variable represents some sort of behavior occuring within the communication system where the outcome is not absolutely known. For instance, the next value produced by a binary information source has the possibility of producing one of two outputs: a binary 1 or a binary 0. Although we know that either value is possible, we do not have definite knowledge that one of these values will be specifically be produced at the output of the information source. Consequently, we model the output values to be produced by the binary information source as a random variable. The reason we call this random variable a discrete random variable is that it produces a single possible output value from a finite number of options.

To mathematically model this discrete random varaible, suppose we define X such that there exists a distinct set of real numbers x_i that it can produce, each with a specific probability of that value occuring:

$$\sum_i P(X = x_i) = 1, \tag{3.1}$$

where $P(X = x_i)$ is the probability that the random variable will produce an output value x_i. Using the law of total probability [1], we know that

$$P(X \in B) = \sum_{i:x_i \in B} P(X = x_i), \tag{3.2}$$

where the set B is composed of a collection of values x_i. A specific form of discrete random variable is the *integer-valued random variable*, where its output values are the integers $x_i = i$; that is,

$$P(X \in B) = \sum_{i \in B} P(X = i). \tag{3.3}$$

Since each output of the random variable X possesses a different probability of occurrence, we usually define the probability of a specific discrete output x_i being generated by X as

$$p_X(x_i) = P(X = x_i), \tag{3.4}$$

where $p_X(x_i)$ is referred to as the *probability mass function* (PMF). Note that the values of the PMF are constrained by

$$0 \leq p_X(x_i) \leq 1 \text{ and } \sum_i p_X(x_i) = 1. \tag{3.5}$$

Several frequently used PMFs are specified in Table 3.1, including uniform, Poisson, and Bernoulli random variables. In particular, Bernoulli random variables are used to generate random outputs for binary information sources, while Poisson random variables are often used to model the delays of routing packets in computer networks.

Table 3.1 Several Frequently Used Probability Mass Functions

Random Variable	*PMF Definition*	*Graphical Representation*
Uniform	$p_X(k) = \begin{cases} \frac{1}{n}, & k = 1, \ldots, n \\ 0, & \text{otherwise} \end{cases}$	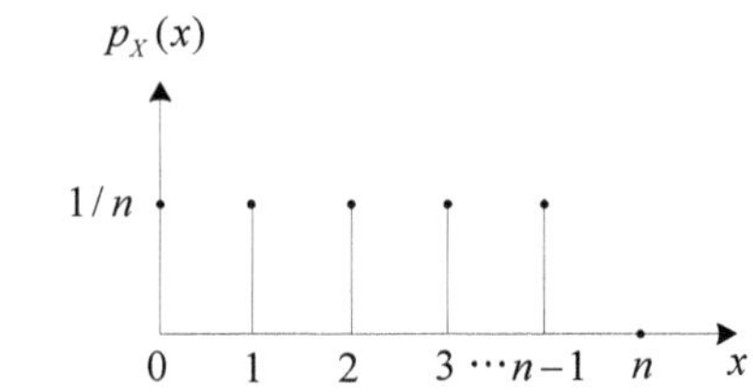
Poisson	$p_X(k) = \frac{\lambda^k e^{-\lambda}}{k!}, \; k = 0, 1, 2, \ldots$	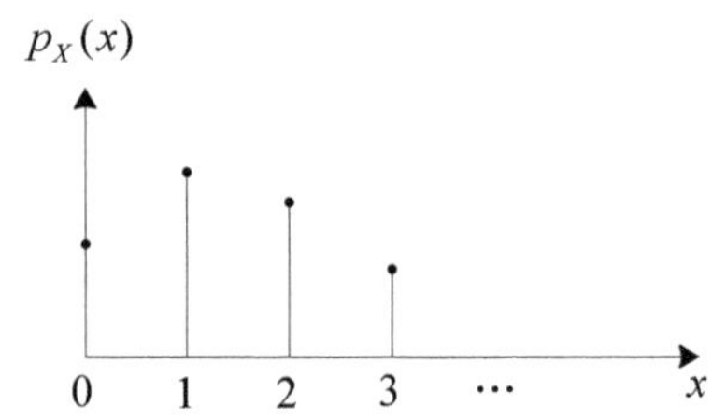
Bernoulli	$p_X(k) = \begin{cases} p, & k = 1 \\ 1 - p, & k = 0 \\ 0, & \text{otherwise} \end{cases}$	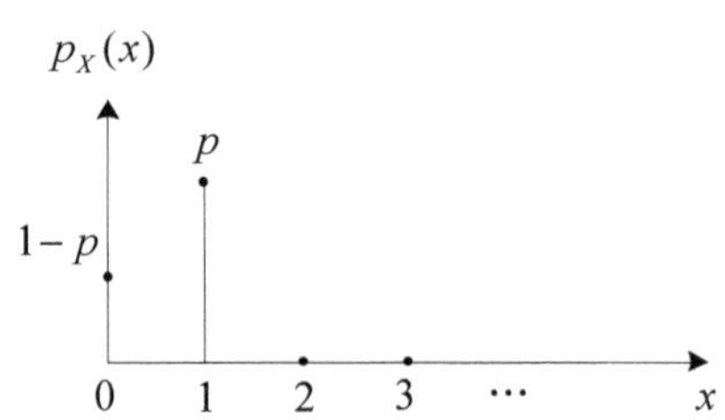

Supposed we now explore how to use these discrete random variables to model an actual component of a communication system; namely, a binary information source. One key characteristic that needs to be incorporated into this tool is the percentage of ones and zeros produced by the *random number generator*, otherwise referred to as an RNG. In the MATLAB script in Code 3.1, a binary RNG is implemented where three vectors are produced. Vector `b1` possesses an equal balance of one and zero values, while vectors `b2` and `b3` generate ratios of 60/40 and 20/80 in terms of ones and zeros, respectively. Note that the MATLAB `rand` uniform RNG function is used to produce the random values between zero and one, which is then rounded to the nearest integer; namely, one or zero.

Code 3.1 Information Source: **`random_example.m`**

```
% Create three binary data streams of length L, each with a different
% percentage of 1 and 0 values
L = 100;
prob1 = 0.5; prob2 = 0.6; prob3 = 0.2; %Probability values for 1 outputs
b1 = round(0.5*rand(1,L)+0.5*prob1); %Even split between 1 and 0 values
b2 = round(0.5*rand(1,L)+0.5*prob2); %Have 60% 1 values and 40% 0 values
b3 = round(0.5*rand(1,L)+0.5*prob3); %Have 20% 1 values and 80% 0 values
```

Manipulating how the output values of the `rand` get rounded to either zero or one by biasing all the values to be closer to one or the other, we can generate binary random values with different percentages of ones and zeros, as shown in Figure 3.1. Using the `stem` command, we can visualize the actual binary values shown in Figure 3.1(a), Figure 3.1(c), and Figure 3.1(e). Although this gives us a general observation about the behavior of these binary RNGs, it is very difficult to distinguish the actual percentages of ones and zeros within the vector. Hence, using a histogram and a very long sequence of randomly generated values, we can characterize these percentages, as shown in Figure 3.1(b), Figure 3.1(d), and Figure 3.1(f). Notice how the histograms accurately show the percentages of ones and zeros in an outputted vector. One important caveat: Since our characterization is dependent on the observation of a random phenomenon, you will need to observe a very substantial amount of data in order to accurately characterize it.

3.1.1 Expectation

Since we are unable to exactly know the output values of these random phenomena that form part of the communication system and its environment, we must instead opt for mathematicallly characterizing the statistical behavior of these random variables. One way to characterize a random variable is by its expected value or mean, which can be quantitatively determined using the expression

$$m_X = E[X] = \sum_i x_i P(X = x_i), \tag{3.6}$$

where the sum of each value is weighted by its probability of occurrence. Consequently, using the definition of the PMF, we can rewrite (3.6) as

$$E[X] = \sum_i x_i p_X(x_i). \tag{3.7}$$

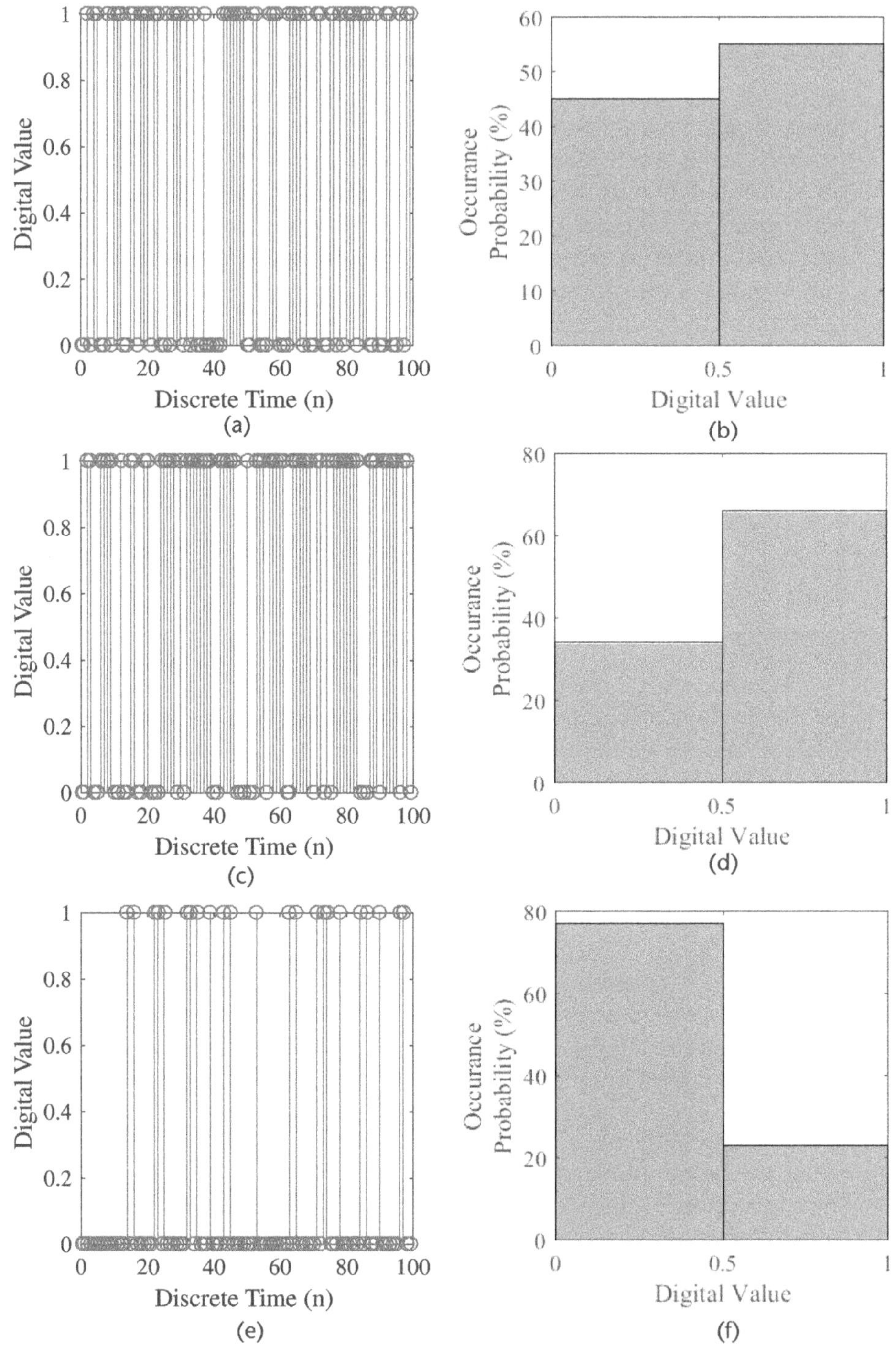

Figure 3.1 Stem plots and histograms of three binary transmissions that each possess different probabilities of ones and zeros being produced. (a) Binary signal (50% ones, 50% zeros), (b) histogram (50% ones, 50% zeros), (c) binary signal (60% ones, 40% zeros), (d) histogram (60% ones, 40% zeros), (e) binary signal (20% ones, 80% zeros), and (f) histogram (20% ones, 80% zeros).

Suppose we have a random variable X and a real-valued function $g(x)$ that maps it to a new random variable Z; that is, $Z = g(X)$. Since X is actually the mapping of points from a sample space Ω to a collection of real numbers, we are thus performing a mapping of a mapping when solving for the random variable Z; that is, $Z(\omega) = g(X[\omega])$. Consequently, in order to compute the expectation of the random variable Z, namely $E[Z]$, we can employ the following expression:

$$E[Z] = E[g(X)] = \sum_i g(x_i) p_X(x_i), \tag{3.8}$$

which is referred to the expectation of a function of a random variable. One of the most important aspects of (3.8) is that the expectation of Z can be solved using the PMF for X rather than having to determine the PMF of Z. Table 3.2 presents several useful properties associated with the expectation operator.

Table 3.2 Several Useful Properties of Expectation

Name	*Definition*
Linearity	If a and b are deterministic constants and X and Y are random variables, then $E[aX+bY] = E[aX]+E[bY] = aE[X]+bE[Y]$.
Moment	The nth moment ($n \geq 1$) of a real-valued random variable X is defined as the expected value of X raised to the nth power; that is, $\text{Moment}_n(X) = E[X^n]$.
Mean	The mean is the first moment of X; that is, $\text{Moment}_1(X) = E[X] = \mu$.
Variance	The second moment of X with its mean subtracted is its variance; that is, $\text{Moment}_2(X-\mu) = E[(X-\mu)^2] = \text{var}(X) = \sigma^2$.
Central moment	The generalization of the variance is the nth order central moment of X; that is, $E[(X-\mu)^n]$. Note that the *skewness* and *kurtosis* of X are the third-order and fourth-order central moments of X, respectively.
Correlation	The correlation between two random variables X and Y is defined to be equal to $E[XY]$.
Covariance	The covariance between X and Y is defined as $\text{cov}(X,Y) = E[(X-\mu_X)(Y-\mu_Y)]$.
Correlation coefficient	The correlation coefficient of two random variables X and Y is given as $\rho_{XY} = E\left[\left(\frac{X-\mu_X}{\sigma_X}\right)\left(\frac{Y-\mu_Y}{\sigma_Y}\right)\right]$.
Markov inequality	If X is a nonnegative random variable, then for any $a > 0$ we have $P(X \geq a) \leq \frac{E[X]}{a}$.
Chebyshev inequality	For any random variable Y and any $a > 0$, we have $P(\lvert Y\rvert \geq a) \leq \frac{E[Y^2]}{a^2}$.
Cauchy-Schwarz inequality	The Cauchy-Schwarz inequality states that $\lvert E[XY]\rvert \leq \sqrt{E[X^2]E[Y^2]}$, which becomes an equality if and only if X and Y are linearly related.
Independence	The random variables are independent if and only if $E[h(X)k(Y)] = E[h(X)]E[k(Y)]$ for all functions $h(x)$ and $k(y)$.

3.2 Binary Communication Channels and Conditional Probability

In a digital communication system, we can view the operations of the receiver as attempting to guess what was sent from a particular transmitter when guessing the value of a transmission, whose values can be modeled as a random variable X, and once it has been observed at the receiver, whose values can be modeled by a random variable Y. Consequently, we need some sort of decision rule to figure out which value of X was originally transmitted. This is illustrated in Figure 3.2, where X can output values of either 0 or 1 while Y is observed at the receiver. Notice how the observation Y can be accidentally interpreted as a value other than the one transmitted; that is, $X \neq Y$.

In order to characterize and understand this random transmission environment, we can use the definition for the conditional probability, which is mathematically expressed as

$$P(X \in B|Y \in C) = P(\{X \in B\}|\{Y \in C\}) = \frac{P(\{X \in B\} \cap \{Y \in C\})}{P(\{Y \in C\})} = \frac{P(X \in B, Y \in C)}{P(Y \in C)}. \tag{3.9}$$

Additionally, the conditional PMF can also be used to mathematically describe this random phenomenon:

$$\begin{aligned} p_{X|Y}(x_i|y_j) = P(X = x_i|Y = y_j) = \frac{P(X = x_i, Y = y_j)}{P(Y = y_j)} = \frac{p_{XY}(x_i, y_j)}{p_Y(y_j)} \\ p_{Y|X}(y_j|x_i) = P(Y = y_j|X = x_i) = \frac{P(X = x_i, Y = y_j)}{P(X = x_i)} = \frac{p_{XY}(x_i, y_j)}{p_X(x_i)} \end{aligned} \tag{3.10}$$

which can be subsequently expressed as

$$p_{XY}(x_i, y_j) = p_{X|Y}(x_i|y_j)p_Y(y_j) = p_{Y|X}(y_j|x_i)p_X(x_i). \tag{3.11}$$

One useful mathematical tool that can be employed to help characterize phenomena described by conditional probabilities and/or conditional PMFs is the law of total probability. For example, suppose that Y is an arbitrary random variable, and we take $A = \{Y \in C\}$, where $C \subset \mathbb{R}$. Consequently, we can define the

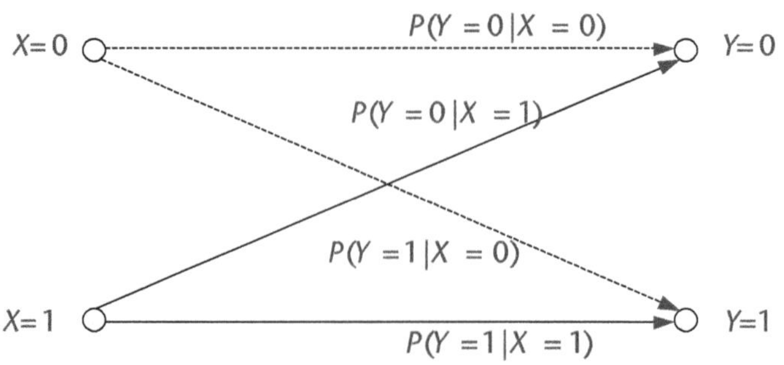

Figure 3.2 An example of a binary channel where transmission values generated by the random variable X are being observed at the receiver as the random variable Y.

Derive the Markov inequality, $P(X \geq a) \leq \frac{E[X^r]}{a^r}$, using the definition for the expectation.

law of total probability as

$$P(Y \in C) = \sum_i P(Y \in C|X = x_i)P(X = x_i). \tag{3.12}$$

Similarly, if Y is a discrete random variable, taking on distinct values y_i and setting $C = \{y_i\}$, then this yields the following law of total probability:

$$\begin{aligned} P(Y = y_j) &= \sum_i P(Y = y_j|X = x_i)P(X = x_i) \\ &= \sum_i p_{Y|X}(y_j)p_X(x_i). \end{aligned} \tag{3.13}$$

Derive the resulting expressions for the law of total probability for expectation and the substitution law.

Returning to our binary communication channel model described by Figure 3.2, suppose we would like to decide on what values were transmitted based on the observation of the intercepted received signal. In this case, we would like to employ the *maximum a posteriori* (MAP) decision rule, where given an observed $Y = j$, the MAP rule states that we should decide on $X = 1$ if

$$P(X = 1|Y = j) \geq P(X = 0|Y = j), \tag{3.14}$$

and to decide on $X = 0$ otherwise. In other words, the MAP rule decides $X = 1$ if the posterior probability of $X = 1$ given the observation $Y = j$ is greater than the posterior probability of $X = 0$ given the observation $Y = j$. Furthermore, we can observe that

$$P(X = i|Y = j) = \frac{P(X = i, Y = j)}{P(Y = j)} = \frac{P(Y = j|X = i)P(X = i)}{P(Y = j)}, \tag{3.15}$$

which we can then use to rewrite (3.14) as

$$\begin{aligned} \frac{P(Y = j|X = 1)P(X = 1)}{P(Y = j)} &\geq \frac{P(Y = j|X = 0)P(X = 0)}{P(Y = j)} \\ P(Y = j|X = 1)P(X = 1) &\geq P(Y = j|X = 0)P(X = 0). \end{aligned} \tag{3.16}$$

If $X = 0$ and $X = 1$ are equally likely to occur, we can simplify this expression such that it yields the maximum likelihood (ML) rule

$$P(Y = j|X = 1) \geq P(Y = j|X = 0). \tag{3.17}$$

Code 3.2 Simulate: **random_example.m**

```
% Define simulation parameters
L = 100000; % Transmission length
prob00 = 0.95; % Probability zero received given zero transmitted
prob11 = 0.99; % Probability one received given one transmitted
prob4 = 0.7; % Have 40% 1 values and 60% 0 values

% Create transmitted binary data stream
b4 = round(0.5*rand(1,L)+0.5*prob4);
                              % Have 40% 1 values and 60% 0 values
b4hat = b4; % Initialize receive binary data stream

% Randomly select 1 and 0 values for flipping
ind_zero = find(b4 == 0); % Find those time instances with zero values
ind_one = find(b4 == 1); % Find those time instances with one values
ind_flip_zero = find(round(0.5*rand(1,length(ind_zero))
                  +0.5*(1-prob00)) == 1); % Flip zero bits to one bits
ind_flip_one = find(round(0.5*rand(1,length(ind_one))
                  +0.5*(1-prob11)) == 1); % Flip one bits to zero bits

% Corrupt received binary data stream
b4hat(ind_zero(ind_flip_zero)) = 1; % Flip 0 to 1
b4hat(ind_one(ind_flip_one)) = 0; % Flip 1 to 0

% Calculate bit error statistics
b4error_total = sum(abs(b4-b4hat))/L;
b4error_1 = sum(abs(b4(ind_one) - b4hat(ind_one)))/length(ind_one);
b4error_0 = sum(abs(b4(ind_zero) - b4hat(ind_zero)))/length(ind_zero);
```

Furthermore, in the general case, we can rearrange (3.16) such that it yields the likelihood ratio; namely:

$$\frac{P(Y=j|X=1)}{P(Y=j|X=0)} \geq \frac{P(X=0)}{P(X=1)} \tag{3.18}$$

where the right-handed side is referred to as the threshold since it does not depend on j.

Given this mathematical formulation, let us now work with this same example via computer simulation. Using the following MATLAB script, we can model a binary channel where we produce `L` binary values with a probability of `prob4` being one values and the rest being zero values. Furthermore, we assume that the binary channel being used is not symmetric, meaning that the probability of one values being flipped into zero values is different than the probability of zero values being flipped into one values. Note that the flipping of bit values is considered to be an error produced by the channel. Consequently, we define the probability of a transmitted one value being received as a one value to be equal to `prob11` while the probability of a transmitted zero value being received as a zero value is equal to `prob00`.

One of fundamental metrics for assessing the performance of any digital communication system is the *probability of bit error*, or the bit error rate (BER). The BER characterizes the amount of bit errors received relative to the total bits transmitted. For various applications, different BER values are considered acceptable while others are considered intolerable. For instance, for typical wireless data transmission applications, a BER of 10^{-5} is considered an acceptable amount of error introduced into the transmission.

In the MATLAB example above involving the binary channel, we would like to characterize the BER values to see if they conform to the parameters we defined. In Figure 3.3, we have the BER for the overall transmission, as well as for only the the transmissions of one values and of zero values. Notice how the BER for the one values only corresponds to the complement of the probability `prob11`, while the same can be observed for the BER of the zero values only and it being the complement of the probability `prob00`. Remember that in order to obtain an accurate statistical assessment of the BER, a very large number of binary values need to be generated.

3.3 Modeling Continuous Random Events in Communication Systems

As we have seen earlier in this chapter, it is possible to mathematically compute the probability of a random event occurring within a communication system described by a discrete random variable; namely,

$$P(a \leq X < b) = \sum_{i=a}^{b-1} p_X(i), \tag{3.19}$$

where X is the discrete random variable, and both a and b are boundaries of a subset belonging to the sample space Ω. However, suppose now that X is a *continuous random variable*, which can take on the entire continuum of values within the interval (a, b). In order to compute the same probability in this case, we can start by realizing this scenario as the summation of an infinite number of points in (a, b) with the space between samples equal to Δx.

There are numerous random elements contained within a communication system where each can produce a continuum of possible output values. As we will discuss later in this chapter, one of these elements represented by a continuous random variable is the noise introduced in a transmission channel. Suppose this noise can be represented by an infinite number of samples such that our Δx becomes

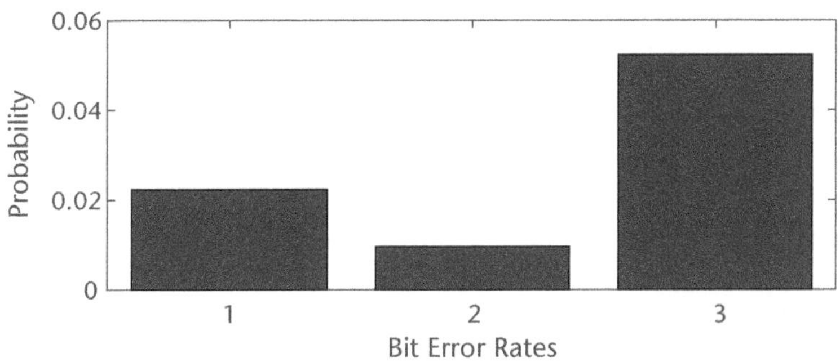

Figure 3.3 Binary channel error probabilities when the probability for a zero being received uncorrupted is 0.95 and the probability for a one being received uncorrupted is 0.99. Note that the transmission consists of 40% ones and 60% zeros. 1 = Total, 2 = one transmitted, 3 = zero transmitted.

so tiny the Δx ultimately converges to dx and the summation in (3.19) becomes an integral expression. Therefore, we can express the probability of a continuous random variable modeling the noise producing output values ranging between a and b using the expression

$$P(a \leq X < b) = \int_{a}^{b} f(t)dt, \tag{3.20}$$

where $f(t)$ is called the *probability density function* (PDF). Note that the PDF is the continuous version of the PMF that we discussed previously in this chapter. Moreover, generally we can express the probability of a continuous random variable using the PDF by the expression

$$P(X \in B) = \int_{B} f(t)dt = \int_{-\infty}^{+\infty} I_B(t)f(t)dt, \tag{3.21}$$

where $I_B(t)$ is an indicator function that produces an output of unity whenever a value of t belongs to the set B and produces an output of zero otherwise. Note that $\int_{-\infty}^{+\infty} f(t)dt = 1$, $f(t)$ is nonnegative, $f(t)$ approximately provides the probability at a point t, and

$$P(a \leq X \leq b) = P(a < X \leq b) = P(a \leq X < b) = P(a < X < b), \tag{3.22}$$

where the end points do not affect the resulting probability measure. Finally, a summary of several commonly used PDFs are presented in Table 3.3, including uniform, Gaussian, and exponential random variables. Note that Gaussian random variables are often used to model the randomness of the noise introduced in a communication channel, while the exponential random variable is used in medium access protocols to help provide random back-off times in the event that two or more wireless systems are attempting to communicate over the same wireless channel via a contention-based framework.

Similar to the expectation of a single discrete random variable, the expectation for a continuous random variable X with PDF $f(x)$ can be computed using the following expression:

$$E[g(X)] = \int_{-\infty}^{+\infty} g(x)f(x)dx, \tag{3.23}$$

where $g(.)$ is some real function that is applied to the random variable X.

Many random variables of practical interest are not independent, where it is often the case that the outcome of one event may depend on or be influenced by the result of another event. Consequently, it is sometimes necessary to compute the *conditional probability* and *conditional expectation*, especially in circumstances where we have to deal with problems involving more than one random variable.

Unlike the conditional probability for a discrete random variable, the conditional probability for a continuous random variable needs to defined in an alternative manner since the probability of a single exact point occurring is zero;

Table 3.3 Several Frequently Used Probability Density Functions

Random Variable	*PDF Definition*	*Graphical Representation*
Uniform	$f(x) = \begin{cases} \frac{1}{(b-a)}, & a \leq x \leq b \\ 0, & \text{otherwise} \end{cases}$	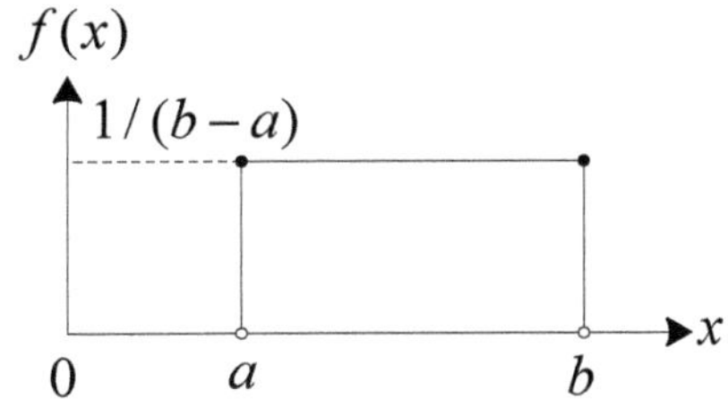
Exponential	$f(x) = \begin{cases} \lambda e^{-\lambda x}, & x \geq 0 \\ 0, & x < 0 \end{cases}$	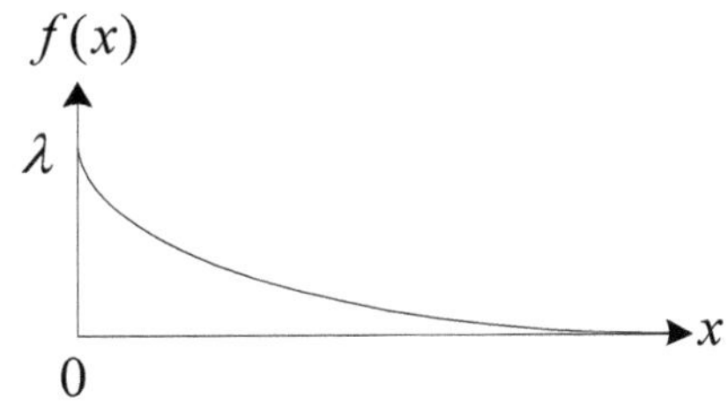
Laplace	$f(x) = \frac{\lambda}{2} e^{-\lambda \lvert x \rvert}$	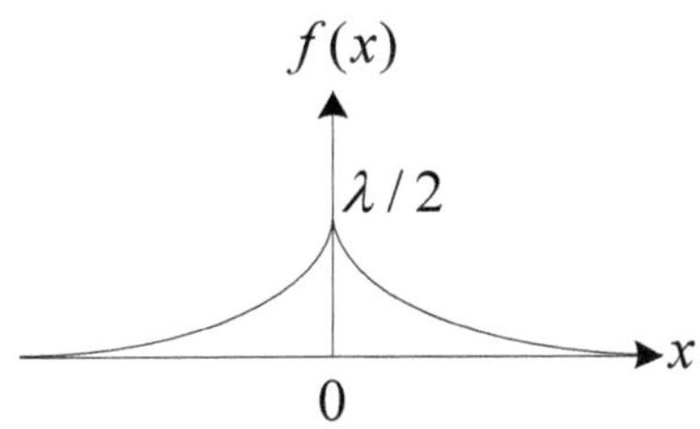
Cauchy	$f(x) = \frac{\lambda/\pi}{\lambda^2 + x^2},\ \lambda > 0$	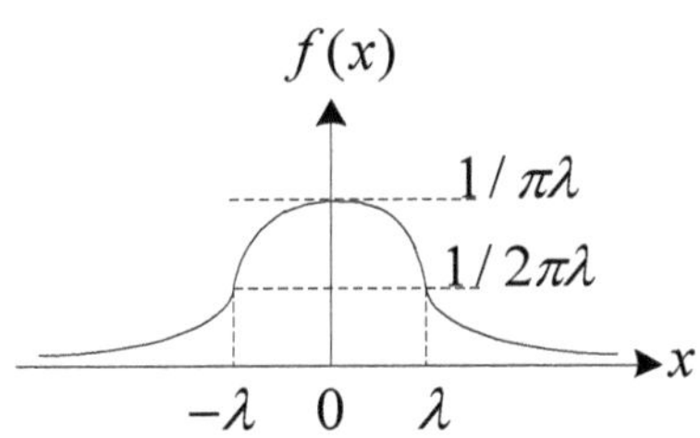
Gaussian	$f(x) = \frac{1}{\sqrt{2\pi\sigma^2}} e^{-0.5((x-\mu)/\sigma)^2}$	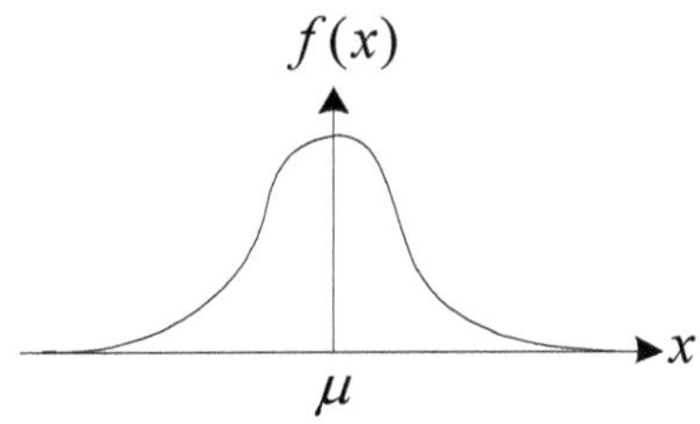

that is, $P(X = x) = 0$. As a result, if we employ the definition for the conditional probability for a discrete random variable from (3.9); namely,

$$P(Y \in C|X = x) = \frac{P(Y \in C, X = x)}{P(X = x)}, \tag{3.24}$$

we observe that the occurrence of $P(X = x) = 0$ would yield a divide-by-zero scenario. Consequently, we need to determine another definition for the conditional probability (and conditional expectation) that would work within the continuous random variable framework.

It can be shown that in order to calculate the conditional probability, one must employ a *conditional density* [1], which can be defined as

$$f_{Y|X}(y|x) = \frac{f_{XY}(x, y)}{f_X(x)}, \tag{3.25}$$

where $f_X(x) > 0$. Thus, leveraging the conditional density, we can now compute the conditional probability without concern of a divide-by-zero scenario by solving for the following expression:

$$P(Y \in C|X = x) = \int_C f_{Y|X}(y|x)dy. \tag{3.26}$$

Furthermore, we can define the law of total probability as the following:

$$P(Y \in C) = \int_{-\infty}^{+\infty} P(Y \in C|X = x)f_X(x)dx, \tag{3.27}$$

where we weigh all the conditional probabilities by the PDF of X before integrating them together to form the overall probability of the event $Y \in C$. Finally, just as in the discrete random variable case, we can employ a form of substitution law for continuous random variables when dealing with conditional probability, which is defined by

$$P((X, Y) \in A|X = x) = P((x, Y) \in A|X = x). \tag{3.28}$$

Note that if X and Y are independent, then the joint density factors, yielding the following expression for the conditional density:

$$\begin{aligned} f_{Y|X}(y|x) &= \frac{f_{XY}(x, y)}{f_X(x)} \\ &= \frac{f_X(x)f_Y(y)}{f_X(x)} \\ &= f_Y(y), \end{aligned} \tag{3.29}$$

which implies that when the two random variables are independent, we do not need to condition one event on the other.

Similarly, the conditional expectation when dealing with continuous random variables is defined as the following expression employing the conditional density;

namely,

$$E[g(Y)|X = x] = \int_{-\infty}^{+\infty} g(y) f_{Y|X}(y|x) dy. \tag{3.30}$$

Furthermore, the law of total probability for a conditional expectation is given as

$$E[g(X, Y)] = \int_{-\infty}^{+\infty} E[g(X, Y)|X = x] f_X(x) dx, \tag{3.31}$$

and the substitution law for a conditional expectation is defined as

$$E[g(X, Y)|X = x] = E[g(x, Y)|X = x]. \tag{3.32}$$

3.3.1 Cumulative Distribution Functions

For both PDFs and PMFs of random variables modeling random elements within a communication system, it is sometimes important to visualize the *cumulative distribution function* or CDF, especially since it provides another perspective on how the random variable behaves probabilistically. Furthermore, the CDF can sometimes be use to solve problems that would otherwise be difficult to access via some other definition.

Mathematically speaking, we can define the CDF by the following expression:

$$F_X(x) = P(X \leq x) = \int_{-\infty}^{x} f(t) dt, \tag{3.33}$$

which describes the probability that the outcome of an experiment described by the random variable X is less than or equal to the dummy variable x.

As an example, suppose that we want to calculate the probability of $P(a \leq X < b)$ using the PDF shown in Figure 3.4(a). One approach for quickly evaluating this probability is to leverage the *tail probabilities* of this distribution; namely, $P(X < a)$ (shown in Figure 3.4[b]) and $P(X < b)$ (shown in Figure 3.4[c]). Notice how the tail probabilities are actually the CDFs of X based on (3.33), where $F_X(a) = P(X < a)$ and $F_X(b) = P(X < b)$. Consequently, given that we are only interested in the region of the PDF where these two tail probabilities do not intersect, we can compute the following probability:

$$P(a \leq X < b) = P(X < b) - P(X < a) = F_X(b) - F_X(a), \tag{3.34}$$

where all we really need are the values for the CDF of X at $x = a$ and $x = b$.

Several fundamental characteristics of the CDF include the fact that $F_X(x)$ is bounded between zero and one, and that $F_X(x)$ is a nondecreasing function; that is, $F_X(x_1) \leq F_X(x_2)$ if $x_1 \leq x_2$. Furthermore, the PDF is the derivative of the CDF in terms of the dummy variable x, which we can define as:

$$f_X(x) = \frac{d}{dx} F_X(x). \tag{3.35}$$

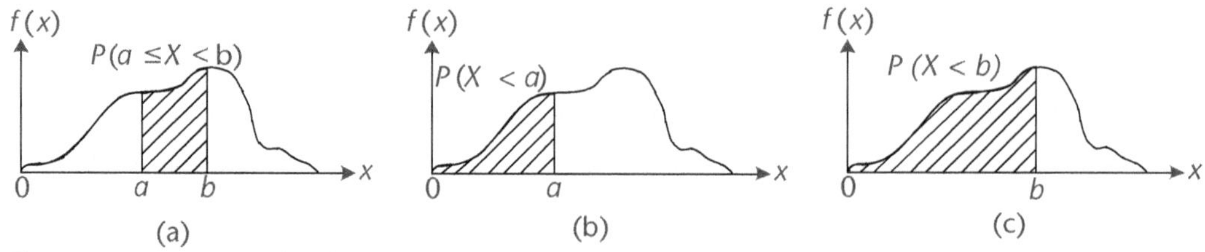

Figure 3.4 An example of how the CDF can be used to obtain the tail probabilities $P(X < a)$ and $P(X < b)$ in order to quickly calculate $P(a \leq X < b)$. (a) The region of the PDF of the random variable X that needs to be integrated in order to yield $P(a \leq X < b)$, (b) the region of the PDF of the random variable X that needs to be integrated in order to yield $P(X < a)$, and (c) the region of the PDF of the random variable X that needs to be integrated in order to yield $P(X < b)$.

> The *Q function* is a convenient way to express right-tail probabilities for Gaussian random variables, $P(X > x)$. Mathematically, this is equivalent to finding the complementary CDF of X; namely [2]:
>
> $$Q(x) = 1 - F_X(x) = 1 - P(X \leq x)$$
> $$= P(X > x) = \frac{1}{\sqrt{2\pi}} \int_{x}^{\infty} e^{-t^2/2} dt,$$
>
> where $F_X(x)$ is the CDF of X.

One important use for CDFs is having them define the exact probabilistic nature of a random element within a communication system. Noise generation, binary outputs of an information source, and random access of a wireless channel by multiple users can all be characterized exactly using CDFs. Consequently, when modeling these phenomena in a computer simulation, we use a RNG that is defined by one of these CDFs. In the MATLAB computer simulation environment, there exists a variety of RNGs that can be used in communication systems experiments, including those based on uniform, Gaussian (normal), and Rayleigh random variables. These random variables can be generated in MATLAB via the `rand` and `randn` functions, as well as their combination to create other random variables. For example, the MATLAB code in Code 3.3 produces three vectors of random values generated in such a way that they possess statistical characteristics equivalent to the uniform, Gaussian, and Rayleigh random variables. Furthermore, using the randomly generated values using these RNGs, it is possible for us to determine the probability densities such that we can then generate their cummulative distribution functions as well as calculate the probability that these random variables produce a value between 0.7 and 1.0.

To obtain values that possess uniform and Gaussian distributions in MATLAB, one can simply use the `rand` and `randn` functions. If a very large number of these random values are generated, it is possible to observe the uniform and Gaussian PDFs, as shown in Figures 3.5(a) and 3.5(c). Since the cumulative distribution function (CDF) is the progressive accumulation of the PDFs from negative infinity to positive infinity, those can also be readily generated from the PDF data, as shown

Code 3.3 Information Source: **random_example.m**

```
%%%%%%%%%%%%%%%%%%%%%%%%%%%%%%%%%%%%%%%%%%%%%%%%%%%%%%%%%%%%%
% Random Variable PDFs and CDFs

% Define simulation parameters
L = 1000000; % Length of random samples
mean_normal = 1; stddev_normal = 2;
                    % Mean and standard deviation of normal RV values
res_hist = 100; % Histogram resolution

% Generate random samples for different distributions
b_unif = rand(1,L); % Uniform random values
b_normal = (stddev_normal*randn(1,L) + mean_normal);
                    % Normal random values
b_rayleigh = (sqrt(randn(1,L).^2 + randn(1,L).^2));
                    % Rayleigh random values

% Obtain cumulative distribution functions
[N_unif, edges_unif] = histcounts(b_unif,res_hist);
N_unif_cum = cumsum(N_unif)./L;
[N_normal, edges_normal] = histcounts(b_normal,res_hist);
N_normal_cum = cumsum(N_normal)./L;
[N_rayl, edges_rayl] = histcounts(b_rayleigh,res_hist);
N_rayl_cum = cumsum(N_rayl)./L;

% Calculate probability of values between 0.7 and 1
x_lower = 0.7;
x_upper = 1.0;
unif_ind_range = find((x_lower <= edges_unif)
                        & (edges_unif < x_upper));
normal_ind_range = find((x_lower <= edges_normal)
                        & (edges_normal < x_upper));
rayl_ind_range = find((x_lower <= edges_rayl)
                        & (edges_rayl < x_upper));
prob_unif = sum(N_unif(unif_ind_range))./L;
prob_normal = sum(N_normal(normal_ind_range))./L;
prob_rayl = sum(N_rayl(rayl_ind_range))./L;
```

for the uniform and Gaussian random values in Figures 3.5(b) and 3.5(d). As for the values produced by a Rayleigh random variable, a quick way of producing these values is to take two *independently and identically distributed* (i.i.d.) Gaussian random variables, take the square of both values, sum them together, and then take their square root. As a result of this operation, and given a very large number of values generated, it is possible to create a PDF and a CDF of a Rayleigh random variable as shown in Figures 3.5(e) and 3.5(f). Note that if one wants to find the probability of a randomly generated value produced by these functions between 0.7 and 1.0, simply either sum up the density values between this range or take the CDF values at these end points and subtract them from each other.

3.4 Time-Varying Randomness in Communication Systems

Until now we have been exploring how to model random phenomena where these probabilistic characteristics remain the same throughout time. Although this

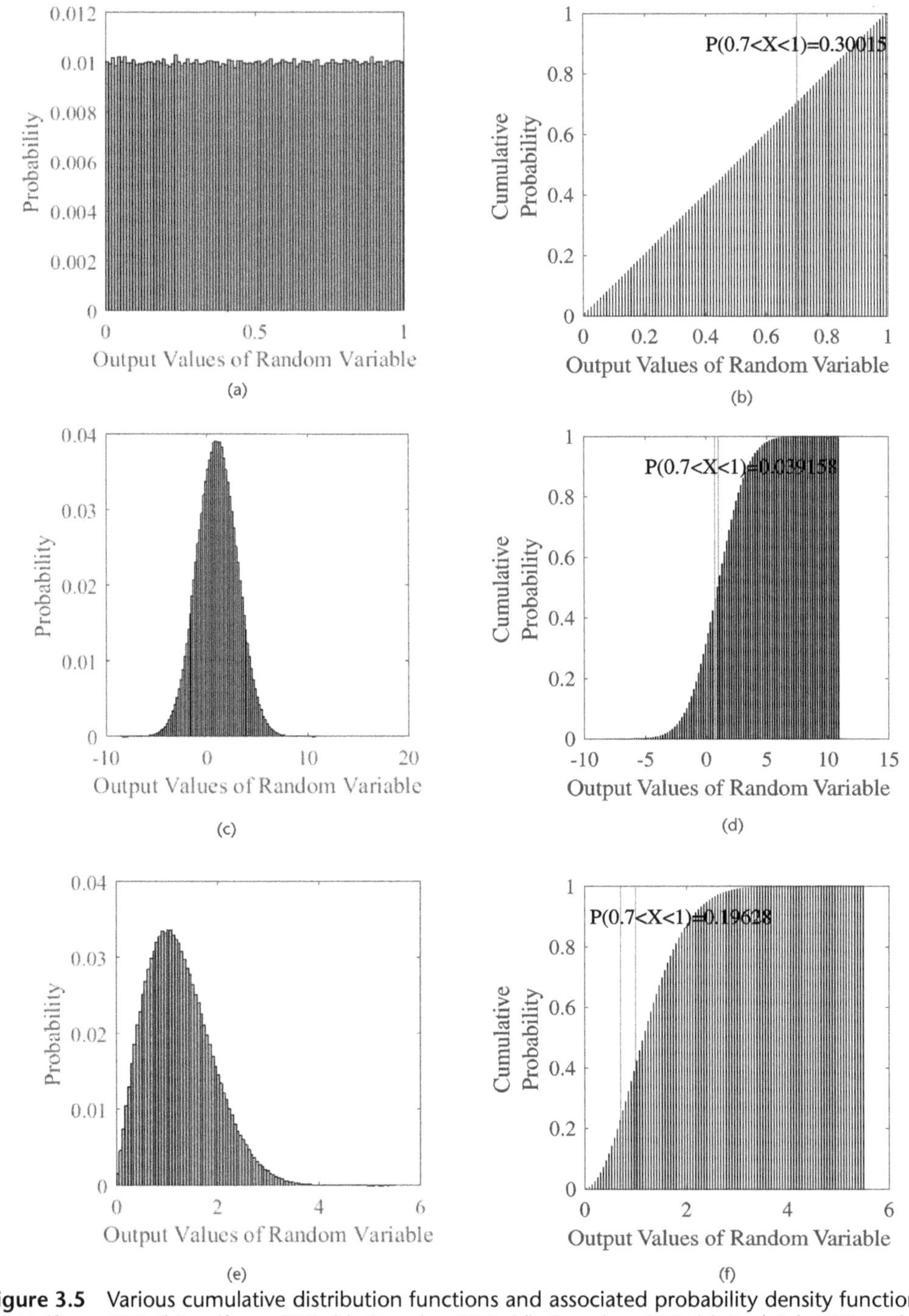

Figure 3.5 Various cumulative distribution functions and associated probability density functions. (a) Uniform PDF, (b) uniform CDF, (c) Gaussian PDF, (d) Gaussian CDF, (e) Rayleigh PDF, and (f) Rayleigh CDF.

simplifies the mathematical derivation of these models, this may not accurately describe the random phenomena. For example, the binary output values from an information source might change over time depending on the real-world data being encoded, such as security camera footage of a dynamic urban environment or the internet traffic of a human user on a computer. Consequently, it is necessary to develop a more comprehensive mathematical representation of these random phenomena that are also functions of time. We refer to these representations as *random processes* or *stochastic processes*. A random process is a family of time domain functions depending on the parameters t and ω; that is, we can define a random process by the function:

$$X(t) = X(t, \omega), \tag{3.36}$$

where the left-handed side is the shortened representation of a random process that implicitly assumes the existence of a mapping of an outcome ω from the sample space Ω to a real-valued number. Note that the domain of ω is Ω while the domain of t is either $\mathbb{R}$ for continuous-time random processes or $\mathbb{Z}$ for discrete-time random processes. An illustration depicting how a random process consists of a family of time domain functions is shown in Figure 3.6.

Suppose we have a random process that is noncountable infinite for each time instant t. Given this situation, we can define its first-order distribution $F(x, t)$ and first-order density $f(x, t)$ as

$$F(x, t) = P(X(t) \leq x) \tag{3.37}$$

and

$$f(x, t) = \frac{\partial F(x, t)}{\partial x}. \tag{3.38}$$

For determining the statistical properties of a random process, knowledge from the function $F(x_1, \ldots, x_n; t_1, \ldots, t_n)$ is required. However, in most communication system applications only certain averages are actually needed. For instance, one of the mostly commonly used statistical characterizations for a random process is the *mean function*, where the mean function $\mu_X(t)$ of a random process $X(t, \omega)$ is the expectation of the random process at a specific time instant t. This can be mathematically represented by

$$\mu_X(t) = E[X(t, \omega)]. \tag{3.39}$$

Another useful statistical characterization tool for a random process $X(t, \omega)$ is the *autocorrelation function* $R_{XX}(t_1, t_2)$, where we evaluate the amount of

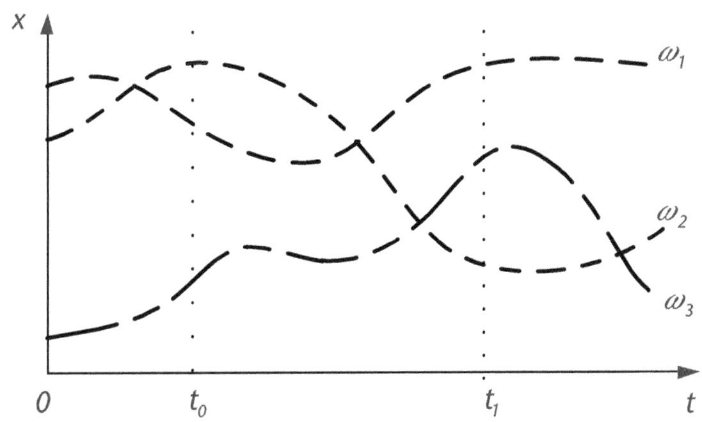

Figure 3.6 Illustration of a random process $X(t, \omega)$.

correlation that the random process $X(t,\omega)$ possesses at two different time instants t_1 and t_2. We can define this mathematically by the expression

$$R_{XX}(t_1,t_2) = E[X(t_1,\omega)X^*(t_2,\omega)] = \int_{-\infty}^{+\infty}\int_{-\infty}^{+\infty} x_1 x_2^* f(x_1,x_2;t_1,t_2)dx_1 dx_2. \quad (3.40)$$

Note that the value of the diagonal for $R_{XX}(t_1,t_2)$ is the average power of $X(t,\omega)$; namely,

$$E[|X(t,\omega)|^2] = R_{XX}(t,t). \quad (3.41)$$

Several other useful properties and observations about the autocorrelation function include the following:

1. Since $R_{XX}(t_1,t_2) = E[X(t_1,\omega)X^*(t_2,\omega)]$, then $R_{XX}(t_2,t_1) = E[X(t_2,\omega)X^*(t_1,\omega)] = R^*_{XX}(t_1,t_2)$.
2. We have $R_{XX}(t,t) = E[|X(t,\omega)|^2] \geq 0$.
3. A random process for which $E[|X(t,\omega)|^2] < \infty$ for all t is called a *second-order process*.
4. For $R_{XX}(t,t) = E[|X(t,\omega)|^2] \geq 0$ and given time instants t_1 and t_2, we have the following inequality:

$$|R_{XX}(t_1,t_2)| \leq \sqrt{E[|X(t_1,\omega)|^2]E[|X(t_2,\omega)|^2]}.$$

5. A normalized process occurs when $X(t,\omega)/\sqrt{C_{XX}(t,t)}$.

An extension of the definition for the autocorrelation function is the *autocovariance function* $C_{XX}(t_1,t_2)$ of $X(t,\omega)$, which is the covariance of the random process $X(t,\omega)$ at time instants t_1 and t_2. Mathematically, we can represent the autocovariance function by the expression.

$$C_{XX}(t_1,t_2) = R_{XX}(t_1,t_2) - \mu_X(t_1)\mu^*_X(t_2). \quad (3.42)$$

Note that for $t_1 = t_2$, the autocovariance function produces the variance of $X(t,\omega)$. Furthermore, we can sometimes represent the autocovariance function of a random process $X(t,\omega)$ using a normalized metric called the correlation coefficient, which we define as

$$\rho_{XX}(t_1,t_2) = \frac{C_{XX}(t_1,t_2)}{\sqrt{C_{XX}(t_1,t_1)C_{XX}(t_2,t_2)}}. \quad (3.43)$$

3.4.1 Stationarity

Although random processes may possess a significant amount variability across time, there does exist a subset of random processes that exhibit the same behavior at any two time instants; that is, the random process is time-invariant. We refer to these types of random processes as *stationary processes*. Two common forms of stationary processes are *strict-sense stationary* (SSS) random processes and *wide-sense stationary* (WSS) random processes. A random process is SSS whenever its statistical properties are invariant to a shift of the origin; that is, the random process

$X(t,\omega)$ and $X(t+c,\omega)$ both possess the same statistics for any time shift c. Therefore, the nth-order density of an SSS random process would be equal to, by definition, the following expression:

$$f(x_1,\ldots,x_n;t_1,\ldots,t_n) = f(x_1,\ldots,x_n;t_1+c,\ldots,t_n+c) \tag{3.44}$$

for any time shift c.

It follows that $f(x;t) = f(x;t+c)$ for any time shift c, which means that the first-order density of $X(t,\omega)$ is independent of the time t; namely,

$$f(x;t) = f(x). \tag{3.45}$$

Furthermore, $f(x_1,x_2;t_1,t_2) = f(x_1,x_2;t_1+c,t_2+c)$ is independent of c for any value of c. Consequently, this means that the density becomes

$$f(x_1,x_2;t_1,t_2) = f(x_1,x_2;\tau), \text{ where } \tau = t_1 - t_2. \tag{3.46}$$

Thus, the joint density of the random process at time instants t and $t+\tau$ is independent of t and is equal to $f(x_1,x_2;\tau)$.

Although SSS random processes can yield mathematically tractable solutions based on their useful time-invariant property, the occurrence of SSS random processes in actual communication systems is not very frequent. On the other hand, the WSS random processes occur more frequently in the analyses of communication systems. A random process $X(t,\omega)$ is considered to be WSS whenever both of the following properties are true:

- The mean function $\mu_X(t)$ does not depend on time t; that is, $\mu_X(t) = E[X(t,\omega)] = \mu_X$.
- The autocorrelation function $R_{XX}(t+\tau,t)$ only depends on the relative difference between t and $t+\tau$; that is, $R_{XX}(t+\tau,t) = R_{XX}(\tau)$.

Several observations about WSS random processes include the following:

- The average power of a WSS random process is independent of time since $E[|X(t,\omega)|^2] = R_{XX}(0)$.
- The autocovariance function of a WSS random process is equal to $C_{XX}(\tau) = R_{XX}(\tau) - |\mu_X|^2$.
- The correlation coefficient of a WSS random process is given by $\rho_{XX}(\tau) = C_{XX}(\tau)/C_{XX}(0)$.
- Two random processes $X(t,\omega)$ and $Y(t,\omega)$ are jointly WSS if each is WSS and their cross-correlation depends on $\tau = t_1 - t_2$.
- If the random process $X(t,\omega)$ is WSS and uncorrelated, then $C_{XX}(\tau) = q\delta(\tau)$, where q is some multiplicative constant.

There exists another form of stationarity characteristic that often occurs in wireless data transmission. A *cyclostationary* random process $Y(t)$ is defined by a mean function $\mu_Y(t)$ that is periodic across time t as well as an autocorrelation function $R_{YY}(\tau+\theta,\theta)$ that is periodic across θ for a fixed value of τ. Consequently, a cyclostationary random process $Y(t)$ with period T_0 can be described mathematically by

$$\bar{R}_{YY}(\tau) = \frac{1}{T_0}\int_0^{T_0} R_{XX}(\tau+\theta,\theta)d\theta. \tag{3.47}$$

In the area of communication systems engineering, cyclostationary random processes are often leveraged in the detection of wireless signals in noisy channel environments, where a target wireless signal will produce a unique characteristic function that will enable its identification assuming that multiple signals are present within the same spatiotemporal-frequency region.

3.5 Gaussian Noise Channels

As mentioned previously, Gaussian; that is, normal, random variables have often been used to model the noise introduced within a communication channel. In fact, many analyses of communication systems and their performance are often conducted assuming that the noisy channel possess Gaussian random behavior. Consequently, this makes the Gaussian random variable one of the most frequently used random variable in the study of communication systems.

We define the univariate Gaussian PDF as

$$f_X(x) = \frac{1}{\sqrt{2\pi\sigma^2}}e^{-(x-\mu)^2/2\sigma^2}, \tag{3.48}$$

where μ is the mean of the Gaussian random variable X, and σ^2 is the variance of X. In the case that $\mu = 0$ and $\sigma^2 = 1$, we refer to X as a standard normal random variable.

Although the univariate Gaussian distribution is frequently used in numerous applications, there are several instances where we must employ the *bivariate Gaussian distribution* in order to characterize a specific application involving two Gaussian random variables possessing some degree of correlation between each other; for example, complex baseband transmission channel with inphase and quadrature signal components. An illustration of an example of a bivariate Gaussian distribution is shown in Figure 3.7.

Mathematically speaking, the general definition for a bivariate Gaussian density with parameters μ_X, μ_Y, σ_X^2, σ_Y^2, and correlation coefficient ρ is given by

$$f_{XY}(x,y) = \frac{\exp\left(\frac{-1}{2(1-\rho^2)}\left(\left(\frac{x-\mu_X}{\sigma_X}\right)^2 - 2\rho\left(\frac{x-\mu_X}{\sigma_X}\right)\left(\frac{y-\mu_Y}{\sigma_Y}\right) + \left(\frac{y-\mu_Y}{\sigma_Y}\right)^2\right)\right)}{2\pi\sigma_X\sigma_Y\sqrt{1-\rho^2}}, \tag{3.49}$$

where the correlation coefficient is defined as

$$\rho = E\left[\left(\frac{x-\mu_X}{\sigma_X}\right)\left(\frac{y-\mu_Y}{\sigma_Y}\right)\right]. \tag{3.50}$$

Suppose we would like to model a complex baseband channel where the inphase and quadrature noise contributions are represented by bivariate Gaussian random variables. In several instances, we would like to have the inphase and quadrature

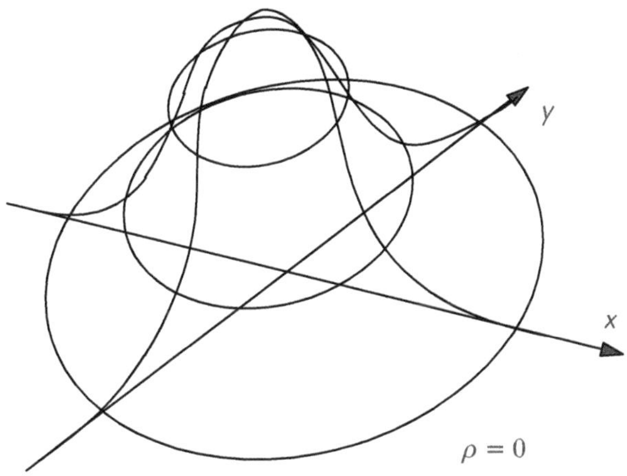

Figure 3.7 Density of a bivariate normal distribution with no correlation; that is, $\rho = 0$.

Code 3.4 Information Source: **random_example.m**

```
% Define simulation parameters
L = 1000000; % Length of data streams
res_hist = 100; % Histogram resolution
std_dev = 5; % Standard deviation of input Gaussian variables

% Create uncorrelated 2D Gaussian random variable
x_normal_1 = std_dev.*randn(1,L);
y_normal_1 = std_dev.*randn(1,L);

% Create correlated 2D Gaussian random data stream
x_normal_2 = x_normal_1+0.1*y_normal_1;
y_normal_2 = y_normal_1+0.9*x_normal_1;
```

components of this noise to be uncorrelated. In other situations, we might want them to be very correlated. As a result, we need to make sure we model these bivariate Gaussian random values accurately. The MATLAB script in Code 3.4 models these types of bivariate Gaussian random variables representing channel noise, where we have employed the function `randn` in order to generate two vectors of length `L` that contain values possessing Gaussian characteristics. Since we have generated these vectors separately, by default they are uncorrelated with each other in this situation. Furthermore, we have made both vectors to be zero mean, and they both possess standard deviations of `std_dev`. From these two vectors, we can readily obtain an uncorrelated bivariate Gaussian distributed data as well as a correlated bivariate Gaussian distributed data.

Based on the two-dimensional density functions shown in Figure 3.8, we can readily observe the difference between uncorrelated and correlated bivariate Gaussian random variables. In the uncorrelated case, the bivariate Gaussian random variables appear to be symmetric about the *x*/*y* plane, which can be seen in Figure 3.8(a) (3-dimensional viewpoint) and Figure 3.8(b) (top-down viewpoint). However, once we introduce some correlation between the inphase and quadrature components of the bivariate Gaussian random variables, this begins to warp the shape of the density, compressing it in one direction while expanding it in another.

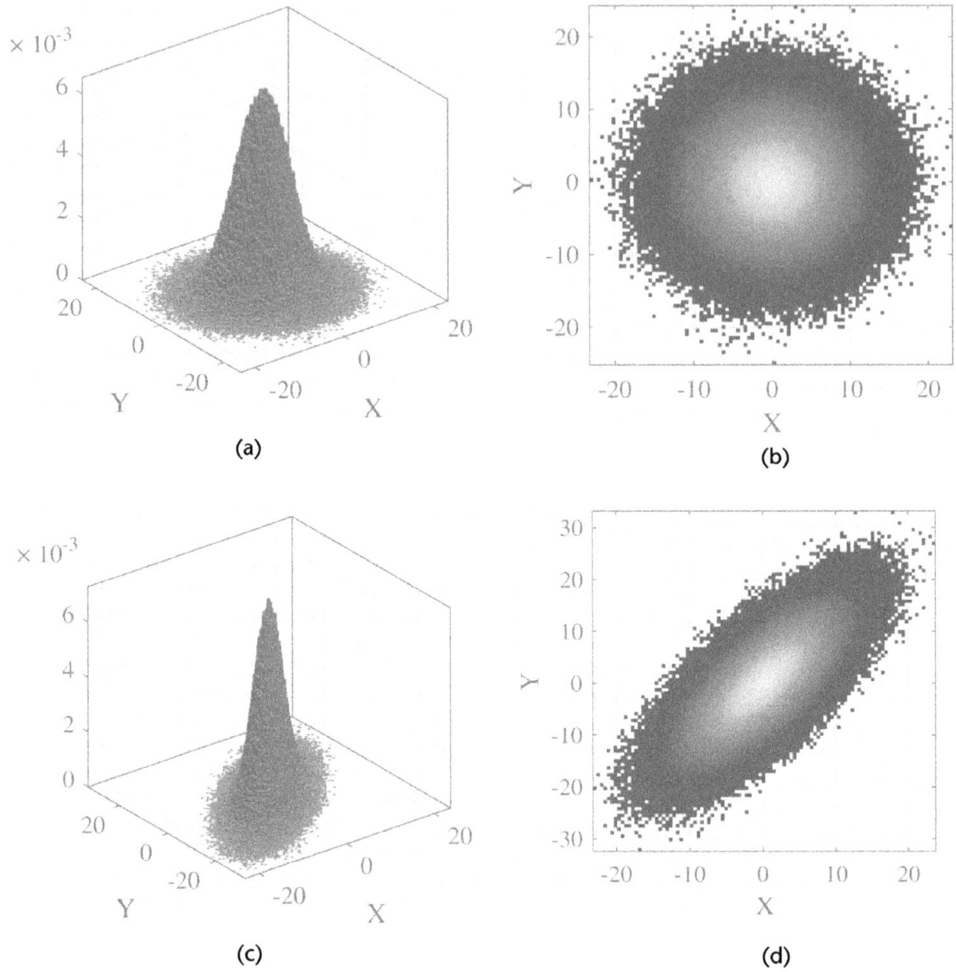

Figure 3.8 Several examples of correlated and uncorrelated bivariate Gaussian densities. (a) Uncorrelated bivariate Gaussian (3-D view), (b) uncorrelated bivariate Gaussian (top-down view), (c) correlated bivariate Gaussian (3-D view), and (d) correlated bivariate Gaussian (top-down view).

This can be observed in Figure 3.8(c) (3-dimensional viewpoint) and Figure 3.8(d) (top-down viewpoint).

3.5.1 Gaussian Processes

As mentioned before, situations exist where the probability characteristics of a random phenomena representing an element of a communication system varies over time (e.g., the properties of a noisy channel represented by a Gaussian random variable). Consequently, we need a mathematical representation that can account for the time-dependent randomness of these phenomena, especially for those events modeled by Gaussian random variables. Refering to Section 3.4, we can model a time-varying Gaussian random variable by a *Gaussian process*, which is a stochastic process whose realizations consist of random values associated with every point in a range of times (or of space) such that each such random variable has a normal distribution. Moreover, every finite collection of those random variables has a multivariate normal distribution.

Gaussian processes are important in statistical modeling because of properties inherited from the normal distribution. For example, if a random process is modeled as a Gaussian process, the distributions of various derived quantities can be obtained explicitly. Such quantities include the average value of the process over a range of times and the error in estimating the average using sample values at a small set of times.

Given the following expression:

$$y = \int_0^T g(t)X(t)\,dt \tag{3.51}$$

we can say that $X(t)$ is a Gaussian process if

- $E(y^2)$ is finite (i.e., does not blow up).
- Y is Gaussian-distributed for every $g(t)$.

Note that the random variable Y has a Gaussian distribution, where its PDF is defined as

$$f_Y(y) = \frac{1}{\sqrt{2\pi\sigma_Y^2}} e^{\frac{-(y-\mu_Y)^2}{2\sigma_Y{}^2}}, \tag{3.52}$$

where μ_Y is the mean and $\sigma_Y{}^2$ is the variance. Such processes are important because they closely match the behavior of numerous physical phenomena, such as *additive white Gaussian noise* (AWGN).

> **Q** Why is an uncorrelated random process referred to as white, such as in the case of additive white Gaussian noise?

3.6 Power Spectral Densities and LTI Systems

To analyze a communication system in the frequency domain, the power spectral density (PSD), $S_{XX}(f)$, is often used to characterize the signal, which is obtained by taking the Fourier transform of the autocorrelation $R_{XX}(\tau)$ of the WSS random process $X(t)$. The PSD and the autocorrelation of a function, $R_{XX}(\tau)$, are mathematically related by the *Einstein-Wiener-Khinchin* (EWK) relations; namely,

$$S_{XX}(f) = \int_{-\infty}^{\infty} R_{XX}(\tau)e^{-j2\pi f\tau}\,d\tau \tag{3.53}$$

$$R_{XX}(f) = \int_{-\infty}^{\infty} S_{XX}(\tau)e^{+j2\pi f\tau}\,df \tag{3.54}$$

A very powerful consequence of the EWK relations is its usefulness when attempting to determine the autocorrelation function or PSD of a WSS random process that is the output of an LTI system whose input is also a WSS random process. Specifically, suppose we denote $H(f)$ as the frequency response of an LTI system $h(t)$. We can then relate the power spectral density of input and output random processes by the following equation:

$$S_{YY}(f) = |H(f)|^2 S_{XX}(f), \tag{3.55}$$

where $S_{XX}(f)$ is the PSD of input random process and $S_{YY}(f)$ is the PSD of output random process. This very useful relationship is illustrated in Figure 3.9.

To understand the impact of an LTI system on the PSD of a random process, we can use the MATLAB script in Code 3.5, where we have a bivariate uniform random number generator producing inphase and quadrature values. To highlight the impact of LTI systems on PSDs, we have designed two filters using the `firls` function. One of the filters has a relatively large passband while the other filter has a relatively small passband. The bivariate uniform random values are then filtered by both systems and we can observe the resulting PSDs in Figure 3.10.

The three dimensional and top-down perspectives of the original bivariate uniform random values are shown in Figures 3.10(a) and 3.10(b). We observe that the density almost appears to be a rectangular block, which is what we are expecting from a bivariate uniform. Once we filter this data using the filter with the narrow passband region and observe the resulting PSD, we can readily notice the effects of the filtering operation, with most of the PSD being filtered away at the peripherals. This is evident in the three-dimensional and top-down perspectives of these filtered bivariate uniform random values shown in Figures 3.10(c) and 3.10(d). When using the filter with the relatively larger passband, we observe that the perimeter of the PSD is not as filtered, as shown in the three-dimensional and top-down perspectives of these filtered bivariate uniform random values in Figures 3.10(e) and 3.10(f). Consequently, the primary take-away point from this example is that the filtering operations of LTI systems can have a significant impact on the PSDs of random processes.

3.7 Narrowband Noise

Now that we have a solid foundation with respect to random variables and random processes, it is time to apply this knowledge to the application of narrowband transmissions for wireless communication systems. In general, most transmissions are designed to be *bandlimited* since there are constraints on the amount of wireless spectrum that any one transmission can use. These constraints are necessary since there is a limited amount of wireless spectrum available and a growing number of wireless applications and users seeking to use this spectrum for their transmissions.

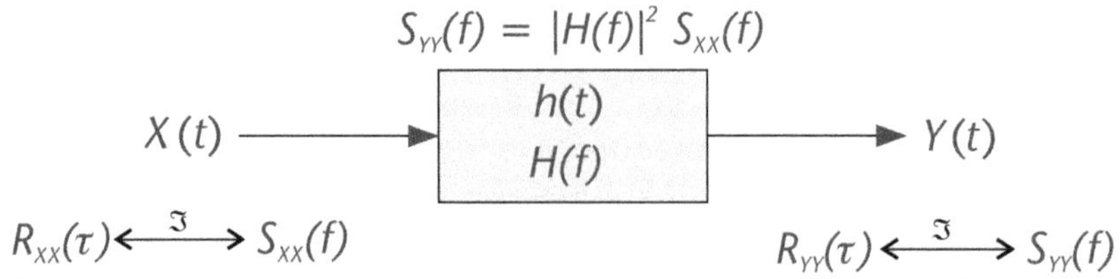

Figure 3.9 An example of how the an LTI system $h(t)$ can transform the PSD between the WSS random process input $X(t)$ and the WSS random process output $Y(t)$.

Code 3.5 Information Source: **random_example.m**

```
% Define simulation parameters
L = 1000000; % Length of data streams
res_hist = 100; % Histogram resolution
cutoff_freq1 = 0.2; % Small passband LPF
cutoff_freq2 = 0.7; % large passband LPF
filt_coeffs1 = firls(13,[0 cutoff_freq1 cutoff_freq1+0.02 1],
                                                    [1 1 0 0]);
filt_coeffs2 = firls(13,[0 cutoff_freq2 cutoff_freq2+0.02 1],
                                                    [1 1 0 0]);

% Create input 2D Gaussian random variable
x_in = rand(1,L);
y_in = rand(1,L);

% Filter input random data stream
filt_output1 = filter(filt_coeffs1,1,(x_in+1j.*y_in));
x_out1 = real(filt_output1);
y_out1 = imag(filt_output1);
filt_output2 = filter(filt_coeffs2,1,(x_in+1j.*y_in));
x_out2 = real(filt_output2);
y_out2 = imag(filt_output2);
```

One of the key elements of a narrowband communication system is the narrowband filters at both the transmitter and receiver, which are designed to only allow only the modulated signals to pass. However, these narrowband filters also allow a portion of the noise intercepted at the receiver to pass through since it is very difficult to separate out the noise from the modulated signals. If it turns out that the noise is white (i.e., uncorrelated), then narrowband noise will take on the shaped of a cosine-modulated bandpass filter response. This is due to the fact that the white noise prior to filtering will have a PSD that is flat and spanning the entire frequency range from negative infinity to positive infinity. When processed by bandpass filters, the resulting narrowband noise PSD will take on the shape of the square of the magnitude response of the bandpass filters since the convolution of the noise with the filters in the time domain translates into the multiplication of the noise PSD with the filter magnitude response (see Section 3.6).

In terms of setting up a convenient mathematical framework to represent narrowband noise, there are two approaches: *in-phase/quadrature representation* and *envelope/phase representation*. Both approaches can describe a complex value x using the definition $x = Ae^{j\phi} = a + jb$, where $x \in \mathbb{C}$ and the envelope A, phase ϕ, inphase component a, and quadrature component b are real numbers $A, \phi, a, b \in \mathbb{R}$. The relationships between the in-phase/quadrature representation and the envelope/phase representation is described by the following:

$$A = \sqrt{a^2 + b^2}\phi = \tan^{-1}(b/a) \tag{3.56}$$

$$a = A\cos(\phi) \quad b = A\sin(\phi) \tag{3.57}$$

Thus, we can describe the in-phase/quadrature representation of narrowband noise in the complex baseband domain via the equation

$$\tilde{n}(t) = n_I(t) + jn_Q(t), \tag{3.58}$$

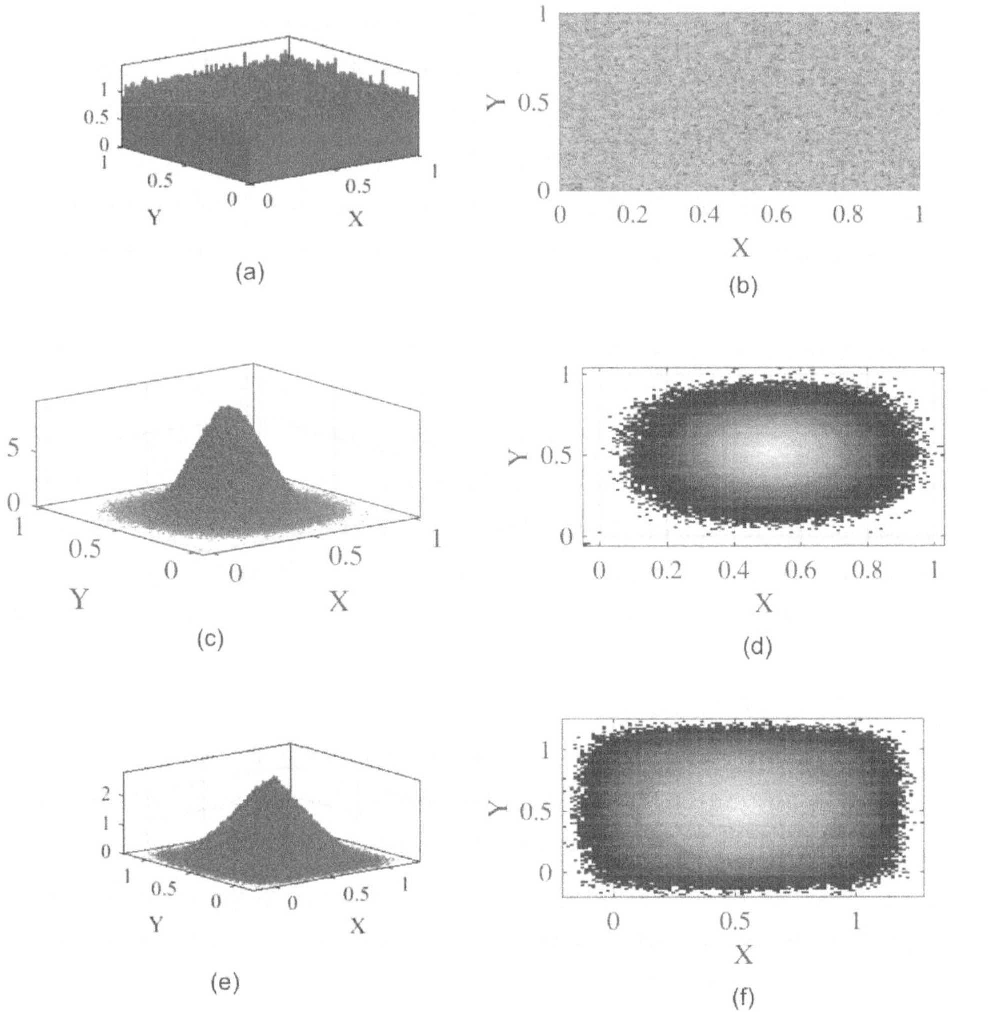

Figure 3.10 Example of how filtering can impact the output power spectral densities of random processes. (a) Power spectral density of input signal (3-D view), (b) power spectral density of input signal (top-down view), (c) power spectral density of output signal using filter coefficients 1 (3-D view), (d) power spectral density of output signal using filter coefficients 1 (top-down view), (e) power spectral density of output signal using filter coefficients 2 (3-D view), and (f) power spectral density of output signal using filter coefficients 2 (top-down view).

which can then be expressed in the bandpass version of a narrowband noise signal as

$$n(t) = \text{Real}\left\{\tilde{n}(t)e^{j2\pi f_c t}\right\}. \tag{3.59}$$

Using Euler's relationship; namely, $e^{j\omega} = \cos(\omega) + j\sin(\omega)$, we get the following expression:

$$n(t) = n_I(t)\cos(2\pi f_c t) - n_Q(t)\sin(2\pi f_c t). \tag{3.60}$$

Several important properties of the in-phase/quadrature representation are presented in Table 3.4.

Table 3.4 Several Important Properties of In-phase/Quadrature Representation

Both $n_I(t)$ and $n_Q(t)$ have zero mean
If $n(t)$ is Gaussian, so are $n_I(t)$ and $n_Q(t)$
If $n(t)$ is stationary, $n_I(t)$ and $n_Q(t)$ are jointly stationary
PSD of $n_I(t)$ and $n_Q(t)$ equal to $S_{N_I}(f) = S_{N_Q}(f) = \begin{cases} S_N(f-f_c)+S_N(f+f_c), & -B \leq f \leq B \\ 0, & \text{otherwise} \end{cases}$
Both $n_I(t)$ and $n_Q(t)$ have the same variance as $n(t)$
If $n(t)$ is Gaussian and its PSD symmetric, then $n_I(t)$ and $n_Q(t)$ are statistically independent
The cross-spectral density between $n_I(t)$ and $n_Q(t)$ is purely imaginary, and for $-B \leq f \leq B$ it is equal to (zero otherwise) $S_{N_IN_Q}(f) = -S_{N_QN_I}(f) = j(S_N(f+f_c) - S_N(f-f_c))$

Similarly, the complex baseband version of the envelope/phase representation of narrowband noise can be written as

$$n(t) = r(t)\cos(2\pi f_c t + \phi(t)) \tag{3.61}$$

where $r(t) = \sqrt{n_I(t) + n_Q(t)}$ is the envelope and $\phi(t) = \tan^{-1}(n_Q(t)/n_I(t))$ is the phase.

In terms of the relationship between the in-phase/quadrature representation and the envelope/phase representation with respect to their joint distributions, the results are very exciting. Suppose we define the joint PDF for $n_I(t)$ and $n_Q(t)$ as a bivariate Gaussian distribution equal to

$$f_{N_IN_Q}(n_I, n_Q) = \frac{1}{2\pi\sigma^2}e^{-\frac{n_I^2+n_Q^2}{2\sigma^2}}. \tag{3.62}$$

It turns out that by using the relationships $n_I = r\cos(\phi)$ and $n_Q = r\sin(\phi)$ as well as a *Jacobian*, we obtain the following distributions:

$$f_{R\Phi}(r,\phi) = \begin{cases} \frac{r}{2\pi\sigma^2}e^{-\frac{r^2}{2\sigma^2}}, & r \geq 0 \text{ and } 0 \leq \phi \leq 2\pi \\ 0, & \text{otherwise} \end{cases} \tag{3.63}$$

which are equivalent to the combination of Rayleigh and uniform PDFs.

3.8 Application of Random Variables: Indoor Channel Model

An excellent example of where random variables are used to model a stochastic phenomenon in the design and implementation of a communication system is the indoor channel model proposed by Saleh and Valenzuela [3]. The premise of this channel model is that the indoor environment generates clusters of multipath components that result from the wireless signals reflecting off of the surrounding environment.

Mathematically speaking, they described these reflections of the transmitted signal intercepted at the receiver by the following expression:

$$h(t) = \sum_{l=0}^{\infty}\sum_{k=0}^{\infty} \beta_{kl}e^{j\theta_{kl}}\delta(t - T_l - \tau_{kl}) \tag{3.64}$$

where β_{kl} is the amplitude level of the kth ray of the lth multipath cluster, θ_{kl} is the phase value of the kth ray of the lth multipath cluster, T_l is the delay of the start of the lth multipath cluster, and τ_{kl} is the delay of the kth ray within the lth multipath cluster. In their work, Saleh and Valenzuela used channel measurements in order to characterize β_{kl} as a Rayleigh random variable, θ_{kl} as a uniform random variable, and both T_l and τ_{kl} as Poisson arrival processes with arrival rates Λ and λ. Graphically, this model can be illustrated using Figure 3.11.

3.9 Chapter Summary

In this chapter, a brief introduction to some of the key mathematical tools for analyzing and modeling random variables and random processes for communication systems has been presented. Of particular importance, the reader should understand how to mathematically manipulate Gaussian random variables, Gaussian random processes, and bivariate normal distributions since they frequently occur in digital communications and wireless data transmission applications. Furthermore, understanding how stationarity works and how to apply the EWK relations to situations involving random processes being filtered by LTI systems is vitally important, especially when dealing with the processing and treatment of received wireless signals by the communication system receiver.

3.10 Additional Readings

Although this chapter attempts to provide the reader with an introduction to some of the key mathematical tools needed to analyze and model random variables and random processes that frequently occur in many digital communication systems, the treatment of these mathematical tools is by no means rigorous or thorough. Consequently, the interested reader is encouraged to consult some of the available books that address this broad area in greater detail. For instance, the gold standard for any textbook on the subject of probability and random processes is by Papoulis and Pillai [1]. On the other hand, those individuals seeking to understand probability and random processes theory within the context of communication networks would potentially find the book by Leon-Garcia to be highly relevant [4].

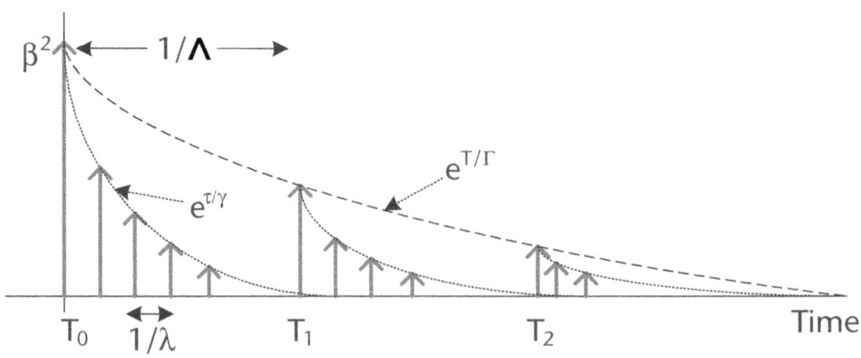

Figure 3.11 Illustration of the Saleh and Valenzuela statistical indoor channel model.

For individuals who are interested in activities involving physical layer digital communication systems and digital signal processing, such as Wiener filtering, the book by Gubner would be a decent option given that many of the examples and problems included in this publication are related to many of the classic problems in communication systems [5]. Regarding books that possess numerous solved examples and explanations, those by Hsu [6] and Krishnan [7] would serve as suitable reading material. Finally, for those individuals who are interested in studying advanced topics and treatments of random processes, the book by Grimmett and Stirzaker would be a suitable publication [8].

References

[1] Papoulis,A., and S. Unnikrishna Pillai, *Probability, Random Variables and Stochastic Processes*, Fourth Edition, New York: McGraw Hill Higher Education, 2002.

[2] Abramowitz, M., and I. Stegun, *Handbook of Mathematical Functions: with Formulas, Graphs, and Mathematical Tables*, New York: Dover Publications, 1965.

[3] Saleh, A. A. M., and R. A. Valenzuela, A Statistical Model for Indoor Multipath Propagation, *IEEE Journal on Selected Areas in Communications*, Vol. 5, No. 2, 1987, pp. 128–137.

[4] Leon-Garcia, A., *Probability, Statistics, and Random Processes for Electrical Engineering*, Third Edition, Upper Saddle River, NJ: Prentice Hall, 2008.

[5] Gubner, J. A., *Probability and Random Processes for Electrical and Computer Engineers*, Cambridge, MA: Cambridge University Press, 2006.

[6] Hsu, H., *Schaum's Outline of Probability, Random Variables, and Random Processes*, New York: McGraw-Hill, 1996.

[7] Krishnan, V., Probability and Random Processes, Hoboken, NJ: Wiley-Interscience, 2006.

[8] Grimmett, G., and D. Stirzaker, *Probability and Random Processes*, Third Edition, Oxford, UK: Oxford University Press, 2001.

CHAPTER 4
Digital Communications Fundamentals

In this chapter, we will provide an overview of several key fundamental concepts employed in the transmission of digital data. Starting with an understanding of how binary data can be used to manipulate the physical characteristics of electromagnetic waveforms, we will then look at the basic anatomy of a digital communication system before concentrating our attention on several different approaches for modulating electromagnetic waveforms using binary data. Once we have established how a digital data transmission process operates, we will then explore one of the key analytical tools for assessing the quantitative performance of such systems: the bit error rate. Finally, this chapter will conclude with an introduction to the design of digital receivers via a signal vector space perspective.

4.1 What Is Digital Transmission?

A digital transceiver is a system composed of a collection of both digital and analog processes that work in concert with each other in order to handle the treatment and manipulation of binary information. The purpose of these processes is to achieve data transmission and reception across some sort of medium, whether it is a twisted pair of copper wires, a fiber optic cable, or a wireless environment. At the core of any digital transceiver system is the *binary digit* or *bit*, which for the purposes of this book is considered to be the fundamental unit of information used by a digital communication system.

Therefore, a digital transceiver is essentially responsible for the translation between a stream of digital data represented by bits and electromagnetic waveforms possessing physical characteristics that uniquely represent those bits. Since electromagnetic waveforms are usually described by sine waves and cosine waves, several physical characteristics of electromagnetic waveforms commonly used to represent digital data per time interval T include the amplitude, phase, and carrier frequency of the waveform, as shown in Figure 4.1. Notice how different combinations of bits represent different amplitude levels or different phase values or different carrier frequency values, where each value uniquely represents a particular binary pattern. Note that in some advanced mapping regimes, binary patterns can potentially be represented by two or more physical quantities.

However, there is much more going on in a digital transceiver than just a mapping between bits and waveforms, as shown in Figure 4.2. In this illustration of the basic anatomy for a digital transceiver, we observe that there are several functional blocks that constitute a communication system. For instance, the mapping between bits and electromagnetic waveform characteristics is represented by the modulation and demodulation blocks. Additionally, there are the source

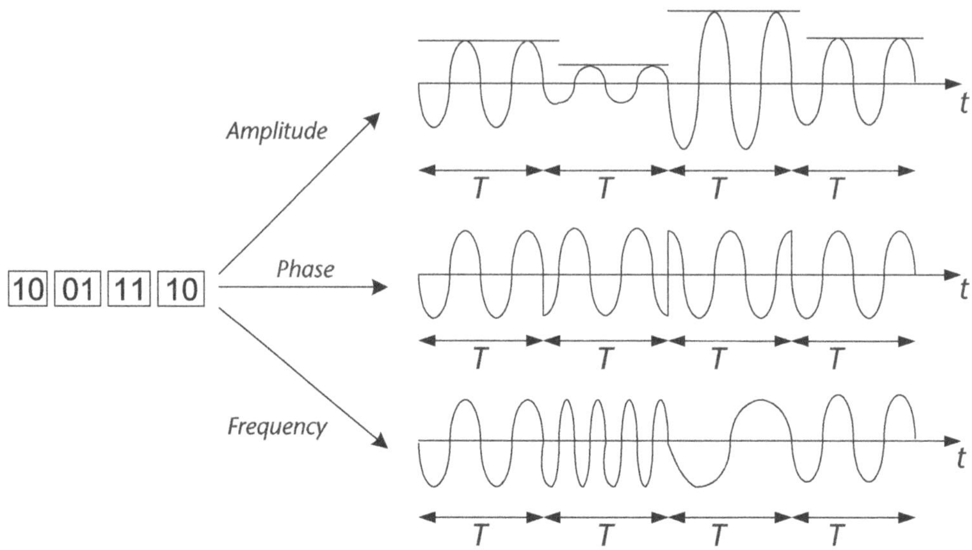

Figure 4.1 Possible mappings of binary information to EM wave properties.

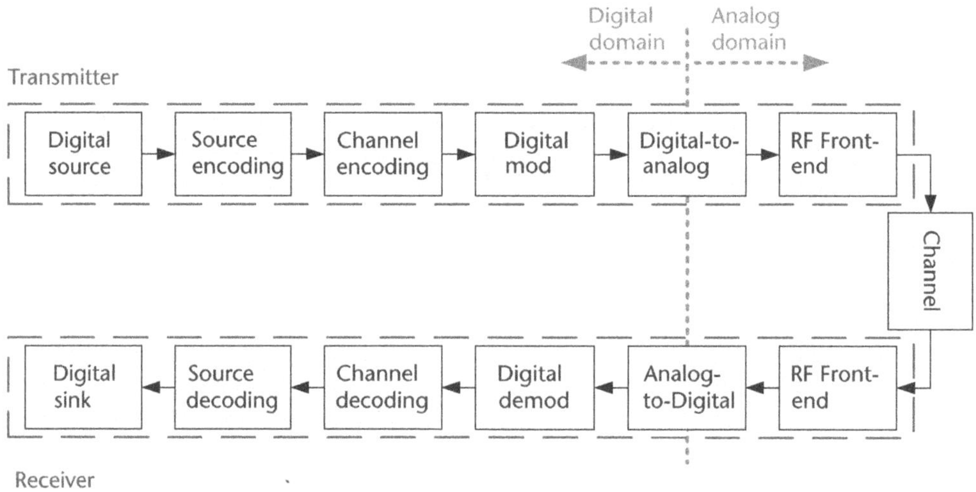

Figure 4.2 Generic representation of a digital communication transceiver.

encoding and source decoding blocks that handle the removal of redundant information from the binary data, channel encoding and channel decoding blocks that introduce a controlled amount of redundant information to protect the transmission for potential errors, and the radio frequency front end (RFFE) blocks that handle the conversation of baseband waveforms to higher carrier frequencies.

One may ask the question, Why do we need all these blocks in our digital communication system? Notice in Figure 4.2 the presence of a channel between the transmitter and the receiver of the digital transmission system. The main reason why the design of a digital communication system tends to be challenging, and that so many blocks are involved in its implementation, is due to this channel. If the channel was an ideal medium where the electromagnetic waveforms from the transmitter are clearly sent to the receiver without any sort of distortion or disturbances, then the design of digital communication systems would be trivial. However, in reality a channel introduces a variety of random impairments to a digital transmission that can potentially affect the correct reception of waveforms intercepted at the receiver. For instance, a channel may introduce some form of noise that can obfuscate some of the waveform characteristics. Furthermore, in many real-world scenarios many of these nonideal effects introduced by the channel are time-varying and thus difficult to deal with, especially if they vary rapidly in time.

Thus, under real-world conditions, the primary goal of any digital communication system is to transmit a binary message $m(t)$ and have the reconstructed version of this binary message $\hat{m}(t)$ at the output of the receiver to be equal to each other. In other words, our goal is to have $P(\hat{m}(t) \neq m(t))$ as small as needed for a particular application. The metric for quantitatively assessing the error performance of a digital communication system is referred to as the probability of error or BER, which we define as $Pe = P(\hat{m}(t) \neq m(t))$. Note that several data transmission applications possess different P_e requirements due in part to the data transmission rate. For instance, for digital voice transmission, a BER of $Pe \sim 10^{-3}$ is considered acceptable, while for an average data transmission application a BER of $Pe \sim 10^{-5} - 10^{-6}$ is deemed sufficient. On the other hand, for a very high data rate application such as those that employ fiber-optic cables, a BER of $Pe \sim 10^{-9}$ is needed since any more errors would flood a receiver given that the data rates can be extremely high.

To help mitigate errors that may occur due to the impairments introduced by the channel, we will briefly study how source encoding and channel encoding works before proceeding with an introduction to modulation.

Hands-On MATLAB Example: Communication systems convey information by manipulating the physical properties of an electromagnetic signal before it is broadcasted across a medium. Signal properties such as the amplitude, phase, and/or frequency are manipulated over time in such a manner that the receiver can interpret the message being conveyed by the transmitter. These electromagnetic broadcasts often use sine wave signals, which makes them relatively straightforward to manipulate for the purposes of conveying information. In the MATLAB script below, we generate three sine wave-based transmissions, where information is embedded in them via their amplitude levels (amplitude shift keying), phase characteristics (phase shift keying), or frequency values (frequency shift keying). In this script, we generate random binary data using the `rand` function and then round it to the nearest integer (one or zero), and then map those binary values to a corresponding amplitude, phase, or frequency value to be used by a sine wave signal. Note that since sine wave signals are continuous waveforms, we have to approximate this situation by using a large number of discrete points to model the sine wave signal as continuous.

The mapping of these random binary values to the physical attributes of a signal wave signal are shown in Figure 4.3, where we can readily observe how the amplitude (Figure 4.3[b]), phase (Figure 4.3[c]), and frequency (Figure 4.3[d]) values change over time in order to represent the binary values being transmitted to the receiver (see Figure 4.3[a]). In all three case, we use the exact same sine wave signal as the basis for communicating this information, but the sine wave characteristics are changing as a function of time.

Code 4.1 Sending Binary Data via Sinusoidal Signal Manipulation: **chapter4.m**

```
%%%%%%%%%%%%%%%%%%%%%%%%%%%%%%%%%%%%%%%%%%%%%%%%%%%%%%%%%%%%%
% Sending binary data via sinusoidal signal manipulation

% Parameters
sig_len = 1000; % Signal length (in samples)
sampl_per_bin = 100; % Samples per binary representation
bin_data_len = sig_len/sampl_per_bin;
        % Length of binary stream is a multiple of signal length
bin_data = round(rand(1,bin_data_len));

% Create sinusoidal carriers
sig_carrier_base = sin(2*pi*(0:(1/sampl_per_bin):
        (1-(1/sampl_per_bin)))); % Baseline carrier
sig_carrier_freq = sin(2*2*pi*(0:(1/sampl_per_bin):
        (1-(1/sampl_per_bin)))); % Double frequency
sig_carrier_phase = sin(2*pi*(0:(1/sampl_per_bin):
        (1-(1/sampl_per_bin)))+(pi/4)); % Phase shifted by 45 degrees

% Modulate sinusoidal carrier via amplitude, phase, and frequency
% manipulations
sig_bin = []; % Binary waveform
sig_ask = []; % Amplitude modulated
sig_psk = []; % Phase modulated
sig_fsk = []; % Frequency modulated
for ind = 1:1:bin_data_len,
    if (bin_data(ind)==1)
        sig_bin = [sig_bin ones(1,sampl_per_bin)];
        sig_ask = [sig_ask sig_carrier_base];
        sig_psk = [sig_psk sig_carrier_base];
        sig_fsk = [sig_fsk sig_carrier_base];
    else
        sig_bin = [sig_bin zeros(1,sampl_per_bin)];
        sig_ask = [sig_ask 0.5*sig_carrier_base];
        sig_psk = [sig_psk sig_carrier_phase];
        sig_fsk = [sig_fsk sig_carrier_freq];
    end;
end;
```

4.1.1 Source Encoding

One of the goals of any communication system is to efficiently and reliably communicate information across a medium from a transmitter to a receiver. As a result, it would be ideal if all the redundant information from a transmission could be removed in order to minimize the amount of information that needs to be sent across the channel, which would ultimately result in a decrease in the amount of time, computational resources, and power being expended on the transmission. Consequently, source encoding is a mechanism designed to remove redundant information in order to facilitate more efficient communications.

The way source encoding operates is by taking a sequence of source symbols $\underline{u}$ and mapping them to a corresponding sequence of source encoded symbols $\underline{v}$, $v_i \in \underline{v}$ as close to random as possible and the components of $\underline{v}$ are uncorrelated (i.e., unrelated). Thus, by performing this source encoding operation we hopefully achieve some level of redundancy minimization in $v_i \in \underline{v}$, thus limiting the amount

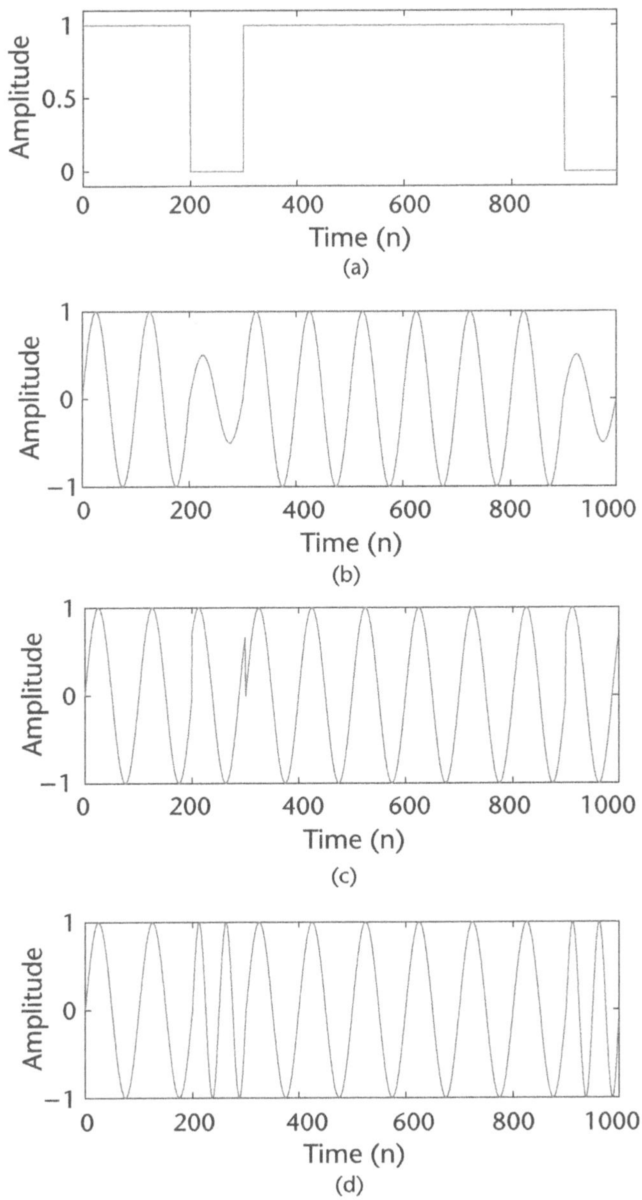

Figure 4.3 Examples of amplitude, phase, and frequency representations of binary transmissions. These sine wave properties are the building blocks for most commonly used modulation schemes used in digital communication systems. (a) Binary signal, (b) amplitude shift keying, (c) phase shift keying, and (d) frequency shift keying.

of wasted radio resources employed in the transmission of otherwise predictable symbols in $\underline{u}$. In other words, a source encoder removes redundant information from the source symbols in order to realize efficient transmission. Note that in order to perform source encoding, the source symbols need to be digital.

Hands-On MATLAB Example: Source coding exploits redundancy in a collection of data by removing it and replacing that redundancy with a short

> A single analog television channel occupies 6 MHz of frequency bandwidth. On the other hand, up to eight digitally encoded television channels can fit within the same frequency bandwidth of 6 MHz.

codeword, thus reducing the overall amount of information. In the following MATLAB script, we will illustrate how source coding works when it is applied to data possessing different amounts of redundancy. Using a combination of the `rand` and `round` functions, we generate two binary data vectors, with one possessing an equal amount of one and zero values while the other vector possesses approximately 90% one values and 10% zero values. To compress the data in these vectors, we use an encoding technique where we take all continuous strings of ones in each vector and replace it with a decimal value representing the length of these strings of one. For example, if a binary vector existed, and 15 one values exist betwen two zero values, we would replace those 15 one values with a decimal value (or equivalent codeword) indicating that 15 one values exist in that part of the binary vector.

Based on the implemented encoding scheme, our intuition would dictate that the binary vector with the 90/10 ratio of one to zero values would be compressed more relative to the binary vector with the 50/50 ratio since the former would have a greater likelihood of having long strings of one values to compress. Referring to Figure 4.4, our intuition is confirmed, with a significant reduction in size for the compressed 90/10 binary data stream. On the other hand, we observe that the 50/50 binary vector does not even compress but rather grow in size. This is due to the fact that the amount of overhead needed to replace every string of one values with a corresponding codeword actually takes up more information than the original binary sequence. Consequently, when performing source coding, it is usually worth our while to compress data streams that possess obvious amounts of redundancy, otherwise we can actually make the situation even more inefficient.

4.1.2 Channel Encoding

To protect a digital transmission from the possibility of its information being corrupted, it is necessary to introduce some level of controlled redundancy in order to reverse the effects of data corruption. Consequently, channel encoding is designed to correct for channel transmission errors by introducing controlled redundancy into the data transmission. As opposed to the redundancy that is removed during the source encoding process, which is random in nature, the redundancy introduced by

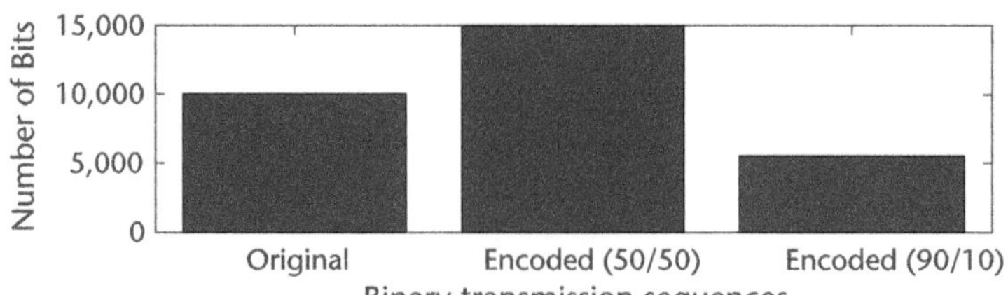

Figure 4.4 Impact of source coding on binary transmissions.

Code 4.2 Reducing Amount of Data Transmission Using Source Coding: **chapter4.m**

```
% Define parameters
len = 10000; % Length of binary data stream

% Have two binary sources, one 50/50 and the other 90/10 in terms of ones
% and zeros
bin1 = round(rand(1,len)); % 50/50 binary
bin2 = round(0.5*rand(1,len)+0.45); %90/10 binary

% Encode strings of ones in terms of the length of these strings
enc_bin1 = [];
enc_bin2 = [];
for ind = 1:1:len,
    if (bin1(ind) == 1)  % Encoding 50/50
        if (ind == 1)
            enc_bin1 = 1;
        else
            enc_bin1(end) = enc_bin1(end)+1;
        end;
    else
        enc_bin1 = [enc_bin1 0];
    end;
    if (bin2(ind) == 1)  % Encoding 90/10
        if (ind == 1)
            enc_bin2 = 1;
        else
            enc_bin2(end) = enc_bin2(end)+1;
        end;
    else
        enc_bin2 = [enc_bin2 0];
    end;
end;

% Find size of encoded binary streams
% (assume all one string length values possess the same number of bits)
ind1 = find(enc_bin1 ~= 0);
ind2 = find(enc_bin2 ~= 0);
[largest_ebin1,ind_largest_ebin1] = max(enc_bin1(ind1));
[largest_ebin2,ind_largest_ebin2] = max(enc_bin2(ind2));
numbits1 = length(dec2bin(largest_ebin1)-'0');
numbits2 = length(dec2bin(largest_ebin2)-'0');
total_size_ebin1 = length(ind1)*numbits1 + length(find(enc_bin1 == 0));
total_size_ebin2 = length(ind2)*numbits2 + length(find(enc_bin2 == 0));
```

a channel encoding is specifically designed to combat the effects of bit errors in the transmission (i.e., the redundancy possesses a specific structure known to both the transmitter and receiver).

In general, channel encoding operates as follows: Each vector of a source encoded output of length K; namely, v_l where $l = 1, 2, ..., 2^K$, is assigned a unique *codeword* such that the vector $v_l = (101010\ldots)$ is assigned a unique codeword $c_l \in \mathbb{C}$ of length N, where $\mathbb{C}$ is a *codebook*. During this process, the channel encoder has introduced $N - K = r$ controlled number of bits to the channel encoding process. The *code rate* of a communications system is equal to the ratio of the number of information bits to the size of the codeword (i.e., the code rate is equal to k/N).

When designing a codebook for a channel encoder, it is necessary to quantitatively assess how well or how poorly the codewords will perform in a situation involving data corruption. Consequently, the *Hamming distance* is often used to determine the effectiveness of a set of codewords contained within a codebook by evaluating the relative difference between any two codewords. Specifically, the Hamming distance $d_H(c_i, c_j)$ between any two codewords, say c_i and c_j, is equal to the number of components in which c_i and c_j are different. When determining the effectiveness of a codebook design, we often are looking for the minimum Hamming distances between codewords; that is,

$$d_{H,\min} = \min_{c_i, c_j \in \mathbb{C},\ i \neq j} d_H(c_i, c_j), \tag{4.1}$$

since our goal is to maximize the minimum Hamming distance in a given codebook to ensure that the probability of accidentally choosing a codeword other than the correct codeword is kept to a minimum. For example, suppose we have a codebook consisting of $\{101, 010\}$. We can readily calculate the minimum Hamming distance to be equal to $d_{H,min} = 3$, which is the best possible result. On the other hand, a codebook consisting of $\{111, 101\}$ possesses a minimum Hamming distance of $d_{H,min} = 1$, which is relatively poor in comparison to the previous codebook example.

In the event that a codeword is corrupted during transmission, *decoding spheres* (also known as *Hamming spheres*) can be employed in order to make decisions on the received information, as shown in Figure 4.5, where codewords that are corrupted during transmission are mapped to the nearest eligible codeword. Note that when designing a codebook, the decoding spheres should not overlap in order to enable the best possible decoding performance at the receiver (i.e., $\rightarrow d_{H,\min} = 2t + 1$).

Hands-On MATLAB Example: Let us apply repetition coding to an actual vector, using a range of repetition rates, and observe its impact when the binary vector is exposed to random bit-flips. The MATLAB script below takes the original binary vector `bin_str` and introduces controlled redundancy into it in the form of repeating these binary values by a factor of N. For example, instead of transmitting `010`, applying a repetition coding scheme with a repetition factor of $N = 3$ would yield an output of `000111000`. The reason this is important is that in the event a bit is flipped from a one to a zero or from a zero to a one, which is considered an error, the other repeated bits could be used to nullify the error at the receiver;

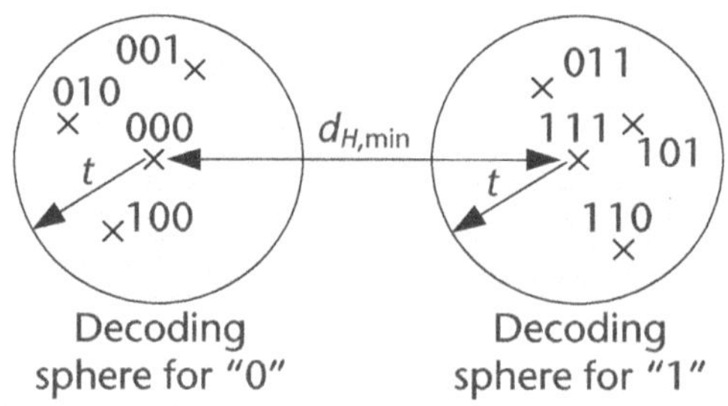

Figure 4.5 Example of decoding spheres.

A rate 1/3 *repetition code* with no source encoding would look like:

$$1 \rightarrow 111 = c_1 \text{ (1st codeword)}$$
$$0 \rightarrow 000 = c_2 \text{ (2nd codeword)}$$
$$\therefore C = \{000, 111\}$$

What are the Hamming distances for the codeword pairs $d_H(111{,}000)$ and $d_H(111{,}101)$?

that is, if a zero value is flipped to a one, the other two zero values will inform the receiver that one of the bits is in error and discard that value when decoding the incoming binary vector. Intuitively, we would assume that with the more repetition applied to a binary vector, the more secure it is from corruption; for example, there are circumstances = where more than one bit is corrupted, which could still yield an error unless there are even more repeated bits to inform the receiver otherwise.

Code 4.3 Protect Data Using Simple Repetition Channel Coding: **`chapter4.m`**

```
% Define parameters
len = 100000; % Length of original binary data stream
N1 = 3; % First repetition factor; should be odd to avoid tie
N2 = 5; % Second repetition factor; should be odd to avoid tie
N3 = 7; % Third repetition factor; should be odd to avoid tie

% Generate binary data stream
bin_str = round(rand(1,len));

% Employ repetition code with repetition factors N1, N2, N3
chcode1_bin_str = zeros(1,N1*len);
chcode2_bin_str = zeros(1,N2*len);
chcode3_bin_str = zeros(1,N3*len);
for ind = 1:1:max([N1 N2 N3]),
    if (ind<=N1)
        chcode1_bin_str(ind:N1:(N1*(len-1)+ind))=bin_str;
    end;
    if (ind<=N2)
        chcode2_bin_str(ind:N2:(N2*(len-1)+ind))=bin_str;
    end;
    if (ind<=N3)
        chcode3_bin_str(ind:N3:(N3*(len-1)+ind))=bin_str;
    end;
end;

% Corrupt both binary strings with zero-mean unit variance Gaussian
% noise followed by rounding (creates "bit flipping" errors)
noisy_bin_str = bin_str + randn(1,len);
rx_bin_str0 = zeros(1,len);
ind0 = find(noisy_bin_str >= 0.5);
rx_bin_str0(ind0) = 1;
noisy_chcode1_bin_str = chcode1_bin_str + randn(1,N1*len);
rx_chcode1_bin_str = zeros(1,N1*len);
ind1 = find(noisy_chcode1_bin_str >= 0.5);
rx_chcode1_bin_str(ind1) = 1;
```

```
noisy_chcode2_bin_str = chcode2_bin_str + randn(1,N2*len);
rx_chcode2_bin_str = zeros(1,N2*len);
ind2 = find(noisy_chcode2_bin_str >= 0.5);
rx_chcode2_bin_str(ind2) = 1;
noisy_chcode3_bin_str = chcode3_bin_str + randn(1,N3*len);
rx_chcode3_bin_str = zeros(1,N3*len);
ind3 = find(noisy_chcode3_bin_str >= 0.5);
rx_chcode3_bin_str(ind3) = 1;

% Decode three encoded binary sequences
dec1_bin = (vec2mat(rx_chcode1_bin_str,N1)).';
dec2_bin = (vec2mat(rx_chcode2_bin_str,N2)).';
dec3_bin = (vec2mat(rx_chcode3_bin_str,N3)).';
ind11 = find(((sum(dec1_bin,1))/N1) >= 0.5);
ind12 = find(((sum(dec2_bin,1))/N2) >= 0.5);
ind13 = find(((sum(dec3_bin,1))/N3) >= 0.5);
rx_bin_str1 = zeros(1,len);
rx_bin_str1(ind11) = 1;
rx_bin_str2 = zeros(1,len);
rx_bin_str2(ind12) = 1;
rx_bin_str3 = zeros(1,len);
rx_bin_str3(ind13) = 1;

% Calculate bit error rate
ber0 = sum(abs(bin_str - rx_bin_str0))/len;
ber1 = sum(abs(bin_str - rx_bin_str1))/len;
ber2 = sum(abs(bin_str - rx_bin_str2))/len;
ber3 = sum(abs(bin_str - rx_bin_str3))/len;
```

Based on the MATLAB script, we can see the impact of repetition coding on a binary vector being corrupted by bit-flipping in an error-prone environment (see Figure 4.6). As we introduce more controlled redundancy in the form of larger repetition rates, the amount of bit errors present in the transmission decreases gradually. This makes sense since as we are introducing more resources into the transmission to make it more reliable in an error-prone environment, the rate at which errors occur will start to decrease. However, one should note that there is a cost-benefit trade-off here since we are introducing more resources into the transmission but this may or may not linearly correlate with the benefits we are obtaining.

4.1.2.1 Shannon's Channel Coding Theorem

In digital communications, it is sometimes necessary to determine the upper limit of the data rate for a specific digital transceiver design. Consequently, in 1949

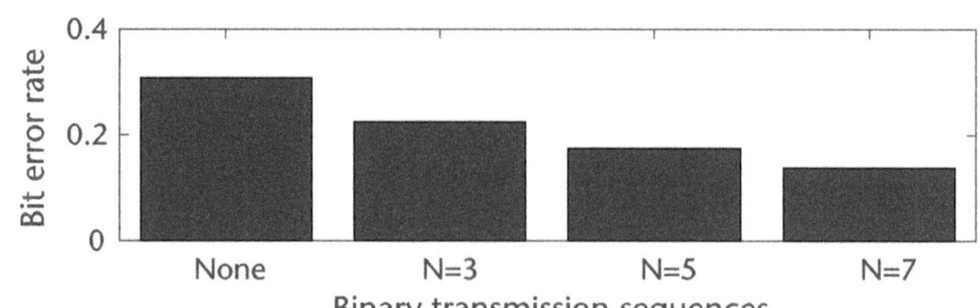

Figure 4.6 Impact of repetition coding on binary transmissions for different repetition factors N.

Claude Shannon published his seminar paper that addressed this topic, entitled "Communication in the Presence of Noise" [1]. In this paper, he defined a quantitative expression that described the limit on the data rate, or capacity, of a digital transceiver in order to achieve error-free transmission.

Suppose one considers a channel with capacity C and we transmit data at a fixed code rate of K/N, which is equal to R_c (a constant). Consequently, if we increase N, then we must increase K in order to keep R_c equal to a constant. What Shannon states is that a code exists such that for $R_c = K/N < C$ and as $N \rightarrow \infty$, we have the probability of error $P_e \rightarrow 0$. Conversely, for $R_c = K/N \geq C$, Shannon indicated that no such code exists. Hence, C is the limit in rate for reliable communications (i.e., C is the *absolute limit* that you cannot go any faster than this amount without causing errors).

So why is the result so important? First, the concept of reliability in digital communications is usually expressed as the probability of bit error, which is measured at the output of the receiver. As a result, it would be convenient to know what this capacity is given the transmission bandwidth, B, the received SNR using mathematical tools rather than empirical measurements. Thus, Shannon derived the information capacity of the channel, which turned out to be equal to

$$C = B\log_2(1 + SNR) \qquad [\text{b/s}], \tag{4.2}$$

where this information capacity tells us the achievable data rate. Note that Shannon only provided us with the theoretical limit for the achievable capacity of a data transmission, but he does not tell us how to build a transceiver to achieve this limit.

Second, the information capacity of the channel is useful since this expression provides us with a bound on the achievable data rate given bandwidth B and received SNR, employed in the ratio $\eta = R/C$, where R is the signaling rate and C is the channel capacity. Thus, as $\eta \rightarrow 1$, the system becomes more efficient. Therefore, the capacity expression provides us with a basis for trade-off analysis between B and SNR, and it can be used for comparing the noise performance of one modulated scheme versus another.

4.2 Digital Modulation

In analog modulation schemes, the analog message signal modulates a continuous wave prior to transmission across a medium. Conversely, digital modulation involves having a digital message signal modulating a continuous waveform. As we have seen earlier in this chapter, this can be accomplished by uniquely manipulating the amplitude and phase information of a signal during each symbol period T based on a specific pattern of bits. However, most digital modulation techniques possess an intermediary step that we will focus on in this section, where collections of b bits forming a binary message m_b are mapped to a symbol, which is then used to define the physical characteristics of a continuous waveform in terms of amplitude and phase. In particular, for each of the possible 2^b values of m_b, we need a unique signal $s_i(t)$, $1 \leq i \leq 2^b$ that can then be used to modulate the continuous waveform, as shown in Figure 4.7.

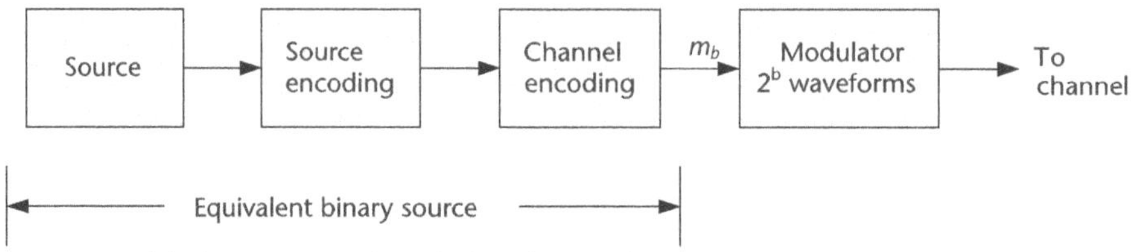

Figure 4.7 Modulation process of equivalent binary data.

In this section, we will study several different families of approaches for mapping binary data into symbols that can then be used to modulate continuous waveforms. These modulation scheme families are defined by which physical characteristic or combination of characteristics are manipulated by the mapping process in order to uniquely represent a specific bit pattern. However, there exist various trade-offs between these different families, including how efficiently a bit is mapped to a symbol in terms of the transmit power expended. Consequently, we will first explore how we assess this trade-off before studying three major families of digital modulation schemes and how they compare to each other.

4.2.1 Power Efficiency

In order to assess the effectiveness of mapping a bit to a symbol in terms of the transmit power expended per symbol, we can employ the *power efficiency* metric. Suppose we define the energy of a symbol $s(t)$ as

$$E_s = \int_0^T s^2(t)dt, \tag{4.3}$$

where T is the period of the symbol. Then, for a modulation scheme consisting of M symbols, we can define the average symbol energy via the following weighted average:

$$\bar{E}_s = P(s_1(t)) \cdot \int_0^T s_1^2(t)dt + \cdots + P(s_M(t)) \cdot \int_0^T s_M^2(t)dt, \tag{4.4}$$

where $P(s_i(t))$ is the probability that the symbol $s_i(t)$ occurs. Furthermore, if we would like to calculate the average energy per bit, we can approximate this using $\bar{E}_s$ and dividing this quantity by $b = \log_2(M)$ bits per symbol, yielding

$$\bar{E}_b = \frac{\bar{E}_s}{b} = \frac{\bar{E}_s}{\log_2(M)}. \tag{4.5}$$

To quantitatively assess the similarity between two symbols in terms of their physical characteristics, we define the *Euclidean distance* as

$$d_{ij}^2 = \int_0^T (s_i(t) - s_j(t))^2 dt = E_{\Delta s_{ij}}, \tag{4.6}$$

where $\Delta s_{ij}(t) = s_i(t) - s_j(t)$. Since we are often interested in the worst-case scenario when assessing the performance of a modulation scheme, we usually compute the minimum Euclidean distance; namely:

$$d_{min}^2 = \min_{s_i(t), s_j(t), i \neq j} \int_0^T (s_i(t) - s_j(t))^2 dt. \tag{4.7}$$

Thus, the power efficiency of a signal set used for modulation is given by the expression

$$\varepsilon_p = \frac{d_{\text{min}}^2}{\bar{E}_b}. \tag{4.8}$$

> Suppose we would like to find the ε_p given the following waveforms:
>
> $$s_1(t) = A \cdot [u(t) - u(t - T)] = s(t)$$
> $$s_2(t) = -A \cdot [u(t) - u(t - T)] = -s(t)$$
>
> where $u(t)$ is the unit step function. Compute the following:
>
> - The minimum Euclidean distance d_{min}^2.
> - The average bit energy $\bar{E}_b$.
> - The power efficiency ε_P.

4.2.2 Pulse Amplitude Modulation

Of the various physical characteristics of a signal waveform that can be manipulated in order to convey digital information, the most obvious choice is the signal amplitude level. Leveraging this physical characteristic, *pulse amplitude modulation* (PAM) is a digital modulation scheme where the message information is encoded in the amplitude of a series of signal pulses. Furthermore, demodulation of a PAM transmission is performed by detecting the amplitude level of the carrier at every symbol period.

The most basic form of PAM is binary PAM (B-PAM), where the individual binary digits are mapped to a waveform $s(t)$ possessing two amplitude levels according to the following modulation rule:

- "1" $\rightarrow s_1(t)$
- "0" $\rightarrow s_2(t)$

where $s_1(t)$ is the waveform $s(t)$ possessing one unique amplitude level while $s_2(t)$ is also based on the waveform $s(t)$ but possesses another unique amplitude level. Note that the waveform $s(t)$ is defined across a time period T and is zero otherwise. Since the duration of the symbols is equivalent to the duration of the bits, the bit rate for a B-PAM transmission is defined as $R_b = 1/T$ bits per second.

The energy of the waveform $s(t)$ is defined as

$$E_s = \int_0^T s^2(t)dt \quad \text{(Joules)}. \tag{4.9}$$

Suppose we define $s(t)$ as a rectangular waveform; namely,

$$s(t) = A \cdot [u(t) - u(t - T)], \tag{4.10}$$

where $u(t)$ is the unit step function and A is the signal amplitude. Furthermore, suppose that the bit "1" is defined by the amplitude A while the bit "0" is defined by the amplitude $-A$. We can subsequently write our modulation rule to be equal to

- "1" $\rightarrow s(t)$
- "0" $\rightarrow -s(t)$

Therefore, the symbol energy is given by $E_s{=}E_{-s}{=}A^2T{=}\frac{A^2}{R_b}$. From this result, we can define the energy per bit for a B-PAM transmission as

$$\bar{E}_b = P(1) \cdot \int_0^T s_1^2(t)dt + P(0) \cdot \int_0^T s_2^2(t)dt, \tag{4.11}$$

where $P(1)$ is the probability that the bit is a "1," and $P(0)$ is the probability that the bit is a "0." Thus, if we define $s_1(t) = s(t)$ and $s_2(t) = -s(t)$, then the average energy per bit is equal to

$$\bar{E}_b = E_s\{P(1) + P(0)\} = E_s = \int_0^T s^2(t)dt = A^2T. \tag{4.12}$$

Calculating the minimum Euclidean distance, we get

$$d_{min}^2 = \int_0^T (s(t) - (-s_(t)))^2 dt = \int_0^T (2s(t))^2 dt = 4A^2T, \tag{4.13}$$

which is then plugged into (4.8) in order to yield a power efficiency result for a B-PAM transmission of

$$\varepsilon_p = \frac{d_{\text{min}}^2}{\bar{E}_b} = \frac{4A^2T}{A^2T} = 4. \tag{4.14}$$

As we will observe throughout the rest of this section, a power efficiency result of 4 is the best possible result that you can obtain for any digital modulation scheme when all possible binary sequences are each mapped to a unique symbol.

Suppose we now generalize the B-PAM results obtained for the average bit energy, the minimum Euclidean distance, and the power efficiency and apply them to the case when we try mapping binary sequences to one of M possible unique signal amplitude levels, referred to as M-ary pulse amplitude modulation (M-PAM). First, let us express the M-PAM waveform as

$$s_i(t) = A_i \cdot p(t), \text{ for } i = 1, 2, \ldots, M/2 \tag{4.15}$$

where $A_i = A(2i - 1)$, $p(t) = u(t) - u(t - T)$, and $u(t)$ is the unit step function. Graphically speaking, the M-PAM modulation scheme looks like the signal constellation shown in Figure 4.8.

In order to compute the power efficieny of M-PAM, $\varepsilon_{p,\text{M-PAM}}$, we select the d_{min}^2 pair $s_1(t) = A \cdot p(t)$ and $s_2(t) = -A \cdot p(t)$ since this pair of signal waveforms are the closest to each other in terms of shape. Thus, using this selection of waveforms, we can solve for the difference between them:

$$\Delta s(t) = 2A \cdot p(t), \tag{4.16}$$

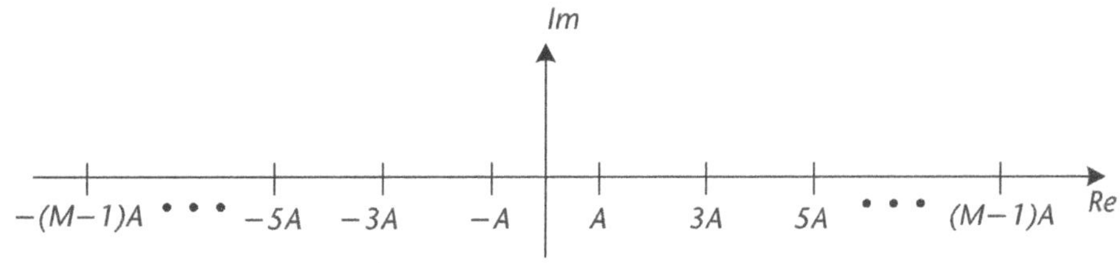

Figure 4.8 M-PAM signal constellation.

which yields a minimum Euclidean distance of

$$d_{\min}^2 = 4A^2T. \tag{4.17}$$

In order to calculate the average symbol energy, $\bar{E}_s$, we can simplify the mathematics by exploiting the symmetry of signal constellation, which yields

$$\begin{aligned}\bar{E}_s &= \frac{2}{M}A^2T\sum_{i=1}^{M/2}(2i-1)^2 \\ &= A^2T\frac{(M^2-1)}{3} \quad \text{which is simplified via tables} \\ \rightarrow \bar{E}_b &= \frac{\bar{E}_s}{\log_2(M)} = \frac{A^2T(2^{2b}-1)}{3b}.\end{aligned} \tag{4.18}$$

Finally, solving for the power efficiency yields

$$\varepsilon_{p,\text{M-PAM}} = \frac{12b}{2^{2b}-1}. \tag{4.19}$$

4.2.3 Quadrature Amplitude Modulation

Similar to PAM, quadrature amplitude modulation (QAM) implies some sort of amplitude modulation. However, QAM modulation is a two-dimensional signal modulation scheme as opposed to PAM modulation. The two dimensions of the QAM modulation; namely, the in-phase and quadrature components, are orthogonal to each other, which implies that one can essentially double the transmission data rate for free. Furthermore, rectangular QAM can be thought of as two orthogonal PAM signals being transmitted simultaneously.

Mathematically, if a rectangular QAM signal constellation consists of M unique waveforms, this could potentially be represented as $\sqrt{M}$-PAM transmissions operating simultaneously in orthogonal dimensions. Note that QAM signal constellations could also take the form of nested circles (called circular QAM), or any other geometric pattern that involves orthogonal modulators. Rectangular QAM is a popular modulation scheme due to its relatively simple receiver structure, as shown in Figure 4.9, where each dimension employs a $\sqrt{M}$-ary PAM detector.

In order to determine the power efficiency of M-QAM, let us first define the mathematical representation of a signal waveform belonging to this form of modulation:

$$s_{ij}(t) = A_i \cdot \cos(\omega_c t) + B_j \cdot \sin(\omega_c t), \tag{4.20}$$

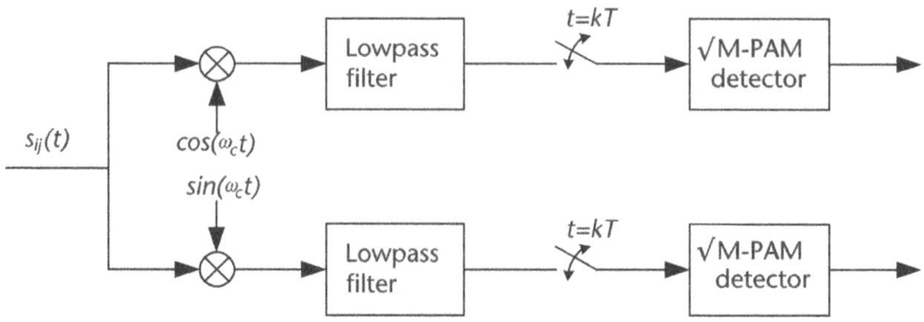

Figure 4.9 M-QAM receiver structure.

where ω_c is the carrier frequency, and A_i and B_j are the in-phase and quadrature amplitude levels. Notice how the cosine and sine functions are used to modulate these amplitude levels in orthogonal dimensions. Visualizing M-QAM as a signal constellation, the signal waveforms will assume positions in both the real and imaginary axes, as shown in Figure 4.10.

To compute the power efficiency of M-QAM, $\varepsilon_{p,\text{M-QAM}}$, we first need to calculate the minimum Euclidean distance, which becomes

$$d^2_{\min} = \int_0^T \Delta s^2(t)dt = 2A^2T, \tag{4.21}$$

where we have selected the following signal waveforms without loss of generality:

$$\begin{aligned} s_1(t) &= A \cdot \cos(\omega_c t) + A \cdot \sin(\omega_c t) \\ s_2(t) &= 3A \cdot \cos(\omega_c t) + A \cdot \sin(\omega_c t). \end{aligned} \tag{4.22}$$

In order to derive the average symbol energy, $\bar{E}_s$, we leverage the expression from M-ary PAM by replacing M with $\sqrt{M}$ such that

$$\bar{E}_s = A^2T\frac{M-1}{3}, \tag{4.23}$$

which can then be used to solve

$$\bar{E}_b = \frac{\bar{E}_s}{\log_2(M)} = A^2T\frac{2^b-1}{3b}. \tag{4.24}$$

Thus, the power efficiency is equal to

$$\varepsilon_{p,\text{M-QAM}} = \frac{3!b}{2^b-1}. \tag{4.25}$$

Hands-On MATLAB Example: QAM modulation is a very useful technique for sending information efficiently within the same symbol period. As mentioned previously, it exploits both the in-phase and quadrature domains in order to transmit information in parallel down both channels orthogonally. In the MATLAB script above, we implement a QAM transmitter and receiver that takes binary vectors `samp_I` and `samp_Q`, modulates them to the in-phase and quadrature channels of a QAM transmission, and then extracts these binary vectors using QAM

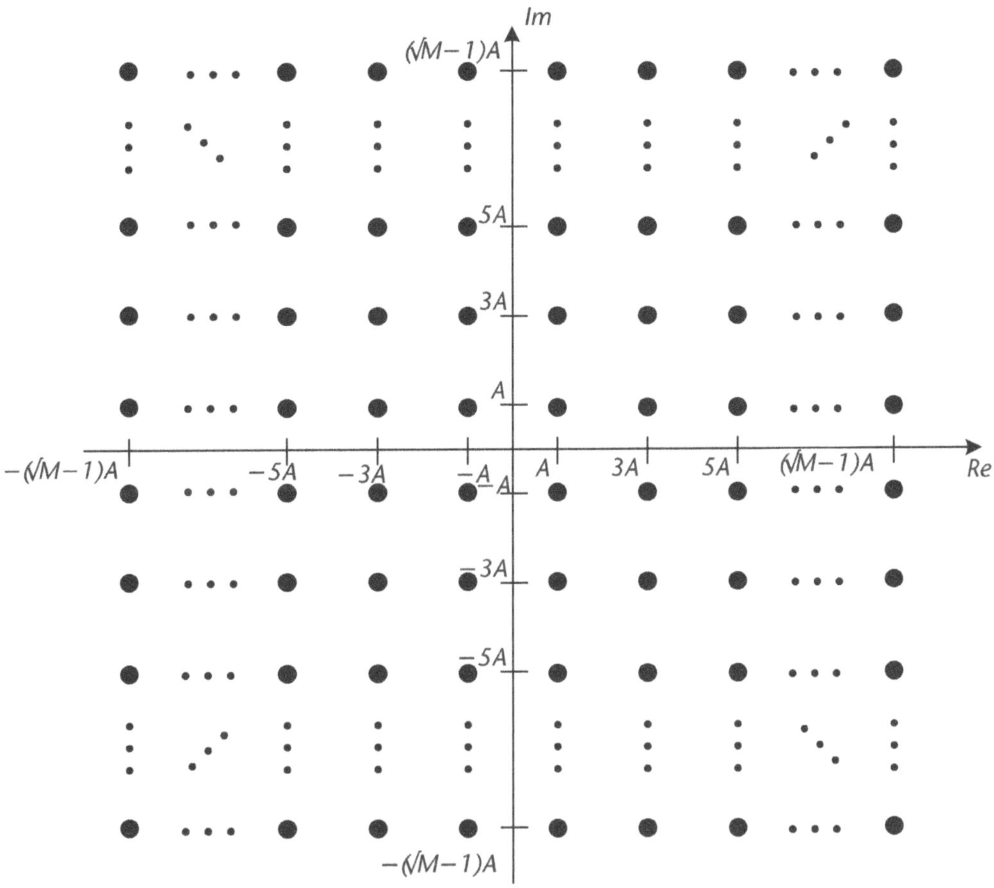

Figure 4.10 M-QAM signal constellation.

demodulation. At the core of QAM modulation are the sine and cosine functions, which are mathematically orthogonal. As we can see in the script, the in-phase and quadrature data is modulated onto the cosine and sine wave signals at the transmitter. At the receiver, the in-phase and quadrature information is separated out of the received signal by exploiting this orthogonality and using trigonometric properties.

In Figure 4.11, we can observe the usage cosine and sine waves as carriers of information, where this information can be transmitted simultaenously and recovered perfectly. By exploiting two dimensions for the transmission of information, we make more efficient use of each symbol period that we use when broadcasting data.

4.2.4 Phase Shift Keying

Phase shift keying (PSK) is a digital modulation scheme that conveys data by changing or modulating the phase of a reference signal (i.e., the carrier wave). Any digital modulation scheme uses a finite number of distinct signals to represent digital data. PSK uses a finite number of phases, each assigned a unique pattern of binary digits. Usually, each phase encodes an equal number of bits. Each pattern of bits forms the symbol that is represented by the particular phase. The demodulator,

Code 4.4 Decoding QAM Waveform Using I/Q Receiver: **chapter4.m**

```
%%%%%%%%%%%%%%%%%%%%%%%%%%%%%%%%%%%%%%%%%%%%%%%%%%%%%%%%%%%%%%%
% Decoding QAM waveform using I/Q receiver

% Define parameters
N_samp = 1000; % Number of samples per symbol
N_symb = 10; % Number of symbols in transmission
cfreq = 1/10; % Carrier frequency of cosine and sine carriers

% Generate inphase and quadrature channels with 2-PAM waveforms
chI = 2*round(rand(1,N_symb))-1;
chQ = 2*round(rand(1,N_symb))-1;
samp_I = [];
samp_Q = [];
for ind = 1:1:N_symb,
    samp_I = [samp_I chI(ind)*ones(1,N_samp)];
    samp_Q = [samp_Q chQ(ind)*ones(1,N_samp)];
end;

% Apply cosine and sine carriers to inphase and quadrature components,
% sum waveforms together into composite transmission
tx_signal = samp_I.*cos(2.*pi.*cfreq.*(1:1:length(samp_I)))
            + samp_Q.*sin(2.*pi.*cfreq.*(1:1:length(samp_Q)));

% Separate out inphase and quadrature components from composite
```

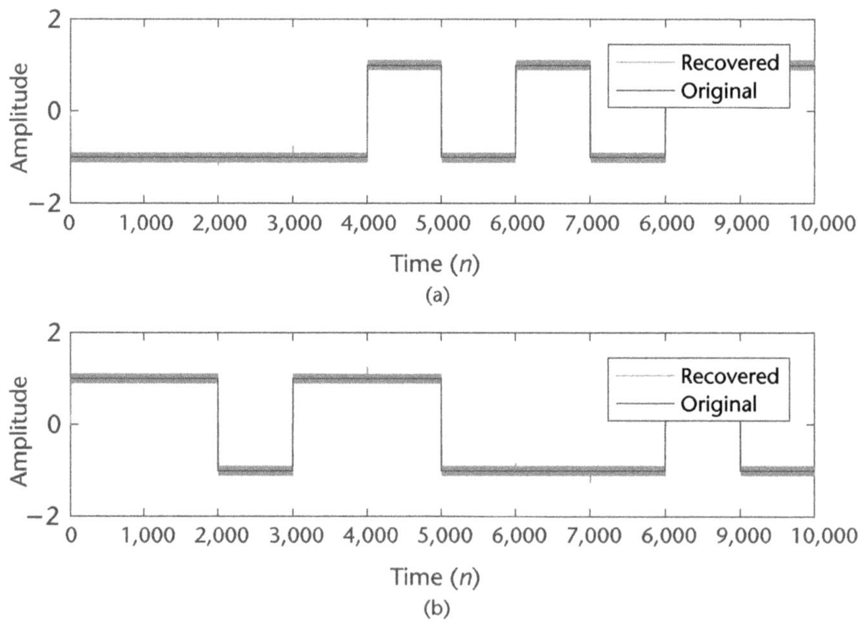

Figure 4.11 Example of quadrature amplitude modulation waveform using an in-phase/quadrature receiver. (a) In-phase signal component, and (b) quadrature signal component.

which is designed specifically for the symbol set used by the modulator, determines the phase of the received signal, and maps it back to the symbol it represents, thus recovering the original data. This requires the receiver to be able to compare the phase of the received signal to a reference signal. Such a system is termed coherent.

PSK characterizes symbols by their phase. Mathematically, a PSK signal waveform is represented by

$$s_i(t) = A\cos\left(2\pi f_c t + (2i-1)\frac{\pi}{m}\right), \quad \text{for} \quad i = 1, ..., \log_2 \text{m}, \tag{4.26}$$

where A is the amplitude, f_c is carrier frequency, and $(2i-1)\frac{\pi}{m}$ is the phase offset of each symbol. PSK presents an interesting set of trade-offs with PAM and QAM. In amplitude modulation schemes, channel equalization is an important part of decoding the correct symbols. In PSK schemes, the phase of the received signal is much more important than the amplitude information.

There are several types of PSK modulation schemes based on the number of M possible phase values a particular PSK waveform can be assigned. One of the most popular and most robust is binary PSK, or B-PSK, modulation, whose signal constellation is illustrated in Figure 4.12. In general, the modulation rule for B-PSK modulation is the following:

$$\begin{aligned} \text{“1”} &\rightarrow s_1(t) = A \cdot \cos(\omega_c t + \theta) \\ \text{“0”} &\rightarrow s_2(t) = -A \cdot \cos(\omega_c t + \theta) \\ &\qquad = A \cdot \cos(\omega_c(t) + \theta + \pi) \\ &\qquad = -s_1(t). \end{aligned} \tag{4.27}$$

In other words, the two signal waveforms that constitute a B-PSK modulation scheme are separated in phase by θ.

In order to derive the power efficiency of a B-PSK modulation scheme, $\varepsilon_{p,\text{BPSK}}$, we first need to compute the minimum Euclidean distance $d^2_{\min}$ by employing the definition and solving for

$$\begin{aligned} d^2_{\min} &= \int_0^T (s_1(t) - s_2(t))^2 dt \\ &= 4A^2 \int_0^T \cos^2(\omega_c t + \theta) dt \end{aligned}$$

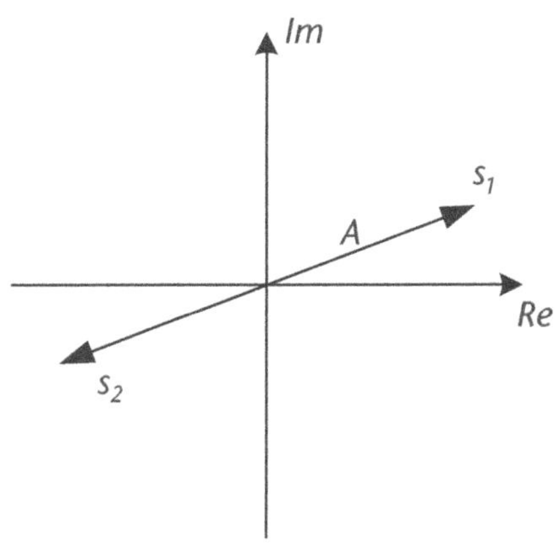

Figure 4.12 BPSK signal constellation.

> Notice how in (4.28) how the second term disappeared from the final result. This is due to the fact that the second term possessed a carrier frequency that was twice that of the original signal. Since the carrier signal is a periodic sinusoidal waveform, integrating such a signal possessing such a high frequency would result in the positive portion of the integration canceling out the negative portion of the integration, yielding an answer than is close to zero. Consequently, we refer to this term as a *double frequency term*. Also note that many communication systems filter their received signals, which means the probability of filtering out the double frequency term is also quite high.

$$
\begin{aligned}
&= \frac{4A^2T}{2} + \frac{4A^2}{2}\int_0^T \cos(2\omega_c t + 2\theta)dt \\
&= 2A^2T.
\end{aligned} \tag{4.28}
$$

Note that another way for computing $d^2_{\min}$ is to use the concept of correlation, which describes the amount of similarity between two different signal waveforms. In this case, we can express the minimum Euclidean distance as

$$
d^2_{\min} = \int_0^T (s_2(t) - s_1(t))^2 dt = E_{s_1} + E_{s_2} - 2\rho_{12} \tag{4.29}
$$

where the symbol energy for symbol i, E_{s_i}, and the correlation between symbols 1 and 2, ρ_{12}, are given by

$$
E_{s_i} = \int_0^T s_i^2(t)dt \quad \text{and} \quad \rho_{12} = \int_0^T s_1(t)s_2(t)dt.
$$

Employing the definition for $\bar{E}_b$, we can now solve for the average bit energy of the B-PSK modulation scheme by first solving for the symbol energies of the two signal waveforms and then averaging them; that is,

$$
\begin{aligned}
E_{s_1} &= \int_0^T s_1^2(t)dt = A^2\int_0^T \cos^2(\omega_c t + \theta)dt \\
&= \frac{A^2T}{2} + \frac{A^2}{2}\int_0^T \cos(2\omega_c t + 2\theta)dt \\
&= \frac{A^2T}{2} \\
E_{s_2} &= \frac{A^2T}{2} \\
\bar{E}_b &= P(0)\cdot E_{s_2} + P(1)\cdot E_{s_1} = \frac{A^2T}{2}.
\end{aligned} \tag{4.30}
$$

Note that since the number of bits represented by a single symbol is equal to one, and both the bit energy and symbol energy are equivalent.

Finally, applying the definition for the power efficiency, we get the following expression:

$$\varepsilon_{p,\mathrm{BPSK}} = \frac{d^2_{\mathrm{min}}}{\bar{E}_b} = 4. \tag{4.31}$$

This is supposed to be the largest possible value for ε_p for a modulation scheme employing all possible signal representations; that is, $M = 2^b$ waveforms. Notice when using the correlation approach to calculate the minimum Euclidean distance, in order to get a large ε_p, we need to maximize d^2_{min}, which means we want $\rho_{12} < 0$. Thus, to achieve this outcome, we need the following situation:

$$E_{s_1} = E_{s_2} = E = A^2T/2, \tag{4.32}$$

which means $d^2_{\mathrm{min}} = 2(E - \rho_{12})$ and consequently $\rho_{12} = -E$.

Q

Show that for the following signal waveforms:

$$s_1(t) = A \cdot \cos(\omega_c t + \theta)$$
$$s_2(t) = 0$$

the power efficiency is equal to $\varepsilon_p = 2$.

Q

Show that for the following signal waveforms:

$$s_1(t) = A \cdot \cos(\omega_c t + \theta)$$
$$s_2(t) = A \cdot \sin(\omega_c t + \theta)$$

the power efficiency is equal to $\varepsilon_p = 2$.

So far we have studied a PSK modulation scheme that consists of only just one of two waveforms. We will now expand our PSK signal constellation repertoire to include four distinct waveforms per modulation scheme. In quadrature PSK (QPSK) modulation, a signal waveform possesses the following representation:

$$s_i(t) = \pm A \cdot \cos(\omega_c t + \theta) \pm A \cdot \sin(\omega_c t + \theta), \tag{4.33}$$

where each signal waveform possesses the same amplitude but one of four possible phase values. This kind of phase modulation is illustrated by the signal constellation diagram shown in Figure 4.13, where each waveform is located at a different phase value.

In order to derive the power efficiency of QPSK, $\varepsilon_{p,\mathrm{QPSK}}$, we first need to solve for the minimum Euclidean distance, d^2_{min}, which is equal to

$$d^2_{\mathrm{min}} = \int_0^T \Delta s^2(t)dt = 2A^2T. \tag{4.34}$$

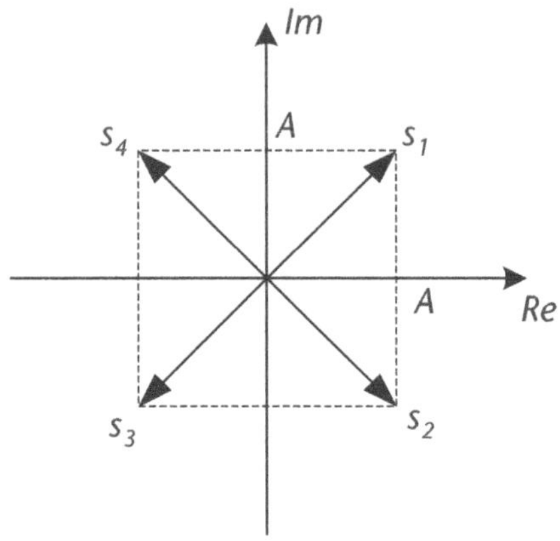

Figure 4.13 QPSK signal constellation.

Next, we would like to find $\bar{E}_b$, which requires us to average over all the signal waveforms. Consequently, this is equal to

$$\bar{E}_b = \frac{(E_{s_1} + E_{s_2} + E_{s_3} + E_{s_4})/4}{\log_2(M)} = \frac{A^2 T}{2}, \tag{4.35}$$

where the symbol energy of all four symbols is equal to $E_{s_1} = E_{s_2} = E_{s_3} = E_{s_4} = A^2 T$. Finally, solving for the power efficiency using (4.8), we get

$$\varepsilon_{p,\text{QPSK}} = \frac{d^2_{\min}}{\bar{E}_b} = 4, \tag{4.36}$$

which is the same as BPSK but with 2 bits per symbol, making this a fantastic result!

Finally, let us study the general case when a PSK modulation scheme has a choice of M possible phase values, where the distance of a signal constellation point to the origin is always a constant and the signal constellation consists of M equally spaced points on a circle. Referred to as M-PSK, a signal waveform can be mathematically represented as

$$s_i(t) = A \cdot \cos\left(\omega_c t + \frac{2\pi i}{M}\right), \text{ for } i = 0, 1, 2, \ldots, M-1. \tag{4.37}$$

Note that there are several advantages and disadvantages with this modulation scheme. For instance, as M increases the spacing between signal constellation points decreases, thus resulting in a decrease in error robustness. Conversely, having the information encoded in the phase results in constant envelope modulation, which is good for nonlinear power amplifiers and makes the transmission robust to amplitude distortion channels.

Regarding the derivation of the power efficiency for an M-PSK modulation scheme, $\varepsilon_{p,\text{M-PSK}}$, suppose we define two adjacent M-PSK signal waveforms as $s_1(t) = A \cdot \cos(\omega_c t)$ and $s_2(t) = A \cdot \cos(\omega_c t + 2\pi/M)$. Calculating the minimum Euclidean distance using

$$d^2_{\min} = E_{s_1} + E_{s_2} - 2\rho_{12} \tag{4.38}$$

where we define the symbol energy as

$$E_{s_i} = \int_0^T s_i^2(t)dt = \frac{A^2T}{2}, \text{ for } i = 1, 2, \tag{4.39}$$

and the correlation between the two signal waveforms as

$$\rho_{12} = \int_0^T s_1(t)s_2(t)dt = \frac{A^2T}{2}\cos\left(\frac{2\pi}{M}\right), \tag{4.40}$$

this yields

$$d_{\min}^2 = A^2T\left(1 - \cos\left(\frac{2\pi}{M}\right)\right). \tag{4.41}$$

The average bit energy $\bar{E}_b$ is equal to $\bar{E}_b = \frac{\bar{E}_s}{\log_2(M)} = \frac{\bar{E}_s}{b}$, where $\bar{E}_s = A^2T/2$. Using the definition for the power efficiency from (4.8), we see that

$$\varepsilon_{p,\text{M-PSK}} = 2b\left(1 - \cos\left(\frac{2\pi}{M}\right)\right) = 4b\sin^2\left(\frac{\pi}{2^b}\right). \tag{4.42}$$

4.2.5 Power Efficiency Summary

After examining the power efficiency performance of several different modulation schemes, it is important to assess the trade-offs between the different schemes such that we can make the appropriate design decisions in the future when implementing a digital communication system. To determine how much power efficiency we are losing relative to $\varepsilon_{p,\text{QPSK}}$, which possesses the best possible result, we use the following expression:

$$\delta\text{SNR} = 10 \cdot \log_{10}\left(\frac{\varepsilon_{p,\text{QPSK}}}{\varepsilon_{p,\text{other}}}\right). \tag{4.43}$$

Using this expression, we created a table of δSNR values for the modulation schemes studied in this chapter, as shown in Table 4.1.

From Table 4.1, notice how the two-dimensional modulation schemes perform better than the one-dimensional modulation schemes. Furthermore, notice how all of the modulation schemes studied are linear modulation schemes, which means they possess a similar level of receiver complexity. Given these insights on the power efficiency performance of these modulation schemes, we now turn our attention to the robustness of a modulation technique in the presence of noise.

Table 4.1 δSNR Values of Various Modulation Schemes

M	*b*	*M-ASK*	*M-PSK*	*M-QAM*
2	1	0	0	0
4	2	4	0	0
8	3	8.45	3.5	—
16	4	13.27	8.17	4.0
32	5	18.34	13.41	—
64	6	24.4	18.4	8.45

Hands-On MATLAB Example: Noise introduced by a transmission medium can potentially result in symbols being decoded in error. In the following MATLAB script, we will examine the behavior of how the introduction of noise can obfuscate the true identity of an intercepted symbol. Specifically, we will compare the originally transmitted symbols and the noisy received symbols using a visual representation referred to as a signal constellation diagram, which plots the locations of symbols across a 2-axis plot with an in-phase axis and a quadrature axis. Notice that we are considering three different waveforms in this example: 4-PAM, 4-QAM, and QPSK. For each of these waveforms, we generated an alphabet of different symbols that each can produce. The randomly generated binary data streams representing in-phase and quadrature information are mapped to these different waveform symbols for each modulation scheme. Then, we introduce Gaussian noise to the three transmissions using the `randn` function.

The before-and-after signal constellation plots for the 4-PAM, 4-QAM, and QPSK modulated transmissions are shown in Figure 4.14. The original symbols are

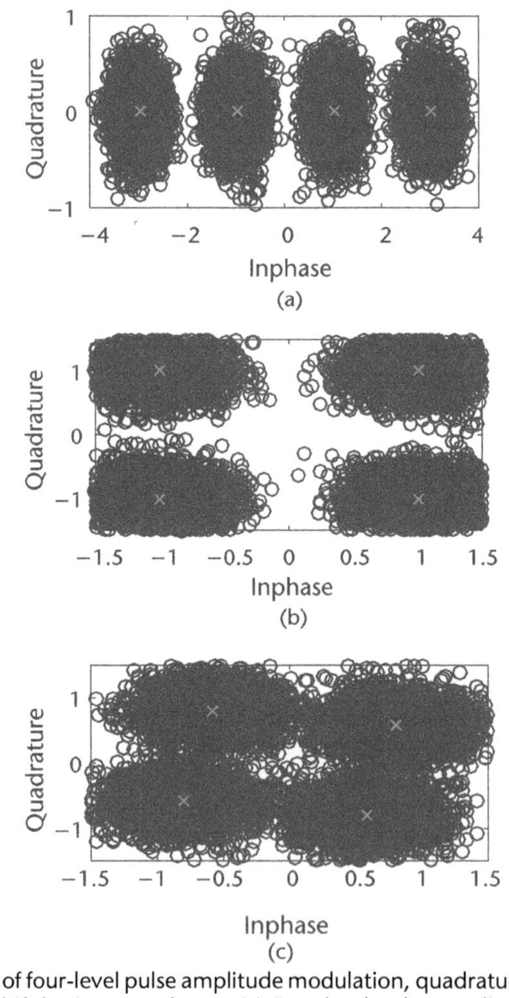

Figure 4.14 Examples of four-level pulse amplitude modulation, quadrature amplitude modulation, and quadrature phase shift keying waveforms. (a) Four-level pulse amplitude modulation, (b) four-level quadrature amplitude modulation, and (c) four-level quadrature phase shift keying.

Code 4.5 Generating Four Level Pulse Amplitude Modulation, Quadrature Amplitude Modulation, and Quadrature Phase Shift Keying Waveforms: **chapter4.m**

```
% Define parameters
len = 10000; % Length of binary string
nvar = 0.15; % Noise variance

% Generate the random bit streams that have already been demultiplexed
% from a single high speed data stream
bin_str1 = round(rand(1,len)); % Inphase data stream
bin_str2 = round(rand(1,len)); % Quadrature data stream

% Perform mapping of binary streams to symbols
ind_wavefm = 2.*bin_str2 + 1.*bin_str1; % Create waveform indices
wavefm_4pam = zeros(1,len); % 4-PAM
wavefm_4qam = zeros(1,len); % 4-QAM
wavefm_qpsk = zeros(1,len); % QPSK
symb_4pam = [-3 -1 3 1];
symb_4qam = [-1+i 1+i -1-i 1-i];
symb_qpsk = [exp(i*(pi/5+pi/2)) exp(i*(pi/5+pi)) exp(i*(pi/5+0))
         exp(i*(pi/5+3*pi/2)) ];
for ind = 1:1:4,
    wavefm_4pam(find(ind_wavefm == (ind-1))) = symb_4pam(ind);
    wavefm_4qam(find(ind_wavefm == (ind-1))) = symb_4qam(ind);
    wavefm_qpsk(find(ind_wavefm == (ind-1))) = symb_qpsk(ind);
end;

% Add complex zero-mean white Gaussian noise
noise_signal = (1/sqrt(2))*sqrt(nvar)*randn(1,len)
         + i*(1/sqrt(2))*sqrt(nvar)*randn(1,len);
rx_wavefm_4pam = wavefm_4pam + noise_signal;
rx_wavefm_4qam = wavefm_4qam + noise_signal;
rx_wavefm_qpsk = wavefm_qpsk + noise_signal;
```

shown as little red cross marks in the center of a cloud of corrupted received symbols after the noise is added to them. This is occurring since whenever a transmission is occurring over a noisy channel, the symbols that are sent over this channel are being displaced from their original coordinates in the in-phase/quadrature plane by the complex Gaussian noise. This displacement is what makes it difficult for the receiver to decode these symbols without any error since the noise might sufficiently displace these symbols closer to another nearby symbol location that is part of the signal constellation. For all of these modulation schemes shown in Figure 4.14(a) (4-PAM), Figure 4.14(b) (4-QAM), and Figure 4.14(c) (QPSK), there is a nonneglible probability that these symbols have been moved closer to another point in the overall signal constellation, which would result in an error in decode that would translate into a bit error.

4.3 Probability of Bit Error

One of the most commonly used quantitative metrics for measuring the performance of a digital communication system is the probability of BER, which is the probability that a bit transmitted will be decoded incorrectly. This metric is very important when assessing whether the design of a digital communication system meets the specific error robustness requirements of the application to be supported (e.g., voice, multimedia, or data). Furthermore, having a metric that quantifies error performance is helpful when comparing one digital communication design with another. Consequently, in this section we will provide a mathematical introduction to the concept of BER.

Suppose that a signal $s_i(t)$, $i = 1, 2$ was transmitted across an AWGN channel with noise signal $n(t)$, and that a receiver intercepts the signal $r(t)$. The objective of the receiver is to determine whether either $s_1(t)$ or $s_2(t)$ was sent by the transmitter. Given that the transmission of either $s_1(t)$ or $s_2(t)$ is a purely random event, the only information that the receiver has about what was sent by the transmitter is the observed intercepted signal $r(t)$, which contains either signal in addition to some noise introduced by the AWGN channel.

Given this situation, we employ the concept of *hypothesis testing* [2] in order to set up a framework by which the receiver can decide on whether $s_1(t)$ or $s_2(t)$ was sent based on the observation of the intercepted signal $r(t)$. Thus, let us employ the following hypothesis testing framework:

$$\mathcal{H}_1 : r(t) = s_1(t) + n(t), \;\; 0 \leq t \leq T$$

$$\mathcal{H}_0 : r(t) = s_2(t) + n(t), \;\; 0 \leq t \leq T$$

where $\mathcal{H}_0$ and $\mathcal{H}_1$ are *Hypothesis 0* and *Hypothesis 1*.

Leveraging this framework, we next want to establish a *decision rule* at the receiver such that it can select which waveform was sent based on the intercept signal. Suppose we assume that $s_1(t)$ was transmitted. In general, we can determine the level of correlation between two signals $x(t)$ and $y(t)$ over the time interval $0 \leq t \leq T$ using the expression

$$\int_0^T x(t)y(t)dt.$$

Consequently, our decision rule on whether $s_1(t)$ or $s_2(t)$ was transmitted given that we observe $r(t)$ is defined as

$$\int_0^T r(t)s_1(t)dt \geq \int_0^T r(t)s_2(t)dt, \tag{4.44}$$

where we assume that $s_1(t)$ was transmitted. Recall that correlation tells us how similar one waveform is to another waveform. Therefore, if the receiver knows the appearance of $s_1(t)$ and $s_2(t)$, we can then determine which of these two waveforms is more correlated to $r(t)$. Since $s_1(t)$ was assumed to be transmitted, ideally the received signal $r(t)$ should be more correlated to $s_1(t)$ than $s_2(t)$.

On the other hand, what happens if some distortion, interference, and/or noise is introduced in the transmission channel such that the transmitted signal waveforms are corrupted? In the situation where a transmitted signal waveform is sufficiently corrupted such that it appears to be more correlated to another possible signal waveform, the receiver could potentially select an incorrect waveform, thus yielding an error event. In other words, assuming $s_1(t)$ was transmitted, an error event occurs when

$$\int_0^T r(t)s_1(t)dt \leq \int_0^T r(t)s_2(t)dt. \quad (4.45)$$

Since $r(t) = s_1(t) + n(t)$, we can substitute this into the error event in order to obtain the decision rule

$$\begin{aligned}\int_0^T s_1^2(t)dt + \int_0^T n(t)s_1(t)dt &\leq \int_0^T s_1(t)s_2(t)dt + \int_0^T n(t)s_2(t)dt \\ E_{s_1} - \rho_{12} &\leq \int_0^T n(t)(s_2(t) - s_1(t))dt \\ E_{s_1} - \rho_{12} &\leq z.\end{aligned}$$

From this expression, we observe that both E_{s_1} and ρ_{12} are deterministic quantities. On the other hand, z is based on the noise introduced by the transmission channel, and thus it is a random quantity that requires some characterization. Since $n(t)$ is a Gaussian random variable, then z is also a Gaussian random variable. This is due to the fact that the process of integration is equivalent to a summation across an infinite number of samples, and since we are summing up Gaussian random variables, the result in is also a Gaussian random variable. With $z \sim \mathcal{N}(0, \sigma^2)$, we now need to calculate the variance of z, σ^2, which can be solved as follows:

$$\begin{aligned}\sigma^2 = E\{z^2\} &= \frac{N_0}{2}\int_0^T (s_1(t) - s_2(t))^2 dt \\ &= \frac{N_0}{2}(E_{s_1 f:ch4_4mods_qpsk} + E_{s_2} - 2\rho_{12}) \rightarrow \text{Assume } E_{s_1} = E_{s_2} = E \\ &= N_0(E - \rho_{12}),\end{aligned}$$

where $E = E_i = \int_0^T s_i^2(t)dt$ and $\rho_{12} = \int_0^T s_1(t)s_2(t)dt$. Note that we are assuming that the channel is introducing zero-mean noise, which means the sum of these noise contributions; that is, z will also be zero-mean.

With both deterministic and random quantities characterized, we can now proceed with the derivation for the probability of bit error. The probability of an error occurring given that a "1" was transmitted; that is, $P(e|1)$ is equal to

$$\begin{aligned}P(z \geq E - \rho_{12}) &= Q\left(\frac{E - \rho_{12}}{\sigma}\right) \rightarrow \begin{array}{l}\text{Since } z \sim \mathcal{N}(0, \sigma^2) \\ \text{and } E - \rho_{12} \text{ is constant}\end{array} \\ &= Q\left(\sqrt{\frac{(E - \rho_{12})^2}{\sigma^2}}\right) \rightarrow \text{Use } \sigma^2 = N_0(E - \rho_{12}) \\ &= Q\left(\sqrt{\frac{E - \rho_{12}}{N_0}}\right),\end{aligned}$$

where the Q-function is defined as

$$Q(x) = \frac{1}{\sqrt{2\pi}} \int_{x}^{\infty} e^{-t^2/2} dt. \tag{4.46}$$

The next step is to optimize the probability of bit error by optimizing the probability of error by minimizing $P(e|1)$, which can be achieved by optimizing the correlation term ρ_{12}. Intuitively, the best choice we can make is when $s_2(t) = -s_1(t)$, which gives us $\rho_{12} = -E$. Consequently, this result yields

$$P(e|1) = Q\left(\sqrt{\frac{2\bar{E}_b}{N_0}}\right). \tag{4.47}$$

Note that when $E_{s_1} \neq E_{s_2}$, we can then use $d^2_{\text{min}} = E_{s_1} + E_{s_2} - 2\rho_{12}$, which yields the following expression for the probability of bit error:

$$P_e = Q\left(\sqrt{\frac{d^2_{\text{min}}}{2N_0}}\right). \tag{4.48}$$

Show that the total probability of bit error is equal to:

$$P_e = P(e|1)P(1) + P(e|0)P(0) = Q\left(\sqrt{\frac{E - \rho_{12}}{N_0}}\right) \tag{4.49}$$

When dealing with a large number of signal waveforms that form a modulation scheme, the resulting probability of error, P_e, is expressed as a sum of pairwise error probabilities; that is, the probability of one received symbol being another specific received symbol. The pairwise error probability of $s_i(t)$ being decoded when $s_j(t)$ was transmitted is given as

$$Q\left(\frac{d^2_{ij}}{2N_0}\right), \tag{4.50}$$

where N_0 is the variance of the noise. An important note here is that we are assuming the noise is AWGN, since Q functions apply specifically to Gaussian random variables. Therefore, the complete expression for P_e can be expressed as

$$Q\left(\frac{d^2_{\text{min}}}{2N_0}\right) \leq P_e \leq Q\left(\frac{d^2_{1j}}{2N_0}\right) + \ldots + Q\left(\frac{d^2_{Mj}}{2N_0}\right), \quad i \neq j, \tag{4.51}$$

where the second half of the relationship is the summation of every single pairwise error probability.

Hands-On MATLAB Example: We have previously observed the impact of a noisy channel on a received signal constellation, where the signal constellation points corresponding to specific symbols of the modulation scheme are potentially shifted to other nearby signal constellation points and risk being incorrectly decoded at the receiver. In the following MATLAB script, we will translate these shifts of signal constellation points into actual BER values in order to quantify the actual severity of the noise being introduced by the channel. The code below employs a nearest neighbor approach for decoding symbols at the receiver, where a received signal constellation point that has been corrupted by noise is decoded by mapping it to the nearest originally transmitted signal constellation point. Using the Euclidean distance, we can calculate the distance between a received signal constellation point with all possible signal constellation point options, map it to the one with the shortest Euclidean distance, and then decode the symbol into the corresponding binary word.

Using this MATLAB script and the Euclidean distance approach for deciding on the nearest neighbors, we can decode the received messages sent across the noisy channel. However, since there are instances where the noise is significant enough that it can move a symbol much closer to another signal constellation point, we should anticipate that there might be several symbols that have been incorrectly decoded. Figure 4.15 presents the BER results for our nearest neighbor decoding scheme for 4-PAM, 4-QAM, and QPSK modulation. Although the first two modulation schemes do not possess a substantial amount of error, the QPSK modulation scheme possesses a large amount of incorrect decisions. This is due to the fact of how the signal constellation points are spaced out, with the QPSK signal constellation points being closer together relative to the other two modulation schemes. As a result, for the same amount of noise, the QPSK modulation will perform noticeably worse compared to the other two schemes.

4.3.1 Error Bounding

Computing each pairwise error probability is not always practical. It is possible to create an upper and lower bound on P_e by computing only the pairwise errors of points that are within one degree of the point of interest. Consider the behavior of the Q function $Q(.)$. As the input to $Q(.)$ increases, the resulting output of the Q function approaches zero. You will find that computing the pairwise error probability of points farther away yields negligible contributions to the total P_e, but can save a significant amount of time as well as cycles. Thus, an accurate estimate of $P(e)$ can be computed from the following bounds.

These upper and lower bounds can be expressed as

$$Q\left(\frac{d_{min}^2}{2N_0}\right) \leq P(e) \leq \sum_{i \in I} Q\left(\frac{d_{ij}^2}{2N_0}\right), \tag{4.52}$$

where I is the set of all signal waveforms within the signal constellation that are immediately adjacent to the signal waveform j. In order to accurately assess the performance of a communications system, it must be simulated until a certain number of symbol errors are confirmed [3]. In most cases, 100 errors will give a 95% confidence interval, which should be employed later on in this book in order to characterize the bit error rate of any digital communication system under evaluation.

Code 4.6 Decode Messages from Previous Example Using Euclidean Distance: **`chapter4.m`**

```
% Go through every received waveform and determine Euclidean distance
% between received waveform and the available waveforms
eucl_dist_4pam = zeros(4,len);
eucl_dist_4qam = zeros(4,len);
eucl_dist_qpsk = zeros(4,len);
for ind = 1:1:4,
    eucl_dist_4pam(ind,1:1:len) = abs(symb_4pam(ind).*ones(1,len)
                                   - rx_wavefm_4pam);
    eucl_dist_4qam(ind,1:1:len) = abs(symb_4qam(ind).*ones(1,len)
                                   - rx_wavefm_4qam);
    eucl_dist_qpsk(ind,1:1:len) = abs(symb_qpsk(ind).*ones(1,len)
                                   - rx_wavefm_qpsk);
end;

% Select shortest Euclidean distances
[mdist_4pam,min_ind_4pam] = min(eucl_dist_4pam);
[mdist_4qam,min_ind_4qam] = min(eucl_dist_4qam);
[mdist_qpsk,min_ind_qpsk] = min(eucl_dist_qpsk);

% Decode into estimated binary streams
bin_str_est_4pam = dec2bin(min_ind_4pam-ones(1,len)).';
bin_str_est_4qam = dec2bin(min_ind_4qam-ones(1,len)).';
bin_str_est_qpsk = dec2bin(min_ind_qpsk-ones(1,len)).';

% Calculate bit error rate
ber_4pam = sum([abs((bin_str_est_4pam(1,:)-'0') - bin_str2) ...
    abs((bin_str_est_4pam(2,:)-'0') - bin_str1)])/(2*len);
ber_4qam = sum([abs((bin_str_est_4qam(1,:)-'0') - bin_str2) ...
    abs((bin_str_est_4qam(2,:)-'0') - bin_str1)])/(2*len);
ber_qpsk = sum([abs((bin_str_est_qpsk(1,:)-'0') - bin_str2) ...
    abs((bin_str_est_qpsk(2,:)-'0') - bin_str1)])/(2*len);
```

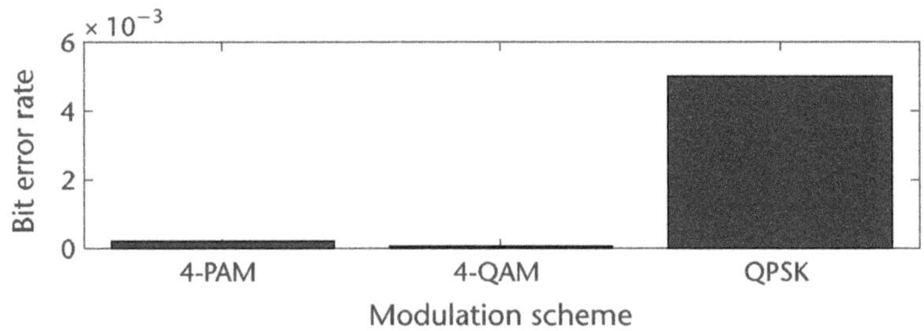

Figure 4.15 Impact of noise on modulation scheme performance.

Hands-On MATLAB Example: We have previosly observed the performance of three simple communication systems using different modulation schemes operating in a noisy environment. Although the results displayed in Figure 4.15 were insightful, we often want to observe how a communication system performs across a broad range of conditions, especially as the intensity of noise varies. In this MATLAB script, we examine the performance of a simple binary communication system across a range of different channel environments possessing varying degrees of noise.

Code 4.7 Create Waterfall Curves for the Bit Error Rate of a Communication System via Monte Carlo: **chapter4.m**

```
% Define parameters
len = 1000; % Length of individual data transmission
N_snr = 9; % Number of SNR values to evaluation
N_tx = 100; % Number of transmissions per SNR
nvar = [(10.^((1:1:N_snr)/10)).^(-1)]; % Noise variance values

ber_data = zeros(N_snr,N_tx);
for ind = 1:1:N_snr, % Different SNR values
  for ind1 = 1:1:N_tx, % Different transmissions for same SNR value

     % Generate BPSK waveform (we will keep this the same for each
     % SNR value for now)
     tx_sig = 2*round(rand(1,len))-1;

     % Create additive noise
     noise_sig = sqrt(nvar(ind))*randn(1,len);

     % Create received (noisy) signal
     rx_sig = tx_sig + noise_sig;

     % Decode received signal back to binary
     decode_bin_str = zeros(1,len);
     decode_bin_str(find(rx_sig >= 0)) = 1;

     % Determine and store bit error rate
     ber_data(ind,ind1) = sum(abs(decode_bin_str - (tx_sig+1)/2))/len;
  end;
end;

% Calculate mean bit error rate and its standard deviation
mean_ber = mean(ber_data,2).';
std_ber = std(ber_data,'',2).';
```

The end result of this analysis, which explores transmission reliability when operating across an AWGN channel, is something referred to as a waterfall curve, as shown in Figure 4.16. Waterfall curves are extensively used by researchers and designers in the digital communications community as a way of characterizing the error robustness of a communication system operating in a noisy environment. The reason why we call these plots waterfall curves is due to the shape they make whenever we generate them using either theoretical analyses or via experimentation (computer simulation or hardware testing). The x-axis describes the SNR of the operating environment and is a gauge of how much noise is present in the channel. The y-axis describes the probability of bit error as a ratio of corrupted bits versus total number of bits transmitted. In Figure 4.16, we not only show the mean BER curve but also the standard deviation above and below the mean in order to establish the degree of confidence we have with respect to the technique we used to generate these curves. Since we are using Monte Carlo techniques for generating the transmission and then corrupting the bits with additive noise, we need to make sure that we conduct this experiment long enough such that the performance results we obtain are reliable (e.g., running an experiment and only obtaining one bit error

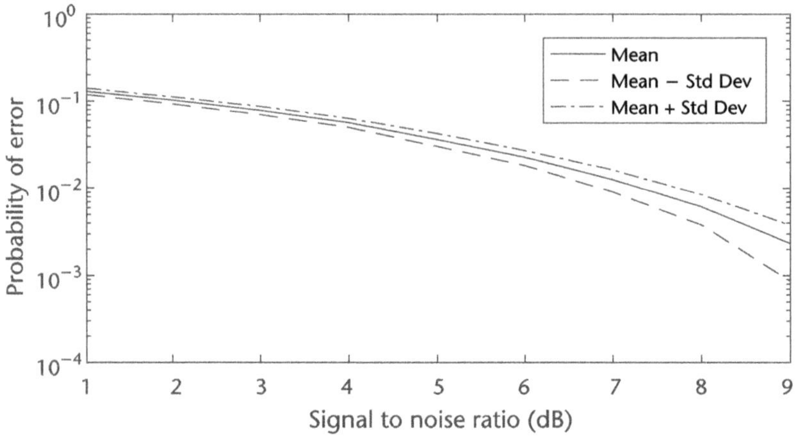

Figure 4.16 Example of waterfall curves for the bit error rate of a communication system employing binary phase shift keying via Monte Carlo simulation techniques.

is not statistically adequate with respect to an accurate assessment of the BER for that communication system). Thus, having the standard deviation curves close to the mean BER curve shows that an adequate number of bits have been used in the experiment, and that a sufficient number of errors have been obtained. Note that for different SNR values, the amount of errors obtained might be different for the same total number of bits transmitted. Consequently, we often have to transmit more bits at higher SNR values in order to obtain an adequate number of bit errors.

4.4 Signal Space Concept

Until this point we have studied digital communication systems from a signal waveform perspective. Leveraging this perspective, we have developed mathematical tools for analyzing the power efficiency and BER of different modulation schemes. However, there are several instances where the use of a signal waveform framework can be tedious or somewhat cumbersome. In this section, we will introduce another perspective on how to characterize and analyze modulation scheme using a different mathematics representation: signal vectors.

Suppose we define $\phi_j(t)$ as an orthonormal set of functions over the time interval $[0, T]$ such that

$$\int_0^T \phi_i(t)\phi_j(t)dt = \begin{cases} 1 & i = j \\ 0 & \text{otherwise} \end{cases}$$

Given that $s_i(t)$ is the ith signal waveform, we would like to represent this waveform as a sum of several orthonormal functions; that is,

$$s_i(t) = \sum_{k=1}^{N} s_{ik}\phi_k(t), \tag{4.53}$$

which can be equivalently represented by the vector

$$\mathbf{s}_i = (s_{i1},\ s_{i2},\ s_{i3}, \ldots s_{iN}), \tag{4.54}$$

where the elements of the vector $\mathbf{s}_i$ define the amplitude scaling of each orthonormal function used to represent the waveform. An illustration of a signal waveform represented by three orthonormal functions is shown in Figure 4.17. Consequently, given this relationship between the signal waveform and the orthonormal functions, where the former can be represented as the weighted sum of the latter, we can readily describe the signal waveform as a simple vector, which we will see next possesses the advantage of enabling us to employ relatively straightforward mathematical operations based on linear algebra.

In order to find the vector elements, s_{il}, we need to solve the expression

$$\int_0^T s_i(t)\phi_l(t)dt = \sum_{k=1}^{N} s_{ik} \int_0^T \phi_k(t)\phi_l(t)dt = s_{il}, \tag{4.55}$$

which is essentially a *dot product* or *projection* of the signal waveform $s_i(t)$ on the orthonormal function $\phi_l(t)$. At the same time, if we perform the vector dot product between the signal waveforms $s_i(t)$ and $s_j(t)$, we get a correlation operation that is equal to

$$\int_0^T s_i(t)s_j(t)dt = \mathbf{s}_i \cdot \mathbf{s}_j = \rho_{ij}, \tag{4.56}$$

while the energy of a signal $s_i(t)$ is equal to

$$E_{s_i} = \int_0^T s_i^2(t)dt = \mathbf{s}_i \cdot \mathbf{s}_i = ||\mathbf{s}_i||^2. \tag{4.57}$$

All of these mathematical operations will be employed when determining the power efficiency of a modulation scheme or deriving the optimal decision rule for a receiver.

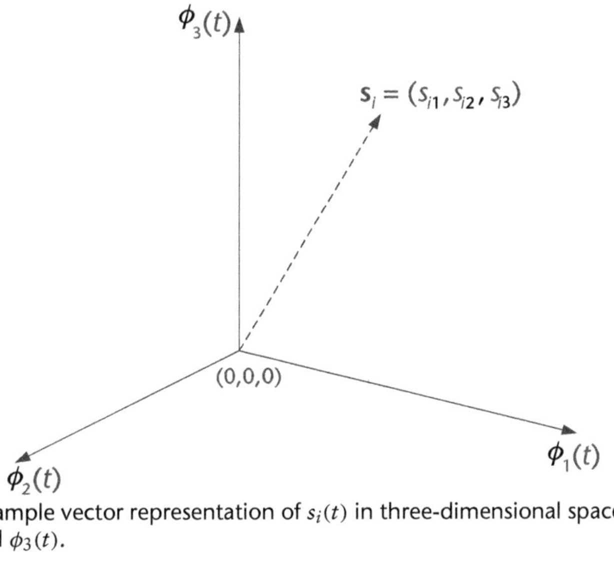

Figure 4.17 Sample vector representation of $s_i(t)$ in three-dimensional space using basis functions $\phi_1(t)$, $\phi_2(t)$, and $\phi_3(t)$.

Suppose we would like to compute the power efficiency for a modulation scheme using a signal vector approach rather than a signal waveform approach. The first step would be to calculate the minimum Euclidean distance, which can be solved using the following:

$$\begin{aligned} d_{\min}^2 &= \int_0^T \Delta s_{ij}^2(t)dt = \int_0^T (s_i(t) - s_j(t))^2 dt \\ &= ||\mathbf{s}_i - \mathbf{s}_j||^2 = (\mathbf{s}_i - \mathbf{s}_j) \cdot (\mathbf{s}_i - \mathbf{s}_j) \\ &= E_{s_i} + E_{s_j} - 2\rho_{ij} \end{aligned}$$

where the correlation term between signal waveforms $s_i(t)$ and $s_j(t)$ is given by

$$\rho_{ij} = \int_0^T s_i(t)s_j(t)dt = \mathbf{s}_i \cdot \mathbf{s}_j. \tag{4.58}$$

In order to solve for the power efficiency, we choose a set of orthonormal basis functions $\phi_i(t)$, $i = 1, 2, \ldots, k$, where k is the dimension of the signal vector space. Given this set of functions, we can now represent the vector $\mathbf{s}_i$, $i = 1, 2, \ldots, M$ where $\mathbf{s}_i = (s_{i1}, s_{i2}, \ldots s_{ik})$ and

$$s_{ij} = \int_0^T s_i(t)\phi_j(t)dt. \tag{4.59}$$

Consequently, using the vector representations for the signals and the orthonormal functions, we can calculate the minimum Euclidean distance:

$$d_{\min}^2 = \min_{i \neq j} ||\mathbf{s}_i - \mathbf{s}_j||^2, \tag{4.60}$$

the average symbol and bit energy values:

$$\begin{aligned} \bar{E}_s &= \frac{1}{M} \sum_{i=1}^{M} ||\mathbf{s}_i||^2 \\ \bar{E}_b &= \bar{E}_s / \log_2(M), \end{aligned} \tag{4.61}$$

and the power efficiency:

$$\varepsilon_p = d_{\min}^2 / \bar{E}_b. \tag{4.62}$$

4.5 Gram-Schmidt Orthogonalization

In mathematics, particularly linear algebra and numerical analysis, the Gram-Schmidt orthogonalization process is a method for creating an orthonormal set of functions in an inner product space such as the Euclidean space $\mathbb{R}^n$. The Gram-Schmidt orthogonalization process takes a finite set of signal waveforms $\{s_1(t), \ldots, s_M(t)\}$ and generates from it an orthogonal set of functions

$\{\phi_1(t), \ldots, \phi_i(t)\}$ that spans the space $\mathbb{R}^n$. Note that an orthonormal function possesses the following property:

$$\int_0^T \phi_i(t)\phi_j(t)dt = \begin{cases} 1 & i = j \\ 0 & \text{otherwise} \end{cases}.$$

Furthermore, it is possible to represent a signal waveform $s_i(t)$ as the weighted sum of these orthonormal basis functions; that is,

$$s_i(t) = \sum_{k=1}^{N} s_{ik}\phi_k(t). \tag{4.63}$$

However, what we need now is an approach to generate the set of orthonormal basis functions $\{\phi_j(t)\}$.

To derive a set of orthogonal basis functions $\{\phi_1(t), \ldots, \phi_i(t)\}$ from a set of signal waveforms denoted by $\{s_1(t), \ldots, s_M(t)\}$, let us first start with $s_1(t)$ and normalize it:

$$\phi_1(t) = \frac{s_1(t)}{\sqrt{E_{s_1}}}$$

where E_{s_1} is the energy of the signal $s_1(t)$. This normalized version of the signal waveform $s_1(t)$ will serve as our first orthonormal basis function from which we will construct the rest of our orthonormal basis function set. In other words, we are effectively bootstrapping a set of orthonormal basis functions based on the existing signal waveforms of the modulation scheme to be used. Note that we can represent $s_1(t)$ as

$$s_1(t) = \sqrt{E_{s_1}}\phi_1(t) = s_{11}\phi_1(t)$$

where the coefficient $s_{11} = \sqrt{E_{s_1}}$ and the orthonormal function $\phi_1(t)$ satisfy the unit energy constraint as required.

Next, we would like to create the second orthonormal function, $\phi_2(t)$. In order to accomplish this task, we use the signal waveform $s_2(t)$ as a starting point. However, $s_2(t)$ may contain elements of $\phi_1(t)$, and thus we need to remove this component before normalizing it in order to transform this waveform into $\phi_2(t)$. To achieve this, we first determine how much of $\phi_1(t)$ is contained within $s_2(t)$ by taking the dot product between these two functions and determining how much $s_2(t)$ projects onto $\phi_1(t)$; that is,

$$s_{21} = \int_0^T s_2(t)\phi_1(t)dt.$$

To help in getting the basis function $\phi_2(t)$, we define the intermediate function:

$$g_2(t) = s_2(t) - s_{21}\phi_1(t),$$

which is orthogonal to $\phi_1(t)$ over the interval $0 \leq t \leq T$ by virtual of the fact that we have removed the $\phi_1(t)$ component from $s_2(t)$. Finally, normalizing $g_2(t)$ yields the basis function $\phi_2(t)$:

$$\phi_2(t) = \frac{g_2(t)}{\sqrt{\int_0^T g_2^2(t)dt}} \tag{4.64}$$

which can be expanded to

$$\phi_2(t) = \frac{s_2(t) - s_{21}\phi_1(t)}{\sqrt{E_{s_2} - s_{21}^2}} \tag{4.65}$$

where E_{s_2} is the energy of the signal $s_2(t)$. A quick sanity check clearly shows that the orthonormal basis function $\phi_2(t)$ satisfies the constraint

$$\int_0^T \phi_2^2(t)dt = 1 \quad \text{and} \quad \int_0^T \phi_1(t)\phi_2(t)dt = 0$$

In general, we can define the following functions that can be employed in an iterative procedure for generating a set of orthonormal basis functions:

$$\begin{aligned} g_i(t) &= s_i(t) - \sum_{j=1}^{i-1} s_{ij}\phi_j(t) \\ s_{ij} &= \int_0^T s_i(t)\phi_j(t)dt, \; j = 1, 2, \ldots, i-1 \\ \phi_i(t) &= \frac{g_i(t)}{\sqrt{\int_0^T g_i^2(t)dt}}, \; i = 1, 2, \ldots, N. \end{aligned} \tag{4.66}$$

We will now work out an example that deals with the Gram-Schmidt orthogonalization process.

An Example: Suppose we want to perform the Gram-Schmidt orthogonalization procedure of the signals shown in Figure 4.18 in the order $s_3(t)$, $s_1(t)$, $s_4(t)$, $s_2(t)$ and obtain a set of orthonormal functions $\{\phi_m(t)\}$. Note that the order in which the signal waveforms are employed to generate the orthonormal basis functions is very important, since each ordering of signal waveforms can yield a potentially different set of orthonormal basis functions.

Starting with $s_3(t)$, we get

$$\phi_1(t) = \frac{s_3(t)}{\sqrt{E_{s_3}}} = \frac{s_3(t)}{\sqrt{3}}. \tag{4.67}$$

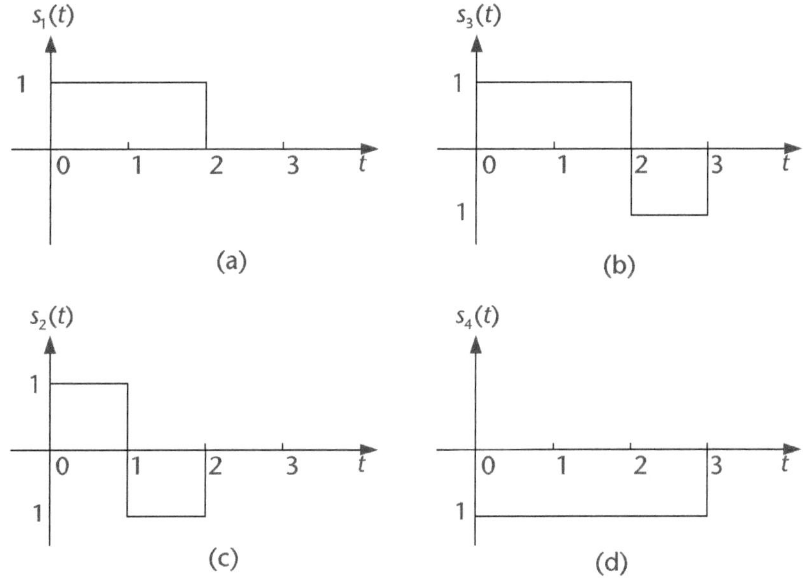

Figure 4.18 Example signal waveforms.

Then, leveraging the result for $\phi_1(t)$, we then derive the orthonormal basis function $\phi_2(t)$ using $s_1(t)$:

$$g_2(t) = s_1(t) - s_{12}\phi_1(t) = s_1(t) - \frac{2}{3}s_3(t) = \begin{cases} 1/3, & 0 \le t < 2 \\ 2/3, & 2 \le t < 3 \\ 0, & t \ge 3 \end{cases}$$

$$\therefore \phi_2(t) = \frac{g_2(t)}{\sqrt{\int\limits_0^T g_2^2(t)dt}} = \begin{cases} 1/\sqrt{6}, & 0 \le t < 2 \\ 2/\sqrt{6}, & 2 \le t < 3 \\ 0, & t \ge 3 \end{cases}. \tag{4.68}$$

We subsequently repeat this operation for $s_4(t)$:

$$g_3(t) = s_4(t) - \sum_{j=1}^{2} s_{4j}\phi_j(t) = 0$$
$$\therefore \phi_3(t) = 0, \tag{4.69}$$

but we notice the resulting $\phi_3(t)$ is equal to zero. This implies that the signal waveform $s_4(t)$ can be entirely characterized by only $\phi_1(t)$ and $\phi_2(t)$. Finally, for $s_2(t)$, we get the following:

$$g_4(t) = s_2(t) - \sum_{j=1}^{3} s_{2j}\phi_j(t) = 0$$
$$\therefore \phi_4(t) = \frac{g_4(t)}{\sqrt{\int\limits_0^T g_4^2(t)dt}} = \frac{s_2(t)}{\sqrt{2}}. \tag{4.70}$$

Consequently, with the orthonormal basis functions $\{\phi_1(t), \phi_2(t), \phi_4(t)\}$ defined, we can now express the four signal waveforms as

- $s_1 = (2/\sqrt{3}, \sqrt{6}/3, 0)$,
- $s_2 = (0, 0, \sqrt{2})$,
- $s_3 = (\sqrt{3}, 0, 0)$,
- $s_4 = (-1/\sqrt{3}, -4/\sqrt{6}, 0)$.

Now let us implement these waveforms via the orthonormal basis functions using the following MATLAB script. In this script, we assume that there are `N_samp` sampling instants per unit time. Consequently, since each waveform is of a duration of 3 seconds, each waveform representation in this MATLAB model is `3*N_samp` long.

Code 4.8 Gram-Schmidt Orthogonalization and Vectorization: **`chapter4.m`**

```
% Define parameters
N_samp = 1000; % Number of samples per time unit

% Create orthonormal basis functions
phi1 = [( 1/sqrt(3))*ones(1,N_samp) ...
     ( 1/sqrt(3))*ones(1,N_samp) ...
     (-1/sqrt(3))*ones(1,N_samp)];
phi2 = [( 1/sqrt(6))*ones(1,N_samp) ...
     ( 1/sqrt(6))*ones(1,N_samp) ...
     ( 2/sqrt(6))*ones(1,N_samp)];
phi3 = [0*ones(1,N_samp) 0*ones(1,N_samp) 0*ones(1,N_samp)];
phi4 = [( 1/sqrt(2))*ones(1,N_samp) ...
     (-1/sqrt(2))*ones(1,N_samp) ...
       0*ones(1,N_samp)];

% Based on these orthonormal basis functions, create the four symbol
% waveforms
sig_s1 = (2/sqrt(3))*phi1 + (sqrt(6)/3)*phi2 + 0*phi3 + 0*phi4;
sig_s2 = 0*phi1 + 0*phi2 + 0*phi3 + sqrt(2)*phi4;
sig_s3 = (sqrt(3))*phi1 + 0*phi2 + 0*phi3 + 0*phi4;
sig_s4 = (-1/sqrt(3))*phi1 + (-4/sqrt(6))*phi2 + 0*phi3 + 0*phi4;
```

Using these orthonormal basis functions, and the results of the Gram-Schmidt orthogonalization process, we are able to produce the same waveforms shown in Figure 4.18 using this MATLAB script, as shown in Figure 4.19.

4.6 Optimal Detection

Detection theory, or signal detection theory, is used in order to discern between signal and noise [2]. Using this theory, we can explain how changing the decision threshold will affect the ability to discern between two or more scenarios, often exposing how adapted the system is to the task, purpose, or goal at which it is aimed.

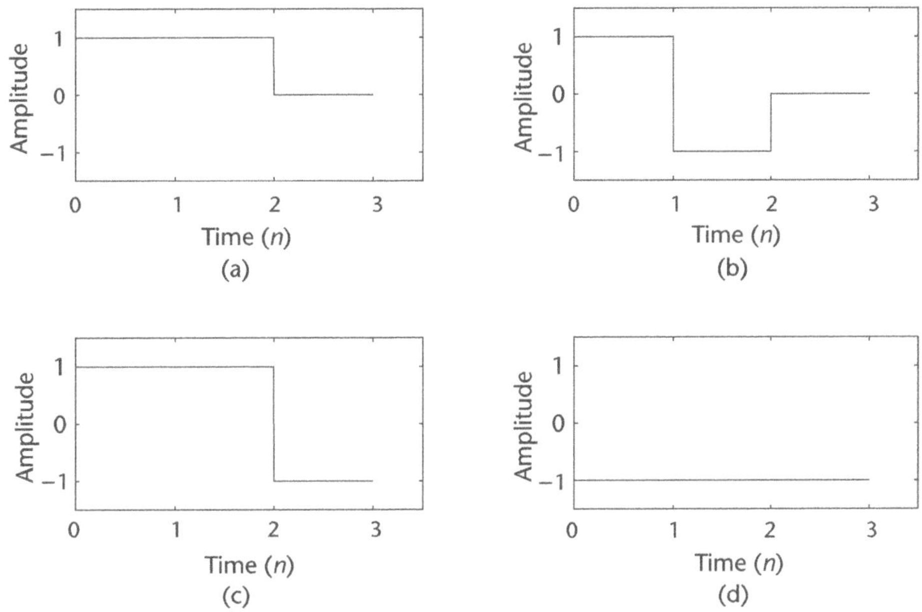

Figure 4.19 Creation of the waveforms (a) $s_1(n)$, (b) $s_2(n)$, (c) $s_3(n)$, and (d) $s_4(n)$ from a collection of orthonormal basis functions.

4.6.1 Signal Vector Framework

Let us assume a simple digital transceiver model as shown in Figure 4.20. As mentioned previously, the receiver only observes the corrupted version of $s_i(t)$ by the noise signal $n(t)$; namely, $r(t)$. The noise signal $n(t)$ usually represents the culmination of all noise sources into a single variable. Therefore, our detection problem in this situation can be summarized as follows: Given $r(t)$ for $0 \leq t \leq T$, determine which $s_i(t)$, $i = 1, 2, \ldots, M$, is present in the intercepted signal $r(t)$.`s1(n)`.

Suppose we decompose the waveforms $s_i(t)$, $n(t)$, and $r(t)$ into a collection of weights applied to a set of orthonormal basis functions; namely,

$$s_i(t) = \sum_{k=1}^{N} s_{ik}\phi_k(t), \quad r(t) = \sum_{k=1}^{N} r_k\phi_k(t), \quad n(t) = \sum_{k=1}^{N} n_k\phi_k(t).$$

Given that all of these signal waveforreliablems use the same orthonormal basis functions, we can rewrite the waveform model expression $r(t) = s_i(t) + n(t)$ into

$$\sum_{k=1}^{N} r_k\phi_k(t) = \sum_{k=1}^{N} s_{ik}\phi_k(t) + \sum_{k=1}^{N} n_k\phi_k(t)$$

$$\mathbf{r} = \mathbf{s}_i + \mathbf{n}.$$

Since $\mathbf{r}$ consists of a combination of the deterministic waveform $\mathbf{s}_i$ and probabilistic signal $\mathbf{n}$, our attention now turns to mathematically characterizing $\mathbf{n}$. Since the noise signal $n(t)$ is assumed to be a Gaussian random variable, we need to determine how the characteristics of this random variable translates into a signal

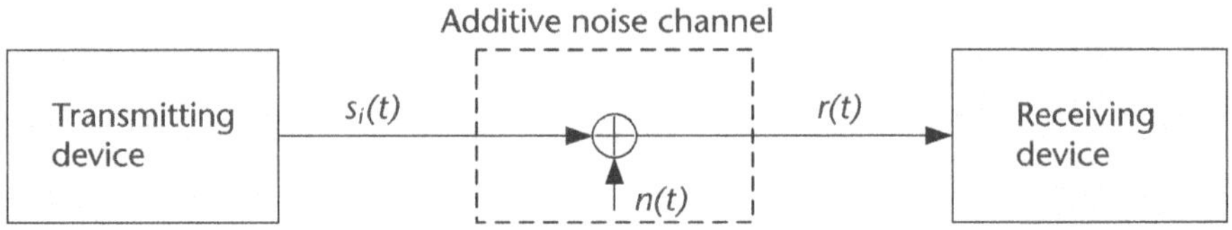

Figure 4.20 Simple digital transceiver model.

vector representation. We know that the noise vector element n_k is equal to

$$n_k = \int_0^T n(t)\phi_k(t)dt, \quad (4.71)$$

which is the projection of the noise signal waveform on the orthonormal basis function $\phi_k(t)$. Since the noise signal $n(t)$ is a Gaussian random variable and the integration process is a linear operation, this means that n_k is a Gaussian random variable as well. Thus, the noise signal vector $\mathbf{n}$ is a Gaussian vector. Let us now proceed with determining the statistical characteristics of $\mathbf{n}$ in order to employ this knowledge in signal waveform detection.

First, we would like to calculate the mean of these vector elements. Thus, by applying the definition for the expectation, this yields

$$\begin{aligned} E\{n_k\} &= E\left\{\int_0^T n(t)\phi_k(t)dt\right\} \\ &= \int_0^T E\{n(t)\}\phi_k(t)dt \\ &= 0 \end{aligned} \quad (4.72)$$

since $E\{n(t)\} = 0$, which ultimately means that the mean of the noise signal vector is $Ereliable\{\mathbf{n}\} = \mathbf{0}$.

The next step is to calculate the variance of these vector elements. Suppose we let $(\mathbf{nn}^T)_{kl} = n_k n_l$ be equal to the (k, l)th element of $\mathbf{nn}^T$. Therefore, in order to determine $E\{n_k n_l\}$, where n_k and n_l are defined by

$$n_k = \int_0^T n(t)\phi_k(t)dt, \quad n_l = \int_0^T n(\rho)\phi_l(\rho)d\rho,$$

we can apply the definition for $E\{n_k n_l\}$, which yields

$$\begin{aligned} E\{n_k n_l\} &= E\left\{\left(\int_0^T n(t)\phi_k(t)dt\right)\left(\int_0^T n(\rho)\phi_l(\rho)d\rho\right)\right\} \\ &= E\left\{\int_0^T\int_0^T n(t)n(\rho)\phi_k(t)\phi_l(t)dtd\rho\right\} \end{aligned}$$

Solving $E\{n_k n_l\}$ yields

$$\begin{aligned} E\{n_k n_l\} &= \int_0^T \int_0^T E\{n(t)n(\rho)\}\phi_k(t)\phi_l(t)dtd\rho \\ &= \int_0^T \int_0^T \frac{N_0}{2}\delta(t-\rho)\phi_k(t)\phi_l(t)dtd\rho \\ &= \frac{N_0}{2}\int_0^T \phi_k(t)\phi_l(t)dt \\ &= \frac{N_0}{2}\delta(k-l), \end{aligned} \tag{4.73}$$

where the integration of the product of the two orthonormal functions $\phi_k(t)$ and $\phi_l(t)$ yields a delta function since only when $k = l$ do these two functions project onto each other. As a result, the matrix equivalent of this outcome is equal to

$$E\{\mathbf{n}\mathbf{n}^T\} = \frac{N_0}{2}\mathbf{I}_{N\times N}. \tag{4.74}$$

Given the vector representation of the Gaussian random variable obtained in (4.74), we need to define the joint probability density function of this representation in order to characterize the individual elements of this vector. Leveraging the assumption that the noise elements are independent to each other, we can express the joint probability density function as the product of the individual probability density functions for each element, yielding

$$\begin{aligned} p(\mathbf{n}) = p(n_1, n_2, \ldots, n_N) &= \frac{1}{(2\pi\sigma^2)^{N/2}} \prod_{i=1}^{N} e^{-n_i^2/2\sigma^2} \\ &= p(n_1)p(n_2)\ldots p(n_N) \end{aligned}$$

where $p(n_i) = \frac{1}{\sigma\sqrt{2\pi}}e^{-n_i^2/2\sigma^2}$ is the probability density function for the vector element n_i. Since we know that $E\{n_k n_l\} = \frac{N_0}{2}\delta(k-l)$, we can then solve $E\{n_k^2\} = \frac{N_0}{2} = \sigma^2$. Additionally, we know that the dot product of a vector can be written as the summation of the squared elements; namely,

$$\sum_{i=1}^{N} n_i^2 = ||\mathbf{n}||^2, \tag{4.75}$$

which can then be used to yield the following expression for the joint probability density function:

$$p(\mathbf{n}) = p(n_1, n_2, \ldots, n_N) = \frac{1}{(2\pi\sigma^2)^{N/2}}e^{-||\mathbf{n}||^2/2\sigma^2}. \tag{4.76}$$

4.6.2 Decision Rules

With the formulation of the joint probability density function derived in (4.76), we can now define a rule for the receiver that can be used to determine which signal waveform is being intercepted given the presence of some noise introduced by the channel. Suppose we define the following criterion for the receiver as

$$\begin{aligned}\text{Minimize } P(\text{error}) &\rightarrow P(\hat{m}_i \neq m_i)\\ \text{Maximize } P(\text{correct}) &\rightarrow P(\hat{m}_i = m_i),\end{aligned} \tag{4.77}$$

where the probability of error is $P(e) = P(\text{error})$, the probability of correct reception is $P(c) = P(\text{correct})$, and $P(e) = 1 - P(c)$ is the complementary relationship between these two probabilities. Then, using the law of total probability, the overall probability of correct detection is equal to

$$P(c) = \int_V P(c|\mathbf{r} = \rho)p(\rho)d\rho, \tag{4.78}$$

where $P(c|\mathbf{r} = \rho) \geq 0$ and $p(\rho) \geq 0$. Therefore, we observe that when $P(c)$ attains a maximum value, this occurs when $P(c|\mathbf{r} = \rho)$ also possesses a maximum value.

In order to maximize $P(c|\mathbf{r} = \rho)$, we use the following decision rule at the receiver:

$$P(\mathbf{s}_k|\rho) \geq P(\mathbf{s}_i|\rho), \text{ for } i = 1, 2, \ldots, M \text{ and } i \neq k, \tag{4.79}$$

for $i = 1, 2, \ldots, M$ and $i \neq k$. Note that for this decision rule we are assuming that $\mathbf{s}_k$ is present in ρ such that

$$\rho = \mathbf{s}_k + \mathbf{n} \rightarrow \hat{m} = m_k. \tag{4.80}$$

Employing a mixed form of *Bayes rule* that is composed of probability density functions and probabilities; namely,

$$P(\mathbf{s}_i|\mathbf{r} = \rho) = \frac{p(\rho|\mathbf{s}_i)P(\mathbf{s}_i)}{p(\rho)}, \tag{4.81}$$

we would like to manipulate this decision rule into a formulation that can be employed by a receiver. Specifically, recall how we wanted to maximize $P(c|\mathbf{r} = \rho)$ earlier in this section. By employing the mixed Bayes rule formulation, the optimal detector can be rewritten such that it is equal to

$$\max_{\mathbf{s}_i} P(\mathbf{s}_i|\mathbf{r} = \rho) = \max_{\mathbf{s}_i} \frac{p(\rho|\mathbf{s}_i)P(\mathbf{s}_i)}{p(\rho)}, \tag{4.82}$$

for $i = 1, 2, \ldots, M$. Since $p(\rho)$ does not depend on $\mathbf{s}_i$, we can simplify the optimal detector expression such that

$$\max_{\mathbf{s}_i} p(\rho|\mathbf{s}_i)P(\mathbf{s}_i), \tag{4.83}$$

for $i = 1, 2, \ldots, M$

Based on our result in (4.83), two types of detectors can be derived based on this expression. The first type of detector is referred to as MAP detector, which can be expressed as

$$P(\mathbf{s}_i|\mathbf{r} = \rho) = \max_{\mathbf{s}_i} p(\rho|\mathbf{s}_i)P(\mathbf{s}_i), \tag{4.84}$$

for $i = 1, 2, \ldots, M$. However, in the event that $P(\mathbf{s}_i) = \frac{1}{M}$, which implies that $P(\mathbf{s}_i)$ does not depend on $\mathbf{s}_i$, we can omit the $P(\mathbf{s}_i)$ term from the optimal detector expression, yielding the second type of detector, referred to as a maximum likelihood (ML) detector:

$$P(\mathbf{s}_i|\mathbf{r} = \rho) = \max_{\mathbf{s}_i} p(\rho|\mathbf{s}_i), \tag{4.85}$$

for $i = 1, 2, \ldots, M$. In the next section, we will mathematically derive the maximum likelihood detector given the optimal decision rule for data transmissions being performed across AWGN channels.

4.6.3 Maximum Likelihood Detection in an AWGN Channel

Maximum likelihood detection is a popular statistical method employed for fitting a statistical model to data, and for identifying model parameters. In general, for a fixed set of data and underlying probability model, a maximum likelihood approach selects values of the model parameters that produce the distribution that are most likely to have resulted in the observed data (i.e., the parameters that maximize the likelihood function).

Suppose that a data transmission is operating across an AWGN channel prior to interception by the receiver. Recall that the transmission model for this scenario is given by

$$\mathbf{r} = \mathbf{s}_i + \mathbf{n}, \tag{4.86}$$

where $\mathbf{s}_i$ is the ith signal waveform sent by the transmitter, $\mathbf{n}$ is the noise introduced to the data transmission by the AWGN channel, and $\mathbf{r}$ is the intercepted signal waveform by the receiver. Given that $\mathbf{s}_i$ is a deterministic quantity, and $\mathbf{n}$ is a random entity that has just been characterized by a joint probability density function, what is needed now is a characterization of $\mathbf{r}$, which can be derived from the characterization of $\mathbf{n}$ coupled with the deterministic properties of $\mathbf{s}_i$.

Suppose we consider the conditional probability of a single element of the received vector $\mathbf{r} = \rho$, say the kth element, given that the signal waveform $\mathbf{s}_i$ was assumed to be transmitted:

$$p(\rho_k|s_{ik}) = \frac{1}{\sqrt{2\pi\sigma^2}} e^{-(\rho_k - s_{ik})^2/2\sigma^2}, \tag{4.87}$$

where the kth element of the noise vector is equal to $n_k = \rho_k - s_{ik}$. Since we assume that the AWGN vector elements are uncorrelated (i.e., independent), we can rewrite this conditional probability expression as

$$p(\rho|\mathbf{s}_i) = \prod_{k=1}^{N} p(\rho_k|s_{ik}), \text{ for } i = 1, 2, \ldots, M. \tag{4.88}$$

Consequently, this product of multiple elemental probability density functions will ultimately yield the following expression:

$$p(\rho|\mathbf{s}_i) = \frac{1}{(2\pi\sigma^2)^{N/2}} e^{-||\rho - \mathbf{s}_i||^2/2\sigma^2}. \tag{4.89}$$

Notice how we now have a formulation for the conditional probability that is entirely represented in terms of $\mathbf{s}_i$, ρ, and their respective elements. Leveraging this expression, we can proceed with mathematically determining the maximum likelihood detector.

Since we would like to solve for $\max_{\mathbf{s}_i} p(\rho|\mathbf{s}_i)$, suppose we take the expression for $p(\rho|\mathbf{s}_i)$, apply it to the detector, and take the natural logarithm of the resulting expression. Performing these operations would yield the following:

$$\ln(p(\rho|\mathbf{s}_i)) = \frac{N}{2}\ln\left(\frac{1}{2\pi\sigma^2}\right) - \frac{||\rho - \mathbf{s}_i||^2}{2\sigma^2}. \tag{4.90}$$

Note that the natural logarithm was employed in order to get rid of the exponential base in the expression, thus yielding a linear expression for the optimal decision rule. Furthermore, since natural logarithms are monotonic functions (i.e., if $x_2 \geq x_1$ then $\ln(x_2) \geq \ln(x_1)$), the decision rule would still remain valid when the inequality is employed.

Solving for this linear decision rule and given the monotonic behavior of the natural logarithm, we can derive the following:

$$\begin{aligned}
\max_{\mathbf{s}_i} \ln(p(\rho|\mathbf{s}_i)) &= \max_{\mathbf{s}_i}\left(\frac{N}{2}\ln\left(\frac{1}{2\pi\sigma^2}\right) - \frac{||\rho - \mathbf{s}_i||^2}{2\sigma^2}\right) \\
&= \max_{\mathbf{s}_i}\left(-\frac{||\rho - \mathbf{s}_i||^2}{2\sigma^2}\right) \\
&= \max_{\mathbf{s}_i}\left(-||\rho - \mathbf{s}_i||^2\right) \\
&= \min_{\mathbf{s}_i} ||\rho - \mathbf{s}_i||.
\end{aligned} \tag{4.91}$$

Since we are interested in the choice of $\mathbf{s}_i$ that yields the maximum value for the decision rule, we can rewrite this decision rule as

$$\mathbf{s}_k = \arg\min_{\mathbf{s}_i} ||\rho - \mathbf{s}_i|| \rightarrow \hat{\mathbf{m}} = \mathbf{m}. \tag{4.92}$$

Note that one of the advantages of employing a vector representation for these decision rules is that the entire scenario can be interpreted in terms of distance. Specifically, the term $||\rho - \mathbf{s}_i||$ actually represents the distance between the heads of two vectors, ρ and $\mathbf{s}_i$, whose tails are located at the origin. Thus, a maximum likelihood detector is the equivalent of a minimum distance detector.

4.7 Basic Receiver Realizations

The fundamental challenge of digital communications is recovering what was transmitted after it has passed through a channel and been corrupted by noise. The first receiver structure we will examine is based on filtering the received signal with a static filter that maximizes the SNR of the channel, which will subsequently minimize the bit error rate. However, one of the disadvantages of a matched filtering

> **Q** Referring to the signal constellation diagram shown in Figure 4.21, implement a simple QPSK transceiver operating in an AWGN channel and implement a maximum likelihood detector. Does this decision rule match the decision regions in Figure 4.21? Does this decision rule adequately declare $\mathbf{s}_i$ as the transmitted signal based on which quadrant ρ appears in? What is the impact of the AWGN channel for different values for the variance given that the noise is zero-mean?

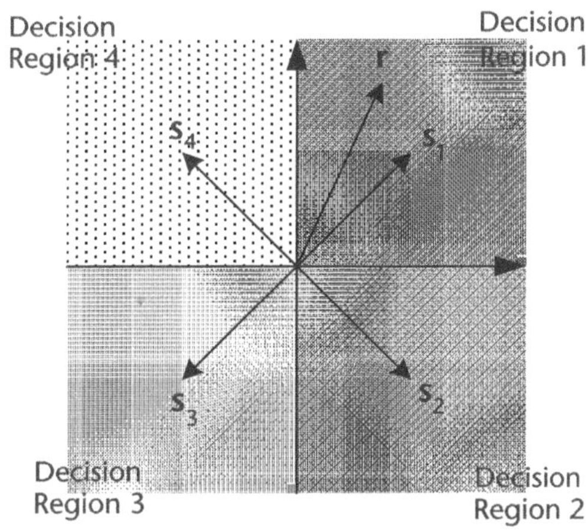

Figure 4.21 Decision regions for QPSK signal constellation.

approach is that it requires a priori knowledge of all the possible signal waveforms sent by the transmitter.

4.7.1 Matched Filter Realization

When designing a receiver, we are interested in a decision rule where the receiver yields the correct result more often than any other decision rule that can be employed by the receiver. In other words, we are interested in detecting a pulse transmitted over a channel corrupted by noise.

Suppose we employ the following transmission model:

$$x(t) = g(t) + w(t), 0 \leq t \leq T, \tag{4.93}$$

where $g(t)$ is a pulse signal, $w(t)$ is a white noise process with mean $\mu = 0$ and power spectral density equal to $\frac{N_0}{2}$, and $x(t)$ is the observed received signal. Assuming the receiver knows all the possible waveforms of $g(t)$ produced by the transmitter, the objective of the receiver is to detect the pulse signal $g(t)$ in an optimum manner based on an observed received signal $x(t)$. Note that the signal $g(t)$ may represent a "1" or a "0" in a digital communication system

In order to enable the receiver to successfully detect the pulse signal $g(t)$ in an optimal manner given the observed received signal $x(t)$, let us filter $x(t)$ the effects of the noise are minimized in some statistical sense such that the probability of correct detection is enhanced. Suppose we filter $x(t)$ using $h(t)$ such that the output of this process yields

$$y(t) = g_0(t) + n(t), \tag{4.94}$$

where $n(t)$ is the result of the noise signal $w(t)$ filtered by $h(t)$ and $g_0(t)$ is the filtered version of $g(t)$ by $h(t)$. The transmission model and filtering operation by $h(t)$ is illustrated in Figure 4.22.

Let us rewrite this filtering operation in the frequency domain, where the time domain convolution operations become frequency domain products. Thus, taking the inverse Fourier transform of $H(f)G(f)$, which is equivalent to a convolution of $h(t)$ and $g(t)$, we get the following expression for the filtered version of $g(t)$:

$$g_0(t) = \int_{-\infty}^{\infty} H(f)G(f)e^{j2\pi ft}\, df, \tag{4.95}$$

where the inverse Fourier transform returns the filtering operation back to the time domain.

Let us now calculate the instantaneous reliable power of the filtered signal $g_0(t)$, which is given as:

$$|g_0(t)|^2 = |\int_{-\infty}^{\infty} H(f)G(f)e^{j2\pi ft}\, df|^2. \tag{4.96}$$

In order to determine a quantitative metric that would indicate when we have achieved the goal of maximizing $g_0(t)$ relative to $n(t)$, let us employ the peak pulse SNR, which is defined as

$$\eta = \frac{|g_0(T)|^2}{E\{n^2(t)\}}, \tag{4.97}$$

where $|g_0(T)|^2$ is the instantaneous power of the output signal at sampling instant T, and $E\{n^2(t)\}$ is the average power of the output noise. Thus, goal of this matched filter realization is to maximize $g_0(t)$ with respect to $n(t)$ using the peak pulse SNR metric, which can be achieved by designing a filter $h(t)$ that can yield the largest possible value for η.

In order to design $h(t)$, we need to mathematically solve for $h(t)$, which consists of evaluating the expression

$$|g_0(t)|^2 = \left| \int_{-\infty}^{\infty} H(f)G(f)e^{j2\pi ft} df \right|^2, \tag{4.98}$$

which is the magnitude squared of the inverse Fourier transform of $H(f)G(f) = \mathcal{F}\{h(t) * g(t)\}$. Since $w(t)$ is a white Gaussian process with power spectral density $\frac{N_0}{2}$, we know from the EWK theorem that the power spectral density of the filtered noise signal $n(t)$ is equal to $S_N(f) = \frac{N_0}{2}|H(f)|^2$. Therefore, applying the definition

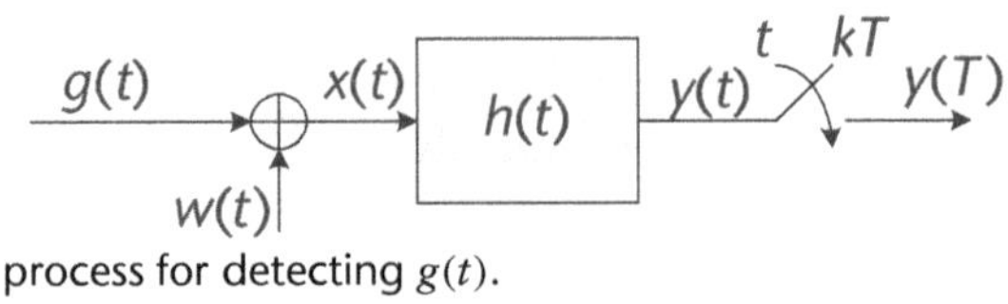

Figure 4.22 Filtering process for detecting $g(t)$.

for η and including these expressions will yield

$$\eta = \frac{|\int_{-\infty}^{\infty} H(f)G(f)e^{j2\pi fT} df|^2}{\frac{N_0}{2}\int_{-\infty}^{\infty} |H(f)|^2 df}. \tag{4.99}$$

From this resulting expression, we see that we need to solve for frequency response $H(f)$ such that it yields the largest possible value for the peak pulse SNR η. In order to obtain a closed-form solution, let us employ Schwarz's inequality. Suppose that we have two complex functions, say $\phi_1(x)$ and $\phi_2(x)$, such that:

$$\int_{-\infty}^{\infty} |\phi_1(x)|^2 dx < \infty \quad \text{and} \quad \int_{-\infty}^{\infty} |\phi_2(x)|^2 dx < \infty. \tag{4.100}$$

Then, by Schwarz's inequality we can rewrite the following integral expression as an inequality:

$$\left|\int_{-\infty}^{\infty} \phi_1(x)\phi_2(x)dx\right|^2 \leq \left(\int_{-\infty}^{\infty} |\phi_1(x)|^2 dx\right) \cdot \left(\int_{-\infty}^{\infty} |\phi_1(x)|^2 dx\right), \tag{4.101}$$

with this expression becoming an equality when $\phi_1(x) = K \cdot \phi_2^*(x)$.

Therefore, leveraging Schwarz's inequality in our expression for the peak pulse SNR, it can be shown that the numerator of (4.99) can be rewritten as:

$$\left|\int_{-\infty}^{\infty} H(f)G(f)e^{j2\pi ft} df\right|^2 \leq \left(\int_{-\infty}^{\infty} |H(f)|^2 df\right) \cdot \left(\int_{-\infty}^{\infty} |G(f)|^2 df\right), \tag{4.102}$$

which then yields the following inequality for η:

$$\eta \leq \frac{2}{N_0}\int_{-\infty}^{\infty} |G(f)|^2 df. \tag{4.103}$$

Thus, in order to make this expression an equality, the optimal value for $H(f)$ should be equal to

$$H_{\text{opt}}(f) = K \cdot G^*(f)e^{-j2\pi fT}, \tag{4.104}$$

whose time domain representation can be mathematically determined using the inverse Fourier transform:

$$h_{\text{opt}}(t) = K \cdot \int_{-\infty}^{\infty} G^*(f)e^{-j2\pi fT}e^{-j2\pi ft} df = K \cdot g(T - t). \tag{4.105}$$

Notice that when we are performing a matched filtering operation, we are convolving the time-flipped and time-shifted version of the transmitted pulse with the transmitted pulse itself in order to maximize the SNR.

> The reason why we call these filters matched filters is due to the fact that when we convolve the time-flipped and time-shifted version of the transmitted pulse signal with itself, the process is SNR maximizing. Consequently, if a receiver intercepts some unknown noise-corrupted signal, it can readily identify which one was sent by a transmitter by matching this intercepted signal to all the available signal waveforms known at the receiver using an implementation illustrated in Figure 4.23.

> Referring to Figure 4.24, suppose we have a signal $g(t)$. Show that $h(t)$ and $g_0(t)$ are the corresponding matched filter and filtered output signals.

4.7.2 Correlator Realization

Recall that a matched filter realization assumes some sort of knowledge regarding the transmitted data. However, if the receiver possesses this information about the reliable transmitter and its signal characteristics, it is also possible to employ a more statistical approach for determining which signal waveforms have been sent, even in the presence of a noisy, corruption-inducing channel. Specifically, we can employ the concept of correlation such that we only need to assume knowledge about the waveforms themselves.[1]

Suppose we start with the decision rule derived at the beginning of this section and expand it such that

$$\begin{aligned}\min_{\mathbf{s}_i} ||\rho - \mathbf{s}_i||^2 &= \min_{\mathbf{s}_i}(\rho - \mathbf{s}_i) \cdot (\rho - \mathbf{s}_i) \\ &= \rho \cdot \rho - 2\rho \cdot \mathbf{s}_i + \mathbf{s}_i \cdot \mathbf{s}_i.\end{aligned} \tag{4.106}$$

Since $\rho \cdot \rho$ is common to all the decision metrics for different values of the signal waveforms $\mathbf{s}_i$, we can conveniently omit it from the expression, thus yielding

$$\min_{\mathbf{s}_i} \left(-2\rho \cdot \mathbf{s}_i + \mathbf{s}_i \cdot \mathbf{s}_i\right) = \max_{\mathbf{s}_i} \left(2\rho \cdot \mathbf{s}_i - \mathbf{s}_i \cdot \mathbf{s}_i\right), \tag{4.107}$$

where $\rho \cdot \mathbf{s}_i$ and $\mathbf{s}_i \cdot \mathbf{s}_i$ are defined by

$$\rho \cdot \mathbf{s}_i = \int_0^T \rho(t) s_i(t) dt \qquad \mathbf{s}_i \cdot \mathbf{s}_i = \int_0^T s_i^2(t) dt = E_{s_i}.$$

We can observe that the waveform representation of $\rho \cdot \mathbf{s}_i$ is equal to the correlation of $r(t) = \rho(t)$ with respect to $s_i(t)$. Thus, when $s_k(t)$ is present in $r(t)$,

1. For a matched filtering implementation, knowledge of both the transmission signal waveforms and the statistical characteristics of the noise introduced by the channel is needed by the receiver.

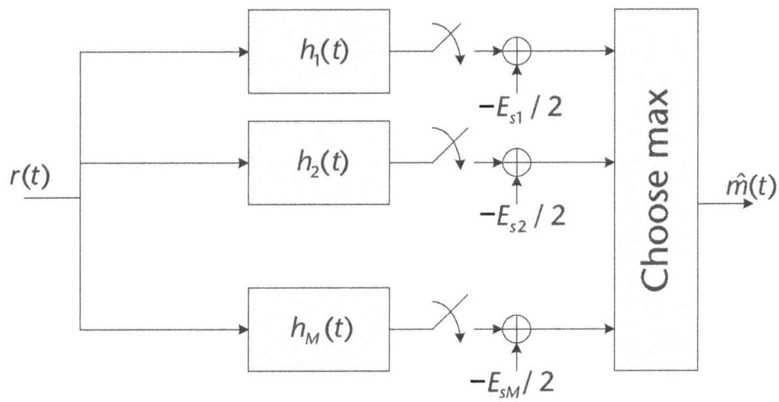

Figure 4.23 Schematic of matched filter realization of receiver structure.

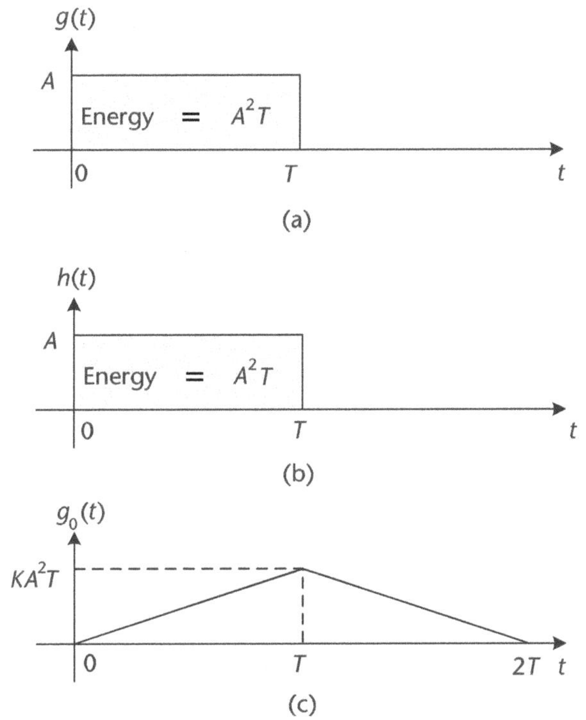

Figure 4.24 An example of the time domain input and output signal waveforms processed by a matched filter. (a) Time domain representation of the input signal to the matched filter, (b) time domain impulse response of the matched filter, (c) time domain representation of the output signal of the matched filter.

the optimal detector is equal to

$$reliables_k = \arg\max_i \left(\int_0^T \rho(t)s_i(t)dt - \frac{E_{s_i}}{2} \right). \tag{4.108}$$

Based on this result, we can design a receiver structure that leverages correlation in order to decide on which signal waveform was sent by the transmitter based on the observed intercepted signal at the receiver. An schematic of a correlation-based implementation is shown in Figure 4.25. Given $r(t) = s_i(t) + n(t)$ and we observe only $r(t) = \rho(t)$ at the input to the receiver, we first correlate $r(t)$ with $s_i(t)$ across all i. Next, we normalize the correlation result by the corresponding signal energy E_{s_i} in order to facilitate a fair comparison. Note that if all energy values are the same for each possible signal waveform, we can dispense with the energy normalization process since this will have no impact on the decision making. Finally, the resulting decision values for each of the branches are compared against each other and the branch with the largest resulting value is selected.

Hands-On MATLAB Example: To highlight the how a correlator receiver structure would work in decoding the intercepted waveforms and translating them in the corrsponding binary output, the following MATLAB script can be used, where we begin by generating a stream of random waveforms consisting of symbols `s1(n)`, `s2(n)`, `s3(n)`, and `s4(n)`. These waveforms are obtained from the MATLAB script shown in Section 4.5. Once this stream of waveforms has been generated, the next step is to vectorize the waveforms into a three-dimensional signal constellation space. Once vectorized, we use the correlation receiver approach in order to find out the Euclidean distance between the received vectorized waveforms and the available signal vectors. Once these distances have been calculated per received waveform, a *decision-making process* is performed in order to find out the closest match between the received and available symbol vectors.

The result of this waveform vectorization and correlator-based receiver design for these examples is shown in Figure 4.26. We can see that both the orignally transmitted and the decoded waveforms are a perfect match. However, in this model we did not include any forms of distortion such as noise. Consequently, it is a perfect exercise to observe how the correlator-based receiver performs when additive white Gaussian noise introduced to the received signal.

4.8 Chapter Summary

A deep and thorough understanding of digital communication theory is vitally essential when designing and evaluating software-defined radio implementations. In this chapter, an overview of several useful topics, including several different types of modulation schemes, the derivation of the probability of error, Gram-Schmidt

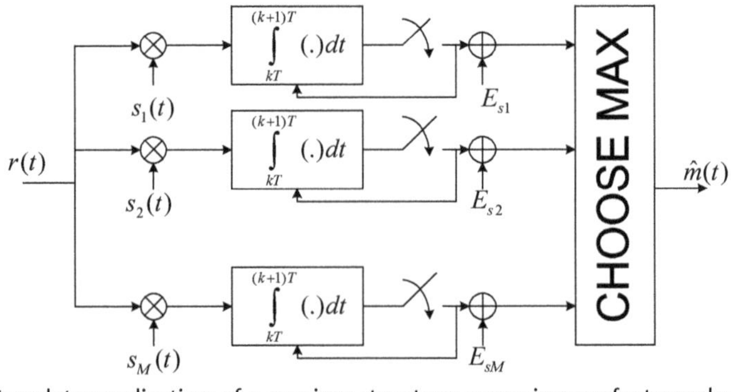

Figure 4.25 Correlator realization of a receiver structure assuming perfect synchronization.

Code 4.9 Correlator-Based Receiver Implementation Using Gram-Schmidt: **chapter4.m**

```
% Define parameters
N_symb = 10; % Number of symbols contained within intercepted signal

% Randomly generate intercepted waveform consisting of s1(n), s2(n),
% s3(n), and s4(n)
rx_sig = [];
orig_msg = [];
for ind = 1:1:N_symb,
    rnd_val = rand(1,1);
    if (rnd_val < 0.25) % Add s1(n) waveform
        rx_sig = [rx_sig sig_s1];
        orig_msg = [orig_msg 1];
    elseif ((rnd_val >= 0.25)&&(rnd_val < 0.5)) % Add s2(n) waveform
        rx_sig = [rx_sig sig_s2];
        orig_msg = [orig_msg 2];
    elseif ((rnd_val >= 0.5)&&(rnd_val < 0.75)) % Add s3(n) waveform
        rx_sig = [rx_sig sig_s3];
        orig_msg = [orig_msg 3];
    else % Add s4(n) waveform
        rx_sig = [rx_sig sig_s4];
        orig_msg = [orig_msg 4];
    end;
end;

% Vectorize the intercepted signal
dim1_comp = [];
dim2_comp = [];
dim4_comp = [];
for ind = 1:1:N_symb,
    dim1_comp = [dim1_comp sum(rx_sig(((ind-1)*3*N_samp+1):1:
                (ind*3*N_samp)).*phi1)];
    dim2_comp = [dim2_comp sum(rx_sig(((ind-1)*3*N_samp+1):1:
                (ind*3*N_samp)).*phi2)];
    dim4_comp = [dim4_comp sum(rx_sig(((ind-1)*3*N_samp+1):1:
                (ind*3*N_samp)).*phi4)];
end;
dim1_comp = dim1_comp/N_samp;
dim2_comp = dim2_comp/N_samp;
dim4_comp = dim4_comp/N_samp;

% Using the correlator receiver approach, we determine the closest
% symbol vector to each vectorized waveform based on Euclidean distance
s1vec = [(2/sqrt(3)) (sqrt(6)/3) 0 0];
s2vec = [0 0 0 sqrt(2)];
s3vec = [(sqrt(3)) 0 0 0];
s4vec = [(-1/sqrt(3)) (-4/sqrt(6)) 0 0];
est_msg = [];
for ind = 1:1:N_symb,
  [val,symb_ind] = min([ ...
  sum((s1vec - [dim1_comp(ind) dim2_comp(ind) 0 dim4_comp(ind)]).^2) ...
  sum((s2vec - [dim1_comp(ind) dim2_comp(ind) 0 dim4_comp(ind)]).^2) ...
  sum((s3vec - [dim1_comp(ind) dim2_comp(ind) 0 dim4_comp(ind)]).^2) ...
  sum((s4vec - [dim1_comp(ind) dim2_comp(ind) 0 dim4_comp(ind)]).^2) ...
    ]);
    est_msg = [est_msg symb_ind];
end;
```

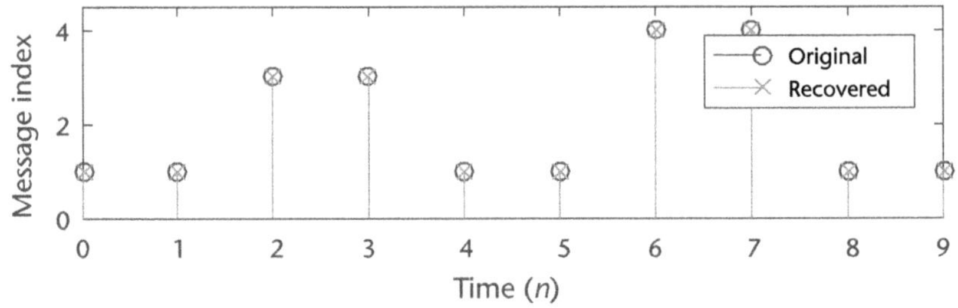

Figure 4.26 Matching correlator receiver output with original transmission.

orthogonalization, formulation of the optimal decision rule, and the presentation of two receiver structures were studied in order to provide the reader with the fundamentals needed in order to master these versatile yet complex systems. In the subsequent chapters, the fundamental knowledge obtained in this chapter, as well as the two previous chapters, will be leveraged extensively when implementing digital communication systems and networks based on software-defined radio technology.

4.9 Additional Readings

Given the introductory nature of this chapter with respect to the topic of digital communications, the interested reader is definitely encouraged to explore the numerous books that provide a substantially more detailed and advanced treatment of this topic. For instance, the latest edition of the seminal digital communications book by Proakis and Salehi [4] provides a rigorous, mathematical treatment of many of the concepts covered in this chapter, in addition to many other topics not presented such as spread spectrum technologies, equalization, and RAKE receiver implementations. To complement this mathematically terse document, the authors also published a companion book that treats digital communications from a more applied perspective, including examples in MATLAB and Simulink [5].

As for introductory textbooks on digital communications, Sklar wrote an excellent document that provides the reader with a balance of mathematical rigor, detailed explanations, and several well-crafted examples [6]. The book by Couch is also in the same category as Sklar, but it treats both analog and digital communications [7], which is well suited for individuals that do not possess a background in the communications field. Rice wrote his introductory book on digital communications from a discrete-time perspective, which is suited for an individual possessing a background in discrete-time signal and systems [8]. The book also provides numerous end-of-chapter problems as well as MATLAB examples available online, providing the reader with many opportunities to practice the theoretical concepts covered within this text.

The classic digital communications book by Barry, Messerschmitt, and Lee [9] is an excellent reference for those individuals who possess some understanding about digital communications but need convenient and immediate access to detailed information. Similarly, the books by Madhow [10] and Pursley [11] both provide readers with a more advanced treatment of digital communication theory. Finally, the book by Hsu [12] is an excellent reference that mostly consists of a popular collection of solved problems.

References

[1] Shannon, C. E., "Communication in the Presence of Noise," in *Proceedings of the Institute of Radio Engineers*, Vol. 37, No. 1, January 1949, p. 1021.

[2] Poor, H. V., *An Introduction to Signal Detection and Estimation*, New York: Springer, 2010.

[3] Wyglinski, A. M., Physical Layer Loading Algorithms for Indoor Wireless Multicarrier Systems, Ph.D. thesis, McGill University, Montreal, 2004.

[4] Proakis, J., and M. Salehi, *Digital Communications*, Fifth Edition, Boston: McGraw-Hill, 2007.

[5] Proakis, J. G., M. Salehi, and G. Bauch, *Contemporary Communication Systems Using MATLAB*, Second Edition, Brooks/Cole, 2003.

[6] Sklar, B., *Digital Communications: Fundamentals and Applications*, Second Edition, Prentice Hall PTR, 2001.

[7] Couch, L. W., Digital and Analog Communication Systems, Seventh Edition, Upper Saddle River: Prentice Hall, 2006.

[8] Rice, M., *Digital Communications: A Discrete-Time Approach*, Third Edition, Pearson/PrenticeHall, 2009.

[9] Barry, J. R., D. G. Messerschmitt, and E.A. Lee, *Digital Communication*, Third Edition, Norwell, MA: Kluwer Academic Publishers, 2003.

[10] Madhow, U., *Fundamentals of Digital Communication*, Cambridge, UK: Cambridge University Press, 2008.

[11] Pursley, M. B., *Introduction to Digital Communications*, Prentice Hall, 2004.

[12] Hsu, H. P., *Schaum's Outline of Theory and Problems of Analog and Digital Communications*, Second Edition, New York: McGraw-Hill, 2002.

CHAPTER 5

Understanding SDR Hardware

In this chapter, we will discuss the real-world implications of using SDR hardware and fundamentals for interacting with the Pluto SDR from MATLAB. Using Pluto SDR as a template we will provide an introduction in the receive and transmit chains, discussing how analog waveforms become digital samples in MATLAB. Once we have a solid grasp on this process a common code templating will be introduced, which will be used throughout the remaining chapters when working with the radio in MATLAB. This templating will provide a simplified workflow that can help alleviate common problems faced when working with SDR's and specifically Pluto SDR. Finally, the chapter will conclude with a small example to make sure the Pluto SDR is configured correctly with MATLAB.

5.1 Components of a Communication System

The software-defined radio described in Section 5.1.1 can constitute a radio node in anything from a point-to-point link to an element in a large ad hoc network of radios. It can be used as an RFFE to a MATLAB script or Simulink model or it can be programmed and used as a complete stand-alone radio. The radio front end, in this case the Pluto SDR, is a single components in a larger communications system, which would also normally include external filters and band-specific antennas. A description of the communication systems, and the block diagram are shown in Figure 5.1(c). The major aspects of that are

- An analog RF section (atennna, RF filters, input mux, LNA, gain, attenuation, mixer);
- An analog baseband section (analog filters, ADC or DAC);
- Some signal processing units (fixed filters inside a transceiver, or user defined inside a FPGA or DSP, or general-purpose processor).

While Pluto SDR provides a great low-cost platform for STEM education and SDR experimentation, it is representive of many SDRs used in commuications systems. Although it is small and low-cost, the Pluto SDR has enough capability to tackle a wide range of SDR applications, such as GPS or weather satellite receiver or ad hoc personal area networks. The Pluto SDR plays the part of the communications systems described above as follows:

- An analog RF section (atennna, RF filters, input mux, LNA, gain, attenuation, mixer)
  - Antenna and RF filters are expected to be done outside the Pluto SDR and are the responsibility of the end user

- The remaining portions of the first first bullet (input mux, LNA, gain, attenuation, mixer), are all implmented in the AD9363, Integrated RF Agile Transceiver
- Analog baseband section (analog filters, ADC or DAC) is implmented in the AD9363, Integrated RF Agile Transceiver
- Signal processing; this is split between
  - Parts of signal processing is implmented in the AD9363, Integrated RF Agile Transceiver. This includes the fixed halfband decimiation and interpolation filters and programmable 128-tap FIR filters.
  - Optional filtering and decimation may be done in the Xilinx Zynq' FPGA fabric.
  - The I/Q data is then passed up to the USB to a host, where MATLAB can continue the signal processing.

To understand the details of these pieces, it is necessary to peel back the plastic on the Pluto SDR and look at the devices that make up the device itself. Being a SDR platform specifically targeted for students, not only are schematics for the Pluto SDR readily available, but also the entire software and HDL stack, so we can examine in detail the makeup of the device at any level from hardware to software.

5.1.1 Components of an SDR

Most modern SDR devices typically share a similar structural design, which makes up the receive and/or transmit chains to translate data from the analog RF domain into analog baseband domain, and into IQ samples, and eventually into a software package such as MATLAB. In the very simplest sense the Pluto SDR (shown in Figure 5.1[b]) is made up of two components, as shown in Figure 5.1(a):

- An analog RF section (which specifies the receive and transmit capabilities);
- The communications mechanism (Ethernet, USB) to get IQ data back to host for processing.

Referring to Figure 5.1(c), the receive, transmit, and communication specifications of the ADALM-PLUTO consist of

- Transmit (SMA connector labeled Tx)
  - 300–3,800 GHz, 200–20,000 kHz channel bandwidth, 65.1–61,440 kSPS
  - 2.4 Hz LO step size, 5 Hz sample rate step size
  - Modulation accuracy (EVM): 40 dB (typical, not measured on every unit)
  - 12-bit DACs
- Receive (SMA connector labeled Rx)
  - 300–3,800 GHz, 200–20,000 kHz channel bandwidth, 65.1–61,440 kSPS
  - 2.4 Hz LO step size, 5 Hz sample rate step size
  - Modulation accuracy (EVM): 40 dB (typical, not measured on every unit)
  - 12-bit ADCs

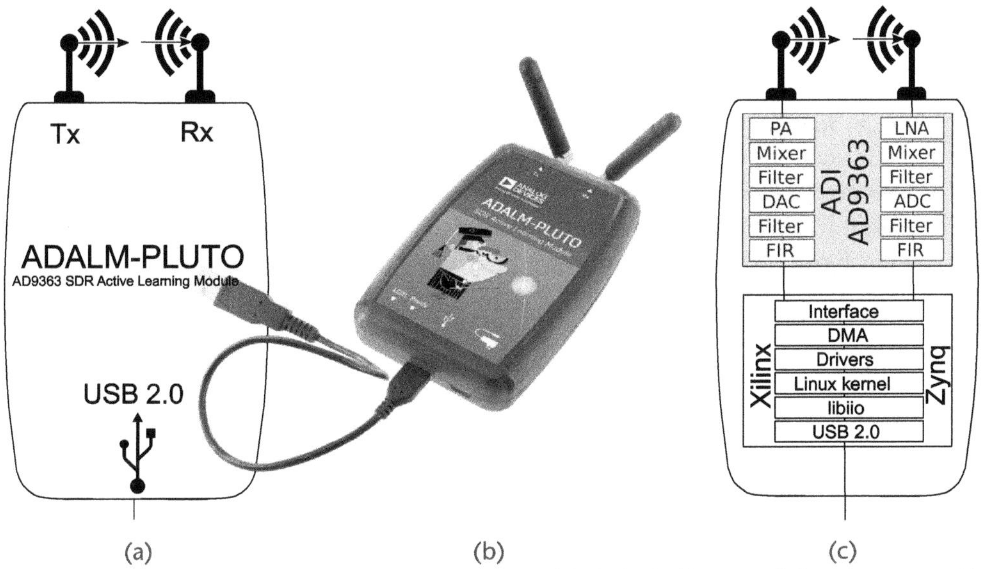

Figure 5.1 Views of the ADALM-PLUTO SDR. (a) Simplified block diagram of the ADALM-PLUTO, (b) photo of the ADALM-PLUTO [1], and (c) I/O on the ADALM-PLUTO.

- USB 2 OTG (480 Mbits/seconds), device mode
  - libiio USB class, for transfering IQ data from/to the RF device to the host
  - Network device, provides access to the Linux on the Pluto device
  - USB serial device, provides access to the Linux console on the Pluto device
  - Mass storage device
- USB 2 OTG (480 Mbits/seconds), host mode
  - Mass storage device, plug in a thumb drive, and capture or playback waveforms
  - Wifi dongle, access the Pluto SDR via WiFi
  - Wired LAN, access the Pluto SDR via wired LAN
- External power, for when using the Pluto SDR in host mode.

It is possible to run the Pluto SDR out of spec and extend the frequency range to 70–6,000 MHz to be able to capture and listen to FM broadcasts (in the 87.5–108.0 MHz bands most places, 76–95 MHz in Japan, and legacy 65.8–74.0 MHz in some Eastern European countries) at the low end, and the all the interesting things happening in 5.8-GHz ISM worldwide bands.

Because of the wide tuning range, 70–6,000 MHz, which is over three orders of magnitude, there are no band-specific receive or transmit filters in the Pluto SDR. What this means is that from a receive side, everything that is broadcasting from 70–6,000 MHz will be picked up, and could affect your signal. This is normally only an issue when you are trying to receive a very low amplitude signal. More about this in Section 5.2.6.

5.1.2 AD9363 Details

At the front of the Pluto SDR is a AD9363 5.2 transceiver from Analog Devices Inc., which is responsible for capturing and digitization of the RF data. This transceiver provides amplification, frequency translation (mixing), digital conversion, and filtering of transmitted and receive signals. In Figure 5.2 we provide a detailed outline of the of the AD9363. While it can look complicated to the beginner, it is has nothing more than the three sections we mentioned before: an analog RF section, an analog baseband section, and some signal processing for both receive and transmit. It is important to understand the physical analog and digital hardware boundary because it will provide the necessary knowledge to configure the device from MATLAB and understand nonidealities experienced in the transmitted and received data. However, an extreme indepth understanding of the device is not required to effectively work with a SDR but some basics are invaluable.

We will discuss the AD9363 from the perspective of the receiver, but logically the same operations just apply in reverse order for the transmitter. At the very front of the AD9363 is a low-noise amplifier (LNA) providing analog gain that is a component of the automatic gain control (AGC) pipeline of the receiver. Following the LNA is the mixer, which is responsible for direct frequency translation. Unlike many traditional heterodyne transceivers, the AD9363 is a direct conversion, or ZeroIF design that does not utilize an intermediate frequency (IF) stage. For more details on the trade-offs between heterodyne and direct-conversion receivers, consider reading Razavi [2].

The mixer in the AD9363 operates from 325 MHz to 3.8 GHz within datasheet specification [2], but software modifications can be made to expand this range, which we will discuss in Section 5.2.6. Prior to this mixing process, the signal is split and fed along two different but identical paths. This process creates the in-phase and quadrature components of our signal through a simple phase rotation of the mixer's clock. Effectively this doubles the effectively bandwidth of the receiver since the in-phase and quadrature signals are orthogonal (bandwidth is $-\frac{f_s}{2}$ to $\frac{f_s}{2}$).

After mixing, the signal is filtered to remove aliasing effects of the now down-mixed signal and to reduce out of band interference and noise. The combined transimpedance amplifier (TIA) and analog filter are configured together to maintain the desired analog bandwidth, which can range from 200 kHz to 20 MHz. The TIA acts as a single pole filter and the analog programmable filter is a third-order Butterworth.

The final stage of the AD9363 is the digital conversion and decimation stage. Here the ADC will typically run at a much higher rate than the desired receive bandwidth, but the ADC itself will not provide all 12 bits defined in the specifications. The additional bits are gained in the halfband filter (HBF) stages, which will allow bit growth. The ADC itself only provides $\sim$ 4.5 bits of resolution. This is a typical design for sigma-delta converters (Σ-Δ ADC), which inherently have low noise and run faster than the alternative successive approximation (SAR) ADCs. Refer to Section 2.5.4 for more information about Σ-Δ ADCs. However, by utilizing a very high speed ADC and associated HBFs the receive signal can be digitized at 12 bits at the desired configured sample rate. Therefore, for the best signal resolution is achieved through large oversampling of the input signal and then followed by several decimation stages.

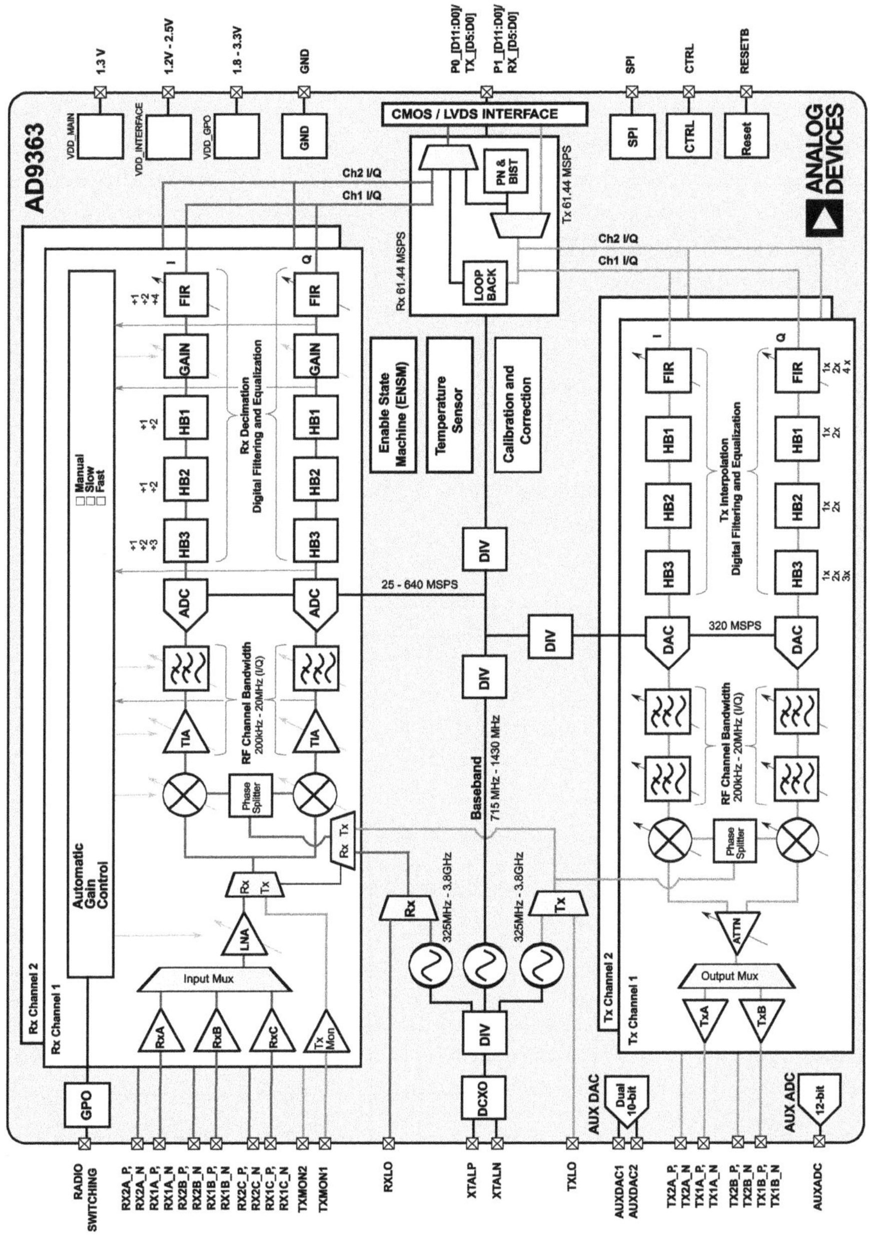

Figure 5.2 Block diagram of the AD9363 [3].

5.1.3 Zynq Details

Once the data is digitized it is passed to the Xilinx Zynq System on Chip (SoC), shown in Figure 5.3. The Zynq-7000 family offers the flexibility and scalability of an FPGA, while providing performance, power, and ease of use typically associated with ASIC and ASSPs. Providing integrated ARM Cortex-A9 based processing system (PS) and programmable logic (PL) in a single device, the Zynq is the used in the Pluto SDR as the main controller.

Having the combination of the programmable logic and a programming subsystem provide some unique advantages. The AD9363 ADC's lowest data conversion rate is 25 MHz. The maximum amount of decimation allows is 48. This provides a lowest sample rate of 520.833 kSPS. An additional divide by 8 decimation filter was put inside the FPGA to extend the lowest sample rate to 65.1042 kSPS. Running Linux on the ARM-A9 inside the Pluto SDR provides some unique advantages. Being able to use the Linux IIO infrastructure allows existing device drivers to be used for the AD9363. Controlling all aspects of the device, from sample rates, to FIR settings, to LO settings, to the additional decimation filters, this proven software did not have to be touched for the creation of the Pluto SDR.

Once the digital receive IQ data from the AD9363, described in Section 5.1.2 is transferred to the FPGA fabric, the AXI DMAC core writes that to the Pluto's external memory. During this step, a change is made from unique continuous samples to *n*-sample buffers.

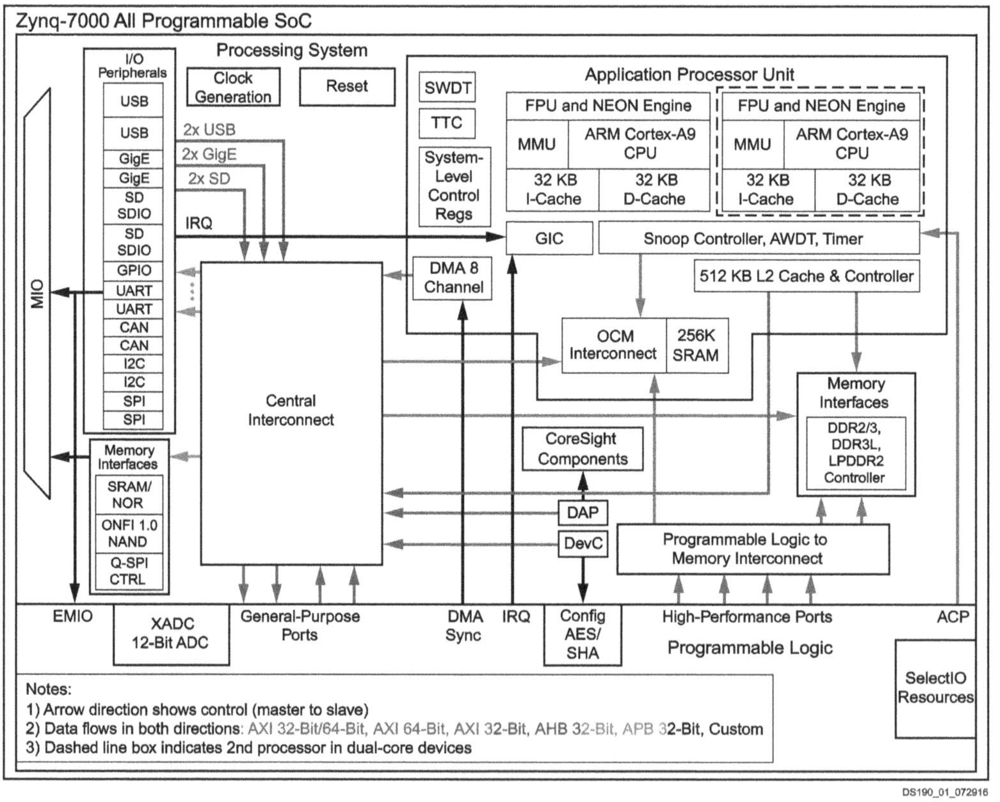

Figure 5.3 Block diagram of the Zynq [4].

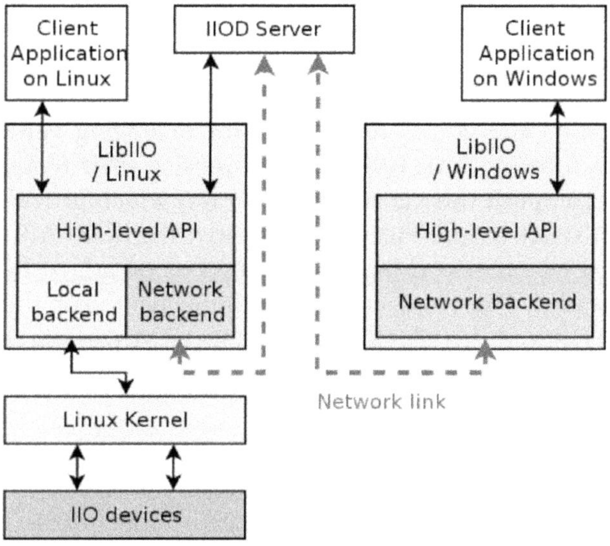

Figure 5.4 Block diagram of libiio and iiod [5].

5.1.4 Linux Industrial Input/Output Details

The industrial input/output (IIO) subsystem inside the Linux kernel is intended to provide support for devices that in some sense are ADCs or DACs, which don't have their own existing subsystems (like audio or video). This is not specific to Pluto nor specific to any SDR implmentation. It is an open-source standard adopted by many different manufactures for providing a common API to deal with a variety of different devices, This includes, but is not limited to, ADCs, accelerometers, gyros, IMUs, capacitance to digital converters (CDCs), pressure sensors, color, light and proximity sensors, temperature sensors, magnetometers, DACs, direct digital synthesis (DDS), phase-locked loops (PLLs), variable/programmable gain amplifiers (VGA, PGA), and integrated RF transceivers, like the AD9363.

There are three main aspects:

- The Linux kernel IIO driver, which runs inside the Linux kernel, in this case in the ARM in the Pluto SDR.
- libiio, the userspace library for accessing local and remote IIO devices, in this case both in the ARM, and on the host.
- iiod, the IIO Daemon, responsible for allowing remote connection to IIO clients, in this case on the ARM inside the Pluto SDR.

libiio is used to interface to the Linux industrial input/output (IIO) subsystem. libiio can be natively used on an embedded Linux target (local mode) or to communicate remotely to that same target from a host Linux, Windows, or MAC over USB, Ethernet, or Serial.

Although libiio was primarily developed by Analog Devices Inc., it is an active open-source library that many people have contributed to. It released under the GNU Lesser General Public License, version 2.1 or later, this open-source license allows anyone to use the library on any vendor's processor/FPGA/SoC that may be controlling any vendor's peripheral device (ADC, DAC, etc.) either locally or remotely. This includes closed- or open-source, commercial or noncommercial applications (subject to the LGPL license freedoms, obligations and restrictions).

Once buffers of Rx data are in external memory, they are passed to iiod, the IIO Daemon. The iiod is responsible for managing various iio clients (normally remote on a network or USB), and translating their requests to local devices. It is able to accomplish this via the libiio library, which provides access to the AD9363 through a series of interfaces. Convinently, the libiio API access to the transceiver is identical whether working on the ARM or on a host PC which also has a libiio driver installed. Therefore, code can be implemented on the host machine connected to Pluto SDR and then deployed onto the ARM with the same code.

5.1.5 MATLAB as an IIO client

MATLAB can be used as a cross-platform IIO client to interface with the Pluto SDR. It includes a Pluto SDR system object interface. A fundamental background on system objects in MATLAB is provided in Appendix B.3. The two system objects provided in the hardware support package (HSP) for Pluto SDR are:

- comm.SDRRxPluto: Pluto SDR Receiver System object
- comm.SDRTxPluto: Pluto SDR Transmitter System object

These objects are typically constructed through the `sdrrx` or `sdrtx` function calls as in Code 5.1.

Code 5.1 Instantiating Pluto SDR System Objects: **`pluto1.m`**

```
rx = sdrrx('Pluto')
tx = sdrtx('Pluto')
```

However, these objects can also be directly instantiated directly. The resulting object of `sdrrx` either way will have the following basic properties, which will be directly printed to the terminal when not using the semicolon as Code 5.2.

Code 5.2 Instantiating Pluto SDR System Objects: **`pluto1.m`**

```
rx = sdrrx('Pluto')
rx =
  comm.SDRRxPluto with properties:
  DeviceName: 'Pluto'
  RadioID: 'usb:0'
  CenterFrequency: 2.4000e+09
  GainSource: 'AGC Slow Attack'
  ChannelMapping: 1
  BasebandSampleRate: 1000000
  OutputDataType: 'int16'
  SamplesPerFrame: 3660
  ShowAdvancedProperties: false
```

Since the Pluto SDR is part of a larger family of SDR devices, it shares the *DeviceName* attribute, which will be defined as `Pluto` for the Pluto SDR. As seen from the usage in Code 5.2, there are various attributes for the System object.

- *RadioID* is related to the enumerate interface that the radio is using. This will either be USB:# where # is a number associated with the number of USB radios connected to MATLAB starting at zero, or ip:<ipaddress>, which utilizes the Pluto SDR over Ethernet.
- *CenterFrequency* defines the RF center frequency in hertz. Note there are separate Rx and Tx LO, on the Pluto SDR, and these are managed separately in the Rx and Tx Objects.
- *BasebandSampleRate* defines the sample rate of the in-phase/quadrature receive chains, respectively. Note, there is only one clock generator for both the ADC and DAC in the AD9363, so these must be set to the same value managing Rx and Tx on the same device.
- *GainSource* has three possible options: Manual, AGC Slow Attack, and AGC Fast Attack. When Manual is selected, another option called Gain will become available. This Gain value is associated with a gain table located within the AD9363. Therefore, when the Gain value changes multiple stages in the receive path shown in Figure 5.2 are updated based on this internal table. Custom gain tables can be used if necessary. However, such settings are considered advanced implementation options and will not be considered in this book. The other GainSource settings enable state machine based gain control within the AD9363 to adapt during operation based on receive signal strength (RSSI).
- The ChannelMapping attribute for the Pluto SDR can only be set to 1. However, on other SDRs in the Analog Devices Inc. family this is used for multichannel (multiple-input and multiple-output, or MIMO) reception.
- *OutputDataType* determines the format data is provided out of the object. Technically, from the AD9363 and libiio, MATLAB can only receive 16-bit complex integers, but we can tell MATLAB to cast them to other data types by default. Typically we will cast them to doubles since they provide the most precision, and working with trigonometric functions will require double or single precision data. As mentioned previously the receive and transmit paths only provide 12 bits of resolution, but since several of the hardware and software layers can only deal with base 8-bit types these 12 bits are provided as a 16-bit integer. For reference, on the receive side the 16-bit samples are sign extended and the transmitter will simply throw away the lowest four significant bits.
- *SamplesPerFrame* determines the number of samples in the buffer or frame that is passed to MATLAB from iiod. This will be the size of the vector provided at a given call to the object. This data will always be continguous as received from the radio unless an overflow has occurerd. However, successive calls to an instantiated object will not guarantee buffer-to-buffer continuity. Therefore, it can be useful to collect large amounts of data at a given time for processing.

The transmitter system object `comm.SDRTxPluto` has nearly identical properties except for *GainSource*, *SamplesPerFrame*, and *OutputDataType*, which do not make sense in the transmitter context.

5.1.6 Not Just for Learning

The architecture combination of the AD936x RF transceiver, the Zynq SoC, and external memory, which is found on the Pluto SDR should not just be thought of as just a learning platform. There are many commercial systems built on a similar architectures that can use the same software (HDL, Linux kernel, and IIO) to communicate with the IIO clients (like MATLAB). For example, Epiq Solutions, located in Schaumburg, Illinois, builds an industrial-grade, commercial solution, shown in Figure 5.5 known as Sidekiq Z2.

Although the Sidekiq Z2 utilizes a similar architecture as Pluto SDR, it does it in a standards-compliant Mini PCIe card form factor measuring $\sim 51 \times 30$ mm. In addition, the Sidekiq Z2 incorporates RF filtering, a high-precision reference clock, a more powerful dual-core Zynq, an extended RF tuning range from 70 – 6,000 MHz using the AD9364, doing so with industrial temperature range (-40° – +70° C) rated components. This allows the Sidekiq Z2 to serve as the basis for real-world flexible RF solutions, even in harsh industrial environments.

By building on the same architecture and using the same software infrastructure, this allows algorithm developers to prototype using a device like the Pluto SDR, and when they are ready for an industrial form factor, or need something embeddable,

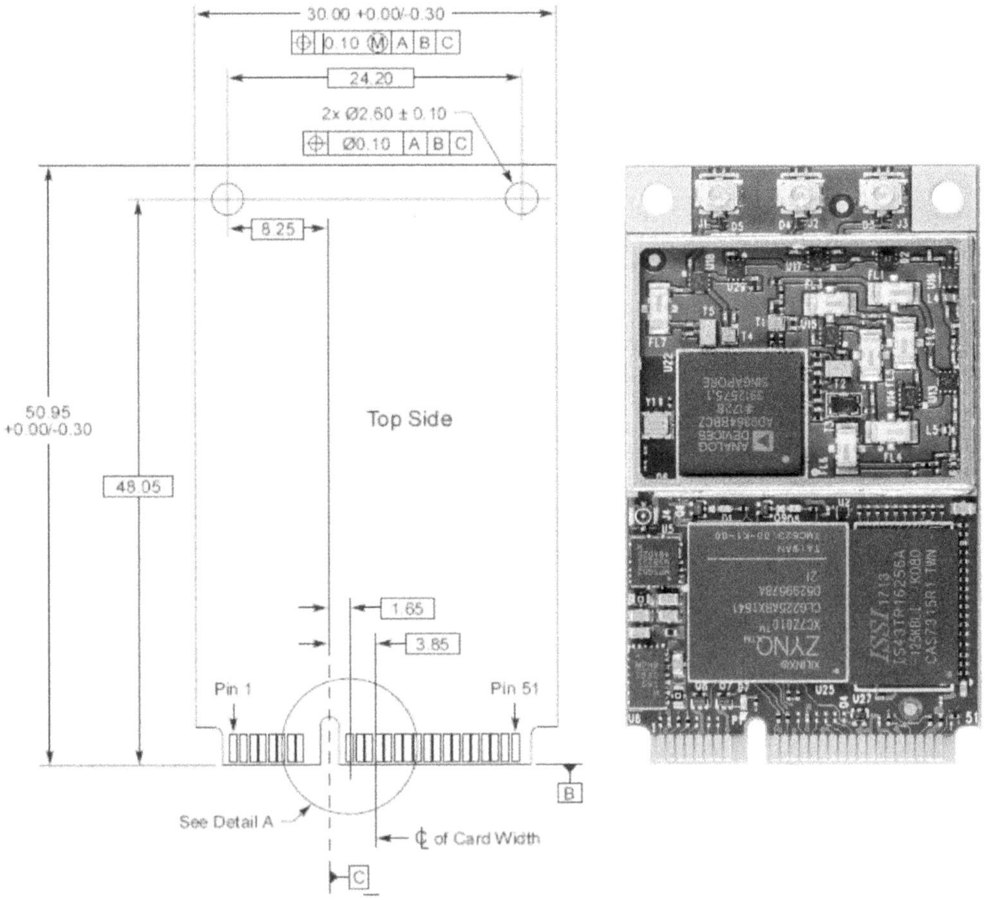

Figure 5.5 *Sidekiq Z2* from Epiq [6]. Not to scale.

they can easily transition over to the Sidekiq Z2. This also allows developers using the Sidekiq Z2 to communicate directly to MATLAB or any of the other IIO client software packages for testing in their specific end product. There are many things about RF that just cannot be simulated, and unless you have a final product with the ability to save and playback waveforms, or to remotely connect to the device, making a final release to production can be difficult to impossible. There is nothing more frustrating for an engineer than to have something that works on the bench fail in the field and have to go there for local troubleshooting. Using the provided open-source, industry-standard frameworks can reduce that.

5.2 Strategies For Development in MATLAB

As we discussed in Section 5.1.5, controlling how data enters MATLAB is very important for consistent operation. In this section we will discuss some strategies for structuring MATLAB code to effectively develop an algorithm. To help guide the development process we have provided templates to show how to appropriately form source around the radio's API. These templates can help progression of designs to real-time or offline work without the radio attached.

5.2.1 Radio I/O Basics

In each of these templates we will assume that a radio has been instantiated as the object `rx`, as in running the code in Code 5.1. Additionally, we assume that the SamplesPerFrame parameter of the object is set to some variable *frameSize*. In the first template presented in Code 5.3 we first collect *framesToCollect* frames of data, where each frame is of *frameSize* samples. The code in Code 5.1 tries to guarantee that we have collect *framesToCollect* × *frameSize* samples of continguous data from the radio with gaps. This is a good technique if more data than 2^{20} samples need to be collected, which is the maximum value you can make the SamplesPerFrame parameter of the Pluto SDR System object. After this data is collected we perform some processing, which in this case is a visualization with `dsp.SpectrumAnalyzer` scope.

Alternatively, if we don't require fresh samples for every run it can be useful to save data to a file so we don't have to worry about clearing data from the workspace. A useful tool for this work is the `comm.BasebandFileWriter`, which saves complex received data with additional metadata like sample rate to a file for off-line processing. We show usage of the `comm.BasebandFileWriter` system object in Code 5.4 with the collected data from Code 5.3.

Utilizing data from a filesource can make testing much more repeatable when debugging issues during algorithm development. It can also be much faster to address a file than to go out to the radio and pull in new data, especially when setting up a transmitter is also required. In Code 5.5 we show use of the complementary System object to the`comm.BasebandFileWriter` called `comm.BasebandFileRead`. The `comm.BasebandFileRead` System object can be configured to provide a specific amount of samples for each call to the object through the SamplesPerFrame parameters to emulate using the Pluto SDR. This is a useful strategy when a radio is not available.

Code 5.3 Template Example: **template1.m**

```
%% Template 1
% Perform data collection then offline processing
data = zeros(frameSize, framesToCollect);
% Collect all frames in continuity
for frame = 1:framesToCollect
    [d,valid,of] = rx();
    % Collect data without overflow and is valid
    if ~valid
        warning('Data invalid')
    elseif of
        warning('Overflow occurred')
    else
        data(:,frame) = d;
    end
end

% Process new live data
sa1 = dsp.SpectrumAnalyzer;
for frame = 1:framesToCollect
    sa1(data(:,frame)); % Algorithm processing
end
```

Code 5.4 Template Example for Saving Data: **template1.m**

```
% Save data for processing
bfw = comm.BasebandFileWriter('PlutoData.bb',...
    rx.BasebandSampleRate,rx.CenterFrequency);
% Save data as a column
bfw(data(:));
bfw.release();
```

Code 5.5 Template Example for Saving Data: **template2.m**

```
%% Template 2
% Load data and perform processing
bfr = comm.BasebandFileReader(bfw.Filename, 'SamplesPerFrame',frameSize);
sa2 = dsp.SpectrumAnalyzer;
% Process each frame from the saved file
for frame = 1:framesToCollect
    sa2(bfr()); % Algorithm processing
end
```

Once an algorithm has been tuned we can place the processing sections within the main loop with the Pluto SDR's System object like in Code 5.6. This type of processing is defined as stream processing in MATLAB [7], where we immediately work on new data. This will limit the amount of information required to be collected and can be useful if logical actions, such as changing channels, need to be applied. As long as the algorithm placed within the loop is able to keep up with the data streaming in from the radio, no overflow warning should occur. This is known as operating in real time. However, since there is some elasticity with the buffers to and from the radio overflows will not always happen immediately. For example, if the algorithm is only slightly slower than the radio's data rate then it may take many loop iterations for the internal radio and operating system buffers to fill to a point of overflow. Therefore, when testing for real-time operation it is useful to run an algorithm for several minutes to check if an overflow will occur.

Code 5.6 Template Example for Saving Data: **`template3.m`**

```
%% Template 3
% Perform stream processing
sa3 = dsp.SpectrumAnalyzer;
% Process each frame immediately
for frame = 1:framesToCollect
    [d,valid,of] = rx();
    % Process data without overflow and is valid
    if ~valid
        warning('Data invalid')
    else
        if of
            warning('Overflow occurred')
        end
        sa3(d); % Algorithm processing
    end
end
```

5.2.2 Continuous Transmit

Anytime the Pluto SDR is powered on, the transceiver is activated and begins to operate even if the user did not intend to. When powered on Pluto SDR will transmit data; this is just how the transceiver was designed. Therefore, when using just the receiver System object (`comm.SDRRxPluto`) data will be transmitted by the actual device. Normally, the transceiver will transmit the last buffer available in the DMA continuously until powered down. If the Tx LO is accedentily tuned to the same value as the RX LO, when communicating between multiple radios or just receiving, this continuous transmit operation can cause significant interference.

Code 5.7 Template Example Transmit Repeat: **`transmitRepeat.m`**

```
% Transmit all zeros
tx = sdrtx('Pluto');
fs = 1e6; fc = 1e4; s = 2*pi*fs*fc*(1:2^14).';
wave = complex(cos(s),sin(s));
tx.transmitRepeat(wave);
```

There are two options to reduce or even remove this interference. The first option is to instantiate a transmitter System object (`comm.SDRTxPluto`) and write a vector of zeros to the object as shown in Code 5.8. This will reduce the transmit energy of the device to a minimal level. However, there will still be leakage into the receiver's data due to Tx LO leakage.

Code 5.8 Template Example Transmit Zeros: **transmitzeros.m**

```
% Transmit all zeros
tx = sdrtx('Pluto');
tx(zeros(1024,1));
```

Alternatively, we can simply shift the LO of the transmitter to a frequency beyond the receive bandwith. We demonstrate this configuration in Code 5.9 where we offset the *CenterFrequency* of the transmitter's System object. This is a better alternative since there LO leakage from the transmitter should not appear in the received data.

Code 5.9 Template Example Transmit Zeros: **transmitoffset.m**

```
% Move transmitter out of receive spectrum
tx = sdrtx('Pluto');
rx = sdrrx('Pluto');
tx.CenterFrequency = rx.CenterFrequency + 100e6;
```

5.2.3 Latency and Data Delays

When working with the Pluto SDR from a host it will soon become obvious that there will delays associated with transmitted and received data, especially when performing loop-back operations from transmitter to receiver. These delays are a function of the internal buffers and FIFOs of the host PC and the Pluto SDR itself. However, there exists both deterministic and elastic/random delay across the various layers illustrated in Figure 5.6. The reason why there is a nondeterministic delay in the transport stacks is due to contention in the kernels of the ARM and host PC. In the worst case the round-trip time should be on the order of tens of milliseconds. In order to guarantee certain delays or minimal delay would require operating directly on the ARM or on the FPGA. Nevertheless, development in these processing regions becomes much more time consuming. However, any radio platform that passes data into MATLAB will have to deal with these delays, but they may have different sources depending on the radio architecture.

One of the complications of looking at Figure 5.6 is that in many places, the transport is defined by bytes, while in other places it convenient to discuss samples. A single I/Q sample (complex) in this circumstance (singe radio channel) is two 16-bit samples, or 4 bytes.

To review Figure 5.6 in detail on the receive side, when an IIO client like MATLAB requests a buffer of 32768 samples at a sample rate of 1 MSPS:

- `iiod` will capture 32768 continuous samples via libiio. This will take 32.768 milliseconds, and consume 131,072 bytes. While `iiod` ensures the packet is contiguous by not sending any data until it has all been captured, it does increase a fixed latency of the sample rate × the number of samples being captured. `iiod` was designed to ensure there are no memory copies after the data has been captured, and simply passes pointers around from the AD9363 driver to `libiio` to `iiod` and then to the USB stack.

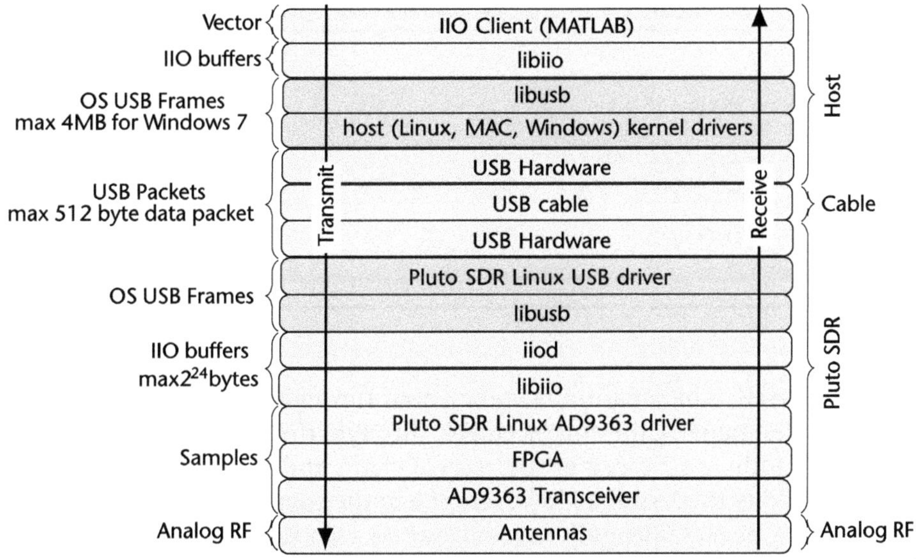

Figure 5.6 Hardware and software stacks transmitted and received packets must traverse from MATLAB to the transceiver throught the various layers in the system.

- This data will then be passed to the Linux kernel on the Pluto where it will be segmented into 512 byte USB packets (per the USB 2.0 spec), were it will be reassembled into a 131,072-byte buffer on the host side. The USB stack will introduce an unknown, and unbounded latency, which will be determined by the host operating system and how much free memory it has available.
- Once the entire buffer is assembled, it is passed to libiio, where it is then passed (untouched) to the iio client, in this case MATLAB, which may do further processing.
- In this case, MATLAB may cast the data from fixed point to floating point or double ± 1.0 data, which also takes some time.

Due to these delays we must design our algorithms and overall system with an understanding of these conditions. For example, if a single frame wanted to be transmitted and received from the same Pluto SDR we can observe a large gap of samples before the frame is observed at the receiver output vectors due to these delays.

It is for these reasons that many systems want to put as much processing as possible as close to the radio as possible. With this architecture, that would be on the FPGA or the ARM processor. However, on the Pluto SDR, the FPGA is small, and there is only a single core ARM, limiting its targeting capabilities.

5.2.4 Receive Spectrum

The receive signals we observe will always will contain noise from the environment and from the radios themselves. To provide perspective on this noise we demonstrate a simple example using the script shown in Code 5.10.

If one employs Code 5.10, we will notice the spectrum is not perfectly flat, containing peaks or spurs in the spectrum. These are actually artifacts of the

Code 5.10 Template Example View Spectrum: **template_rt.m**

```
% View some spectrum
rx = sdrrx('Pluto');
rx.SamplesPerFrame = 2^15;
sa = dsp.SpectrumAnalyzer;
sa.SampleRate = rx.BasebandSampleRate;
for k=1:1e3
  sa(rx());
end
```

radio itself, which naturally come about through the complex receive chain, which provides signal gain enhancement through the AGC stages. Depending on the bandwidths we choose in the receive chain and the AGC configuration itself, we can modify these spurs. However, due to the complexity of the AD9363 transceiver this can be a challenging task since it contain thousands of registers for the AGC itself for configuration. Therefore, since these spurs can be minor relatively to the signal of interest itself we can simply ignore them or perform filtering in software to reduce their affects. Nonetheless, when working with Pluto SDR we should always be aware of these nonidealities in the hardware.

Fortunately, Pluto SDR does maintain many built-in calibrations to help reduce self-induced problems. These include RF DC tracking, quadrature tracking, and baseband DC tracking. To access these features, Pluto SDR enabled the parameter *ShowAdvancedProperties*, which will then display these features. Since the AD9363 is a direction conversion receiver, a common problem with such devices is the presence of a tone or energy at DC or near the zeroith frequencies. This is due to the radio itself. The DC tracking components, RF and baseband, both work to reduce these effects.

The last advanced feature is quadrature tracking. The quadrature tracking reduces and in-phase and quadrature (IQ) imbalance that may occur, which may be difficult to spot in a frequency domain plot. An imbalance would typically appear as a spur reflection in the frequency domain. When enabling quadrature tracking, these image spurs should be reduced significantly. However, when working with a constellation plot IQ imbalances become more noticable. There will always be some residual imbalance, but corrections are performed during initialization so it will not be improved over time necessarily.

5.2.5 Automatic Gain Control

One of the most complex pieces of the AD9363 is the AGC. The AGC is actually spread out through the receive chain apply gain at different stages and sensing the amplitude of the received signals. From the simplistic API of MATLAB we have three options: Manual, AGC Slow Attack, and AGC Fast Attack. Under Manual the receiver will simply fix the input gain to a specific value based on an internal gain table. Essentially the manual gain acts as a single index into the internal table. This Manual setting is useful when using cabling or when the transmitter is at a fixed known distance. Under the Manual setting it can make receiver algorithms easier to debug since the AD9363's state will remain the same.

In practice the amplitude of the receive signal will be unknown and even change across multiple transmissions. Therefore, we can simply utilize the AGC Slow Attack and AGC Fast Attack modes. If the received transmission is very short or rapidly changes in amplitude AGC Fast Attack will be the best option. However, AGC Slow Attack will be prefered when received data has a relatively maintained amplitude. An obvious question would be, why not always used AGC Fast Attack? The main disadvantage of AGC Fast Attack is that it can create large discontinuities in the amplitude, which may distort the received signal. For illustration we provide a comparison of a system setup in loopback with a single Pluto SDR transmitted a QPSK signal. We provide a time plot of the received signal to demonstrate both the delay of the received signal and the changes in amplitude when using the different gain modes. As we can observe there are rapid changes in gain for the AGC Fast Attack mode, but the gain is more gradual over time for the AGC Slow Attack mode. The determination of the correct mode is not always obvious. Therefore, time series testing and observation can be useful during algorithm development.

5.2.6 Common Issues

The way that various signals mix can also be an issue. As described in Section 2.3.1, the mixer accepts a single-ended local oscillator (LO).

5.3 Example: Loopback with Real Data

Now that we have a solid understanding of the system object that controls the Pluto SDR and some coding structures for interacting with Pluto, we can explore a simple loopback example. Loopback means that the waveform is both transmitted and received by the same Pluto SDR, which is a common strategy for algorithm debugging and general hardware debugging.

Starting from the transmitter (tx), in this example you will first notice we have set the gain to −30, which is 20 dB down from the default. The reasoning behind reducing the gain is to prevent clipping or saturation at the receiver. Since the transmit and receive antennae are about 12 mm from one another the received signal will be rather loud. The sinewave used here was simply generated by the `dsp.SineWave` system object for convenience, which will provide a complex output for the Pluto SDR. A special method called `transmitRepeat` was used, which will continuously transmit the passed vector. This will prevent any gaps in the transmission.

In the received waveform in Figure 5.7 we can observe both the complex and real components of the signal over time, which are $\frac{\pi}{2}$ radians out of phase with one another as expected. At the start of the signal we can observe a larger amplitude than future samples. This is a direct result of the AGC settling, and since the AGC starts from a high gain by default at configuration or setup time. In MATLAB this setup time only occurs on the first call to the receiver object (rx), not the construction time. This is also known as the first Step method call, which will call an internal method of the system object called setupImpl.

Code 5.11 Loopback Example: **loopback.m**

```
% Setup Receiver
rx=sdrrx('Pluto','OutputDataType','double','SamplesPerFrame',2^15);
% Setup Transmitter
tx = sdrtx('Pluto','Gain',-30);
% Transmit sinewave
sine = dsp.SineWave('Frequency',300,...
                    'SampleRate',rx.BasebandSampleRate,...
                    'SamplesPerFrame', 2^12,...
                    'ComplexOutput', true);
tx.transmitRepeat(sine()); % Transmit continuously
% Setup Scope
samplesPerStep = rx.SamplesPerFrame/rx.BasebandSampleRate;
steps = 3;
ts = dsp.TimeScope('SampleRate', rx.BasebandSampleRate,...
                    'TimeSpan', samplesPerStep*steps,...
                    'BufferLength', rx.SamplesPerFrame*steps);
% Receive and view sine
for k=1:steps
  ts(rx());
end
```

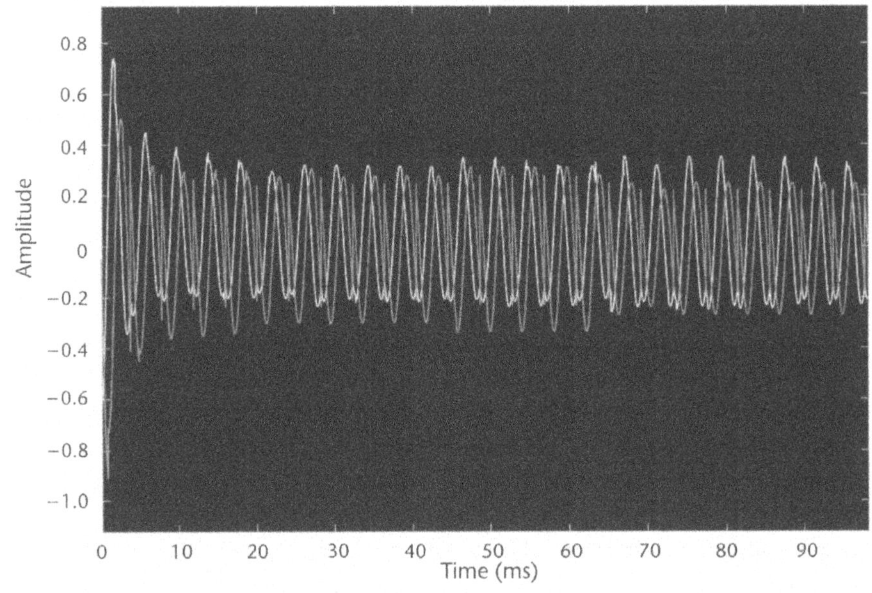

Figure 5.7 Loopback sinewave of 300 Hz from Pluto SDR generated by Code 5.11.

Looking closer on the sine wave, discontinuities can be observed that result from small gaps between received frames. This is actually a result of the transmitter that is repeating the same waveform over and over. Since the transmitted waveform is not cyclic when relating the start and end of the data passed to the transmitter we observe these discontinuities. To avoid these we would need to make sure a period ends at the end of the passed frame and started exactly at the beginning of the passed frame.

5.4 Noise Figure

With the receiver pipeline discussed in the previous section, the AD9363 is able to achieve a noise figure (NF) of only 2 dB at 800 MHz. NF is a common metric to compare receivers, which is a measure of the internal or thermal noise of the electrical components of the device. NF is calculated based on the SNR ratio of output to input in dB as

$$NF = 10\, log_{10} \frac{SNR_{INPUT}}{SNR_{OUTPUT}}, \tag{5.1}$$

where NF is in dB, and both SNR_{INPUT} and SNR_{OUTPUT} are in linear scale [8]. For comparison, another popular SDR the RTL-SDR has a NF of 4.5 dB, which is almost double the NF of the Pluto SDR. NF is important because it will affect the eventual sensitivity of the receiver, and the lower the better. The easiest way to measure NF is with a noise figure analyzer that generates noise into the receive path of a system, which is then fed back out. The output noise is then measured and compared with the input noise to create an NF measurement. Figure 5.8 demonstrates an example set up to evaluate NF using a noise figure analyzer where the device under test (DUT) will operate in loopback mode transmitted out the received noise. When measuring NF it can be useful to use a very stable LO to drive the device to limit internal noise, since generally the noise of a DUT needs to be measured, not the source oscillator driving it. Furthermore, NF is always based on frequency and the measured NF of a DUT such as an SDR will typically be based on the noisiest part of the device.

NF is a common metric that hardware manufacturers typically use but it can be difficult to relate to a communications problem, since communications engineers tend to measure further down the receive chain after the filter stages. NF also requires specific and usually expensive instruments to measure. However, it is important to understand its meaning when specified on a datasheet, since it can give a rough estimate on the low bound for error vector magnitude measurements.

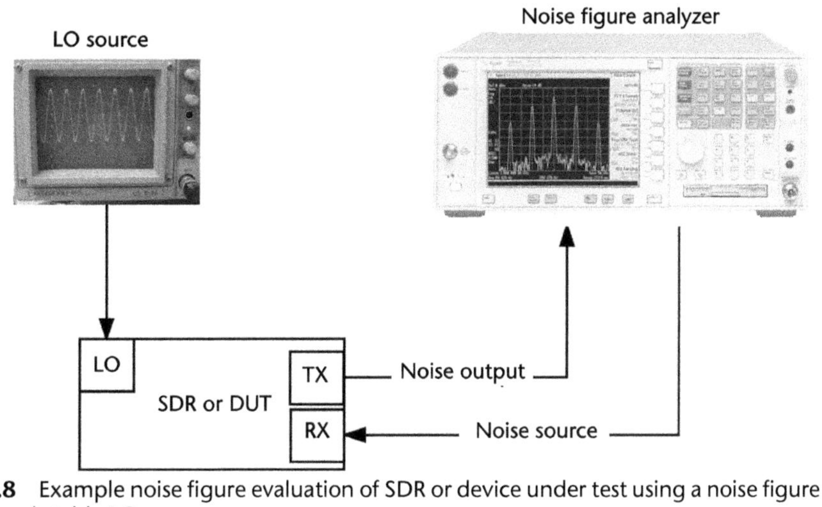

Figure 5.8 Example noise figure evaluation of SDR or device under test using a noise figure analyzer and external stable LO source.

References

[1] Analog Devices ADALM-PLUTO Software-Defined Radio Active Learning Module, http://www.analog.com/plutosdr 2017.

[2] Razavi, B., "Design considerations for direct-conversion receivers," *IEEE Transactions on Circuits and Systems II: Analog and Digital Signal Processing*, Vol. 44, No. 6, June 1997, pp. 428–435.

[3] Analog Devices AD9363 Datasheet, http://www.analog.com/AD9363 2015.

[4] Xilinx, *Zynq-7000 All Programmable SoC Overview*, [Online], 2017, https://www.xilinx.com/support/documentation/data_sheets/ds190-Zynq-7000-Overview.pdf.

[5] Analog Devices, What is libiio?, [Online], 2018, https://wiki.analog.com/resources/tools-software/linux-software/libiio.

[6] Epiq Solutions, *Sidekiq Embeddable Software Defined Radio 70MHz–6GHz*, [Online], https://epiqsolutions.com/sidekiq/.

[7] The Math Works, Inc., Stream Processing in MATLAB: Process Streaming Signals and Large Data with System Objects, https://www.mathworks.com/discovery/stream-processing.html.

[8] Friis, H. T., "Noise Figures of Radio Receivers," *Proceedings of the IRE*, Vol. 32, No. 7, July 1944.

CHAPTER 6

Timing Synchronization

In the next series of chapters we will be introducing receiver synchronization and signal recovery concepts, which are necessary for wireless communications between multiple devices. In this chapter will introduce the concept of timing recovery and approach this topic first for two primary reasons. First the downstream recovery methods used in this book for coherent modulations are sensitive to timing offset and must be compensated for first. The second reason is for prototyping with the Pluto SDR. We will be relying heavily on the loopback features of the radio, which will allow for control of nonidealities to some degree. However, since signals must travel a distance between the transmitting DAC and receiving ADC there will be a fixed but random time offset between the chains. This is where timing recovery is used to correct for this offset. With that said, a receiver can be designed in many different ways but the specific ordering of chapters here relates to the successive application of algorithms to be used: First timing recovery, then carrier phase correction, and finally frame synchronization. Once these three major pieces are handled we will then move on to more advanced topics including equalization and coding. Blocks in Figure 6.1 will be highlighted at the start of each relevant chapter to outline the progress of the overall receiver design and show how they fit with one another. In this chapter, matched filtering and timing recovery are highlighted.

In this chapter, the concept of timing recovery will be broken down into five primary sections. A brief overview of transmit filters will be first discussed, which is necessary to understand how we algorithmically perform timing recovery. Then we will move on to a simple model to demonstrate timing error, which will include Pluto SDR as well for realistic data. Finally, several recovery techniques will be discussed that adaptively handle correction of timing problems. Debugging methodology will be provided to understand how to tune these techniques for your own datasets. In regard to the algorithms discussed, an analytic analysis or derivations will not be provided explicitly. However, these algorithms will instead be treated as tools used to build a receiver component by component, where only a top-level understanding is necessary. Alternative sources will be referenced for deeper analysis, but in this work we will focus on implementations and specifically implementations with SDRs. Our goal here is to motivate the construction of a receiver initially from existing works, and readers can explore further literature if they wish to extract more performance from their implementation.

6.1 Matched Filtering

In digital communications theory when matched filtering is discussed it is typically called pulse-shaping at the transmitter and matched filtering at the receiver for

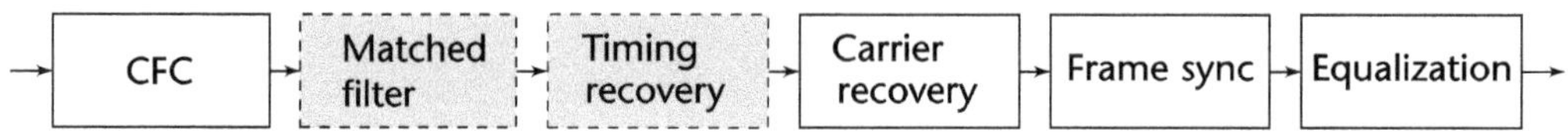

Figure 6.1 Receiver block diagram.

reference. The goal of these techniques is threefold: first, to make the signal suitable to be transmitted through the communication channel by limiting its effective bandwidth, second, increase the SNR of the received waveform, and third, to reduce intersymbol interference (ISI) from multipath channels and nonlinearities. Pulse-shaping was discussed in Section 2.7.5, but we will revisit the topic here from a slightly different approach that focuses on more practical aspects of these filters.

By filtering a symbol, sharp phase and frequency transitions are reduced resulting in a more compact and spectrally efficient signal. Figure 6.2 provides a simple example of a DBPSK signal's frequency representation before and after filtering with a transmit filter. As we can see the effective bandwidth of the signal is reduced, primarily from the upsampling/interpolation that is applied at the transmitter. Since time and frequency are inversely related we get this reduction in bandwidth. These filter stage implementations will typically upsample and downsample signals, which reduce their effective bandwidth. However, upsampling inherently increases the so-called surface area of a symbol, making it easier to determine, since we will have multiple copies of it at the receiver. Therefore, we are trading recoverability for bandwidth since data will be produced at the same rate from the transmitter but will not utilize the entire bandwidth available. These operations of rate transitions (upsampling/downsampling) are performed during the matched filtering stages since it is efficient to utilize a single filter to perform both operations.

The filter used to generate this figure was a square-root raised cosine (SRRC) filter, which is a common filter used in communication systems. We provided the raised cosine (RC) filter in Section 2.7.5, but the more common SRRC has the impulse response:

$$h(t) = \begin{cases} \dfrac{1}{\sqrt{T_s}}\left(1 - \beta + 4\dfrac{\beta}{\pi}\right), & t = 0 \\ \dfrac{\beta}{\sqrt{2T_s}}\left[\left(1+\dfrac{2}{\pi}\right)\sin\left(\dfrac{\pi}{4\beta}\right) + \left(1-\dfrac{2}{\pi}\right)\cos\left(\dfrac{\pi}{4\beta}\right)\right], & t = \pm\dfrac{T_s}{4\beta} \\ \dfrac{1}{\sqrt{T_s}}\dfrac{\sin\left[\pi\dfrac{t}{T_s}(1-\beta)\right] + 4\beta\dfrac{t}{T_s}\cos\left[\pi\dfrac{t}{T_s}(1+\beta)\right]}{\pi\dfrac{t}{T_s}\left[1-\left(4\beta\dfrac{t}{T_s}\right)^2\right]}, & \text{otherwise} \end{cases} \tag{6.1}$$

where T_s is the symbol period and $\beta \in [0, 1]$ is the roll-off factor. In practice these filters will be arranged in two ways as outlined in Figure 6.3. First we can place a single RC filter at the transmitter or place a SRRC at both the transmitter and receiver. Both options produce Nyquist filter arrangements of signals to reduce or eliminate ISI. Details on how to select β and T_s will be discussed later in Section 8.2. However, we examine β through use of an eye diagram in Figure 6.4, and we can easily see the time domain effects for very different roll-offs of $\beta = [0.3, 0.99]$. For

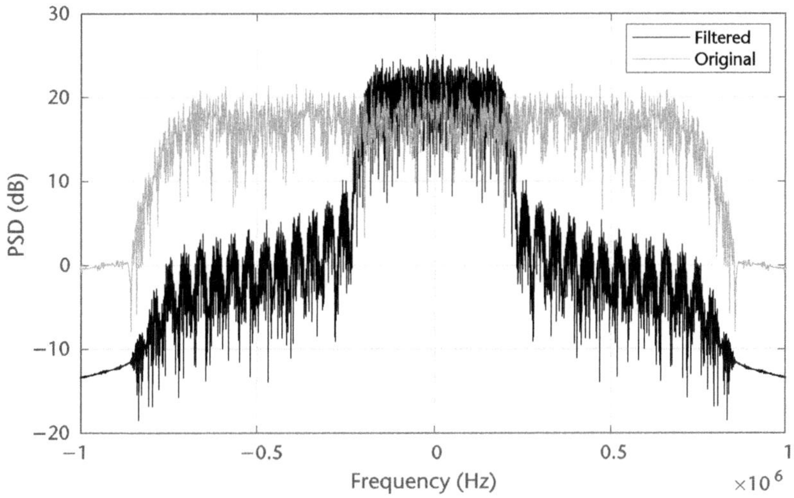

Figure 6.2 Frequency spectrum of PSK signal before and after pulse-shaping filter.

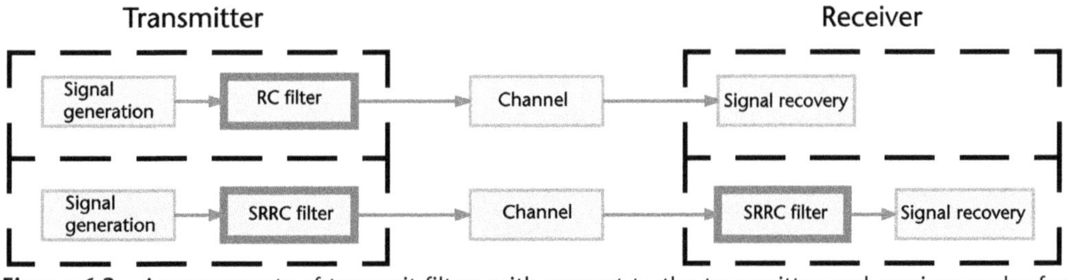

Figure 6.3 Arrangements of transmit filters with respect to the transmitter and receiver nodes for raised cosine and root-raised cosine filters.

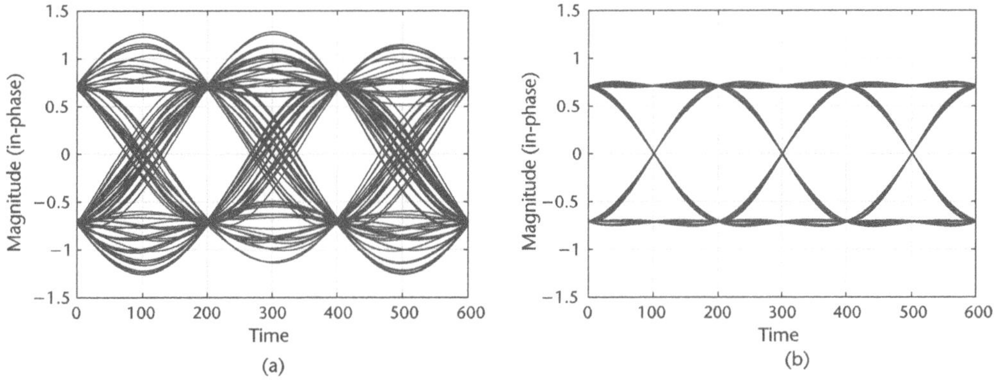

Figure 6.4 Eye diagrams of in-phase portion of QPSK signal after being passed through SRRC filters with different β values. (a) $\beta = 0.3$, and (b) $\beta = 0.99$.

these results the signal was passed through a SRRC filter at the transmitter, AWGN channel with an SNR of 27 dB, and SRRC filter at the receiver. A high SNR was used to keep the eyes open and remove noise as the limiting factor for eventual decisions. At the decisions times 200 and 400 the symbols are well defined in both cases but with $\beta = 0.99$ the transitions are less noisy. However, with a $\beta = 0.99$ the frequency domain outer bands contain more energy as seen in Figure 2.52, which may be undesirable.

Alternatively, we can examine the impulse response with respect to β in Figure 6.5, which compares the RC and SRRC responses. You will notice that unlike the RC filter, the impulse response is not zero at the intervals of T_s. However, the composite response of the SRRC filters at the transmitter and receiver will have zeros at intervals of T_s like the RC filter. The SRRC is used since it is a Nyquist type filter, which produces zero ISI when sampled correctly [1] as discussed in Chapter 2. We can demonstrate the effect of ISI by introducing a simple nonlinearity into the channel and consult the resulting eye diagrams that were introduced in Section 2.4.1. Nonlinearities cause amplitude and phase distortions, which can happen when we clip or operate at the limits of our transmit amplifiers. For more details on the model used, consult [2], but other models exists such as in [3]. In Figure 6.6 we observe the effects of ISI as the eye becomes compressed and correct sampling becomes difficult to determine. We will revisit ISI effects again when equalization is discussed in Chapter 9.

As mentioned previously, rate conversion will typically occur in these transmit or receive filters. Therefore, a polyphase filter can be used where the taps of the SRRC filter are used within the individual arms of the polyphase filter. This is a very efficient implementation since the taps will be applied at the lower rates,

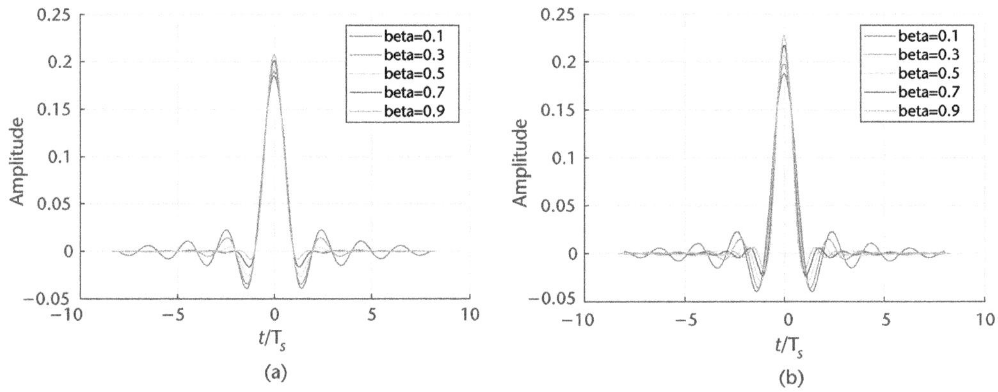

Figure 6.5 Impulse response comparison between raised-cosine and square-root raised-cosine filters. (a) RC impulse response, and (b) SRRC impulse response.

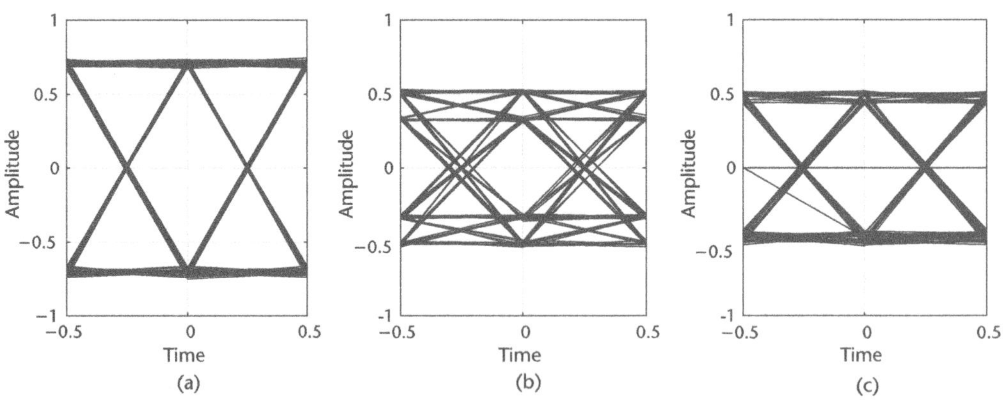

Figure 6.6 Eye diagrams of QPSK signal affected by nonlinearity causing ISI, which is reduced by SRRC matched filtering. (a) Original signal at transmitter, (b) passed through nonlinearity without pulse-shaping, and (c) SRRC filters used at transmitter and receiver with nonlinearity.

before interpolation or after decimation within the polyphase design, reducing the number of multiplies required.

The final aspect of the matched filters we want to discuss and provide insight into is SNR maximization. This argument logically comes out of the concept of correlation. Since the pulsed-shaped/filtered signal is correlated with the pulse-shaped filter and not the noise, matched filtering will have the effect of SNR maximizing the signal, creating peaks at central positions of receive pulses. We demonstrate this effect in Figure 6.7, where we present data transmitted with and without pulse-shaping under AWGN. In the middle plot of Figure 6.7(b) we observe a signal closely related to the originally transmitted sequence, even under high noise. However, without pulse-shaping even visually the evaluation of the transmitted pulse becomes difficult. We even observe demodulation errors in this third plot of Figure 6.7(c) without any timing offset introduced.

6.2 Timing Error

Timing error between transmitter and receiver is a simple concept to understand, but can be difficult to visualize and debug in a communication system. In the most basic sense the purpose of symbol timing synchronization is to align the clocking signals or sampling instances of two disjointed communicating devices. In Figure 6.8(a) we consider a simple example where we overlap the clocking signals of the transmit and receiver nodes and the input signal to be sampled. The sampling occurs at the rising clock edges, and the optimal timing is the transmitter's clock. A small delay τ, less than a single clock cycle, is introduced at the receiver. This is known as a fractional delay since it is less than a sample. Due to this delay the signal is sampled at the wrong positions and the eventual demodulated signal is incorrect. Figure 6.8(b) shows a comparison of the correct and receiver's demodulated symbols with an obvious error occurring at the second transition.

Mathematically we can model this offset received signal r as

$$r(t) = \sum_{n} x(n)h(t - \tau(t) - nT_s) + v(t), \tag{6.2}$$

where x is the transmitted symbol, h is the pulse shape of the transmit filter, τ is the fractional offset, T_s is the sampling period, n is the sample index, and v is the additive channel noise. After reception the r is passed through the receive matched filter and the relation of the source symbols will be

$$y(t) = \sum_{n} x(n)h_A(t - \tau(t) - nT_s) + v_h(t), \tag{6.3}$$

where $h_A = h(t) * \bar{h}(-t)$ is the autocorrelation of the transmit filter and its conjugate used on the source data x, v_h is the shaped noise, and y is the output symbols. This demonstrated our notion of matched filtering and correlation. You will notice that the delay τ is indexed as well, since this delay will change since the oscillator at the transmitter and receiver will not have identical frequencies. Therefore, over time this timing difference will drift. However, changes in τ from symbol to symbol should be small relative to the sample rate of the device in a practical RF communication system.

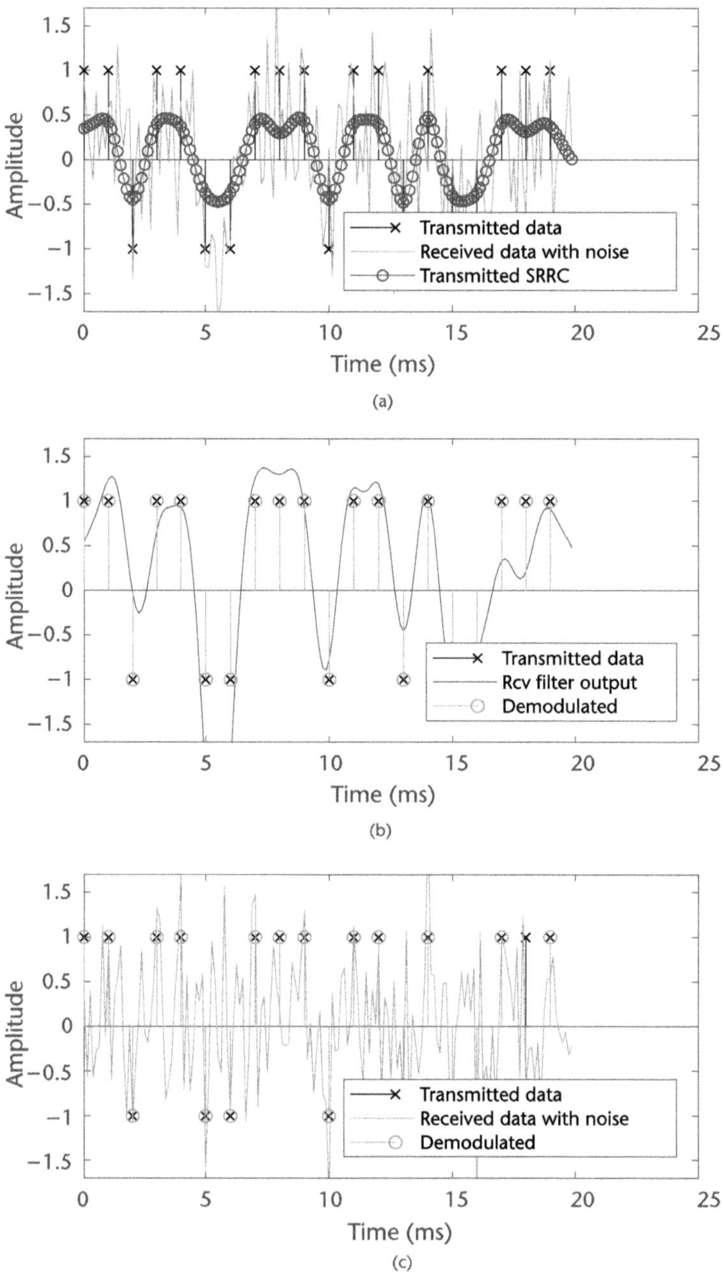

Figure 6.7 Comparison of pulse-shaped and nonpulse-shaped received signals after an AWGN channel. An obvious advantage is visible in the receiver signal when using matched filtering. (a) Transmitted SRRC filtered signal, (b) received SRRC filtered signal, and (c) received signal without SRRC filtering at the receiver.

As discussed in Section 6.1 we will interpolate the signal to be transmitted at the transmit filter stage before actually sending the signal. Although this reduces the throughput of our system it provides the receiver more data to perform decisions without having to oversample the signal itself. In MATLAB we will use `comm.RaisedCosineTransmitFilter`, which first uses a polyphase interpolator to upsample the signal and applies the necessary RC or SRRC taps.

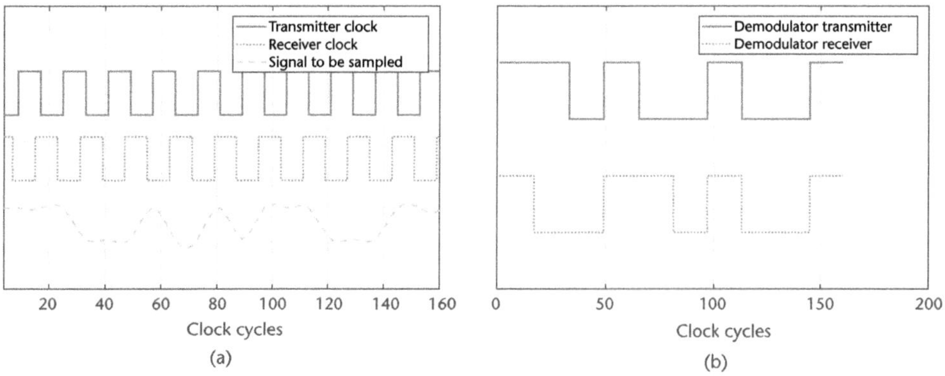

Figure 6.8 Comparison of different clock timings associated with a analog waveform and the resulting sampled and demodulated output. (a) Transmitter and receiver clocking signals with analog waveform to be sampled, and (b) demodulator outputs of receiver and transmitter according to their sampling times.

The upsampling factor N, also known as sample per symbol, will be chosen based on the recovery algorithms used and the desired data rate of the system. In general increasing N can improve the recovery process at the receiver to a point, but again this will reduce our useful bandwidth, forcing hardware to run at higher rates to achieve the same throughput.

Next if we consider timing error from the perspective of the constellation diagram we will observe clustering or scattering of the symbols. In Figure 6.9(a), we provide a simulated timing offsets (τ) of $0.2N$ and $0.5N$, where N is the samples per symbol. An offset of $0.5N$ is the worst case because we are exactly between two symbols. In Figure 6.9(b) we provide a QPSK single transmitted through loopback of a single Pluto SDR. We can clearly observe a similar clustering and some rotation from the transmit and receive chains lack of phase synchronization. This clustering happens because the receiver is actually sampling the transitions between samples. For example, if symbols $y(n)$ and $y(n+1)$ are $[1+i]$ and $[-1-i]$, respectively. Then if they are sampled at time $n+0.5$, the resulting point will be close to zero.

> **Q** In the case of the Pluto SDR constellation plot in Figure 6.10(b) why does the constellation appear rotated? It may be helpful to refer to Chapter 5.

At the receiver the unknown delay τ must be estimated to provide correct demodulation downstream. A crude but simple way we can illustrate a correction for the offset is to fractionally resample the signal with use of a polyphase filter. We will utilize the `dsp.VariableFractionalDelay` in the script below, which implements a polyphase filter for a given delay provided. We can use this with Pluto SDR to demonstrate different delays we should provide to correct for the offset. At the correct value of $\hat{\tau}$, where $\hat{\tau}+\tau = kT_s$ and $k = \mathbb{Z}_{\geq 0}$, the constellation will have four distinct points.

In Figure 6.10, four example delays are used as estimates to correct for the timing missmatch during loopback on a single Pluto SDR. These represent four instances from the above MATLAB script.

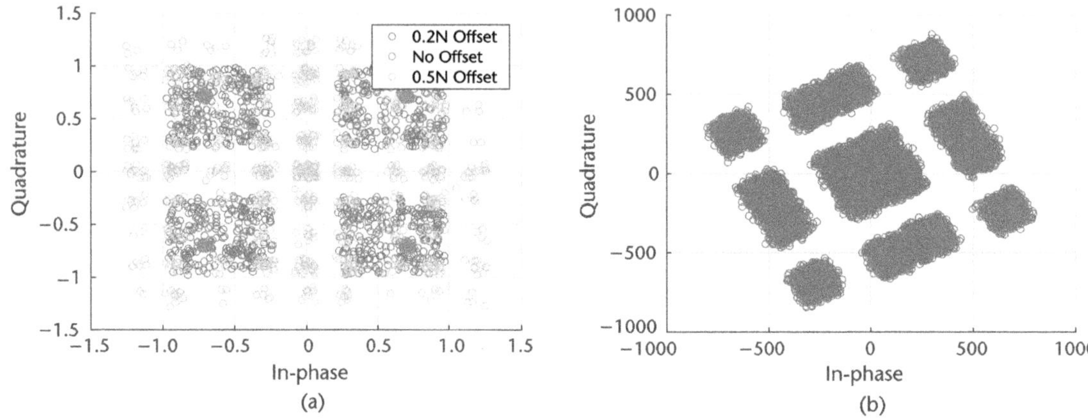

Figure 6.9 Comparison of simulation versus hardware timing offset. (a) Simulation-only example of several timing offsets with a QPSK received signal, and (b) hardware example created with a QPSK signal transmitted through Pluto SDR using a loopback cable.

Code 6.1 Loopback Pluto Example: `plutoLoopback.m`

```
% User tunable (samplesPerSymbol>=decimation)
samplesPerSymbol = 12; decimation = 4;
% Set up radio
tx = sdrtx('Pluto','Gain',-20);
rx = sdrrx('Pluto','SamplesPerFrame',1e6,'OutputDataType','double');
% Create binary data
data = randi([0 1],2^15,1);
% Create a QPSK modulator System object and modulate data
qpskMod = comm.QPSKModulator('BitInput',true); modData = qpskMod(data);
% Set up filters
rctFilt = comm.RaisedCosineTransmitFilter( ...
    'OutputSamplesPerSymbol', samplesPerSymbol);
rcrFilt = comm.RaisedCosineReceiveFilter( ...
    'InputSamplesPerSymbol',  samplesPerSymbol, ...
    'DecimationFactor',       decimation);
% Pass data through radio
tx.transmitRepeat(rctFilt(modData)); data = rcrFilt(rx());
% Set up visualization and delay objects
VFD = dsp.VariableFractionalDelay; cd = comm.ConstellationDiagram;
% Process received data for timing offset
remainingSPS = samplesPerSymbol/decimation;
% Grab end of data where AGC has converged
data = data(end-remainingSPS*1000+1:end);
for index = 0:300
    % Delay signal
    tau_hat = index/50;delayedsig = VFD(data, tau_hat);
    % Linear interpolation
    o = sum(reshape(delayedsig,remainingSPS,...
      length(delayedsig)/remainingSPS).',2)./remainingSPS;
    % Visualize constellation
    cd(o); pause(0.1);
end
```

6.3 Symbol Timing Compensation

There are many ways to perform correction for symbol timing mismatches between transmitters and receivers. However, in this chapter we will examine three digital

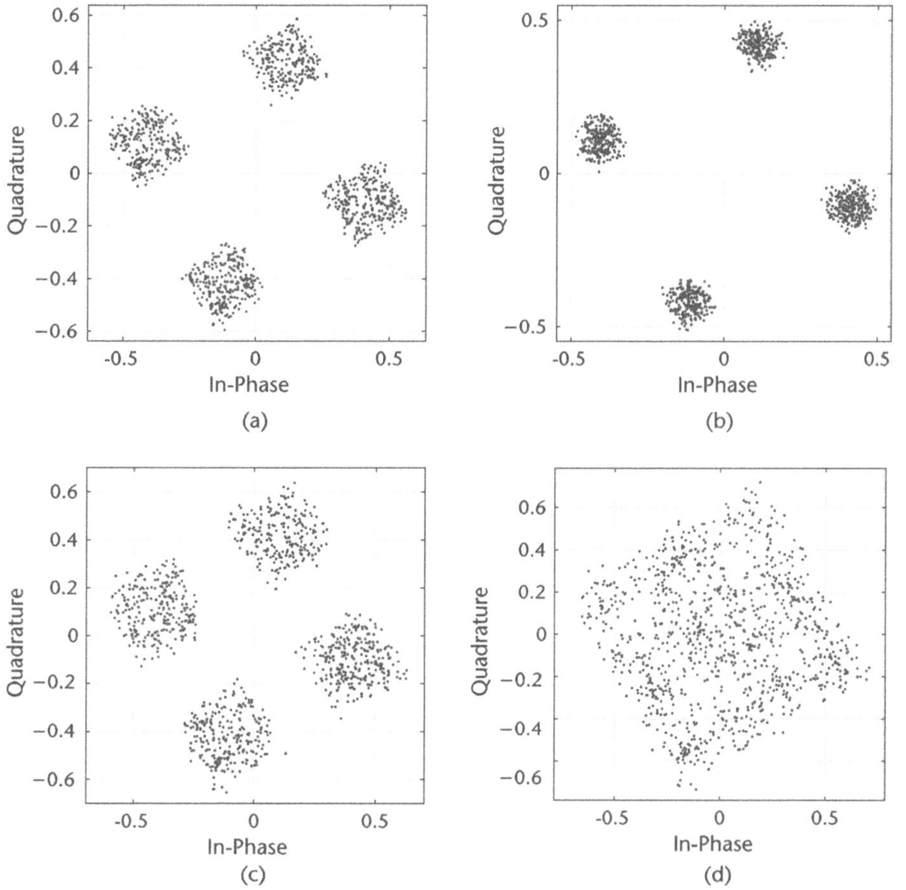

Figure 6.10 Resulting constellations of Pluto SDR loopback data after different fractional delays $\hat{\tau}$. (a) $\hat{\tau} = 0.1$, (b) $\hat{\tau} = 0.5$, (c) $\hat{\tau} = 1$, and (d) $\hat{\tau} = 1.5$.

> **Q** Using the above MATLAB code verify the timing offset observed. Is this a fixed offset? Change the frequency of both transmitter and receiver to 900 MHz, then explain the observation. Change she sampling rate of both the transmitter and receiver to 2 MHz, then explain your observation.

PLL strategies that will also share the same methodology as in Chapter 7 for our carrier recovery implementations. This type of timing recovery was chosen because it can be integrated with our future recovery solutions, can be robust, and is not overly complex algorithmicly. A basic PLL structure will be introduced first, which will be used to derive our feedback timing correction design. This will frame the discussion around Figure 6.11, where we will replace the individual blocks leading to our eventual design in Figure 6.12. During this process we will provide an overview conceptually how timing is synchronized and then move into each individual block, explaining their design. The specific detectors discussed will be Zero-Crossing, Müller/Mueller, and Gardner. However, more can be found in the literature from Mengali [4] and Oerder [5] among others. Rice [6] provides a more indepth analysis

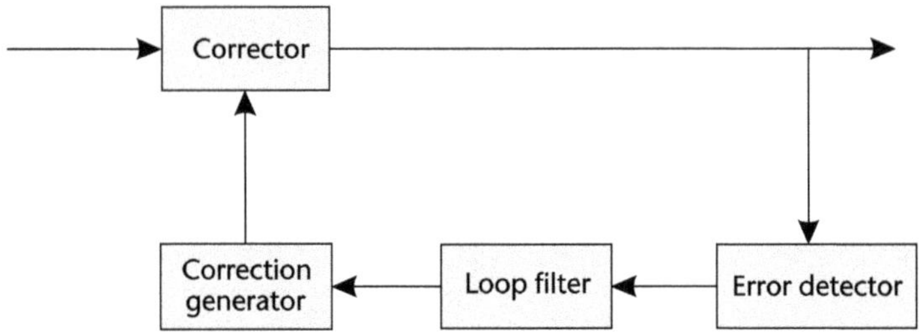

Figure 6.11 Basic PLL structure with four main component blocks.

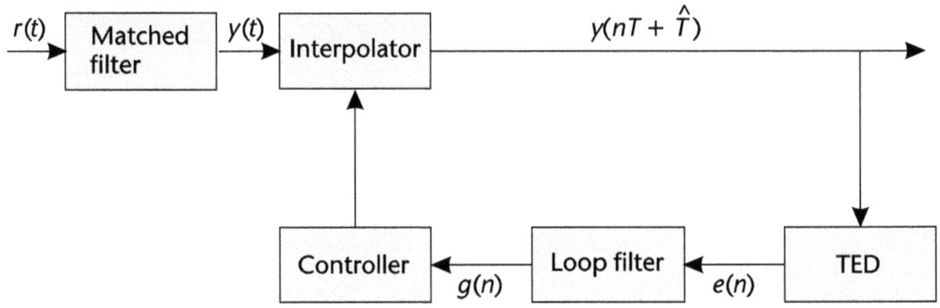

Figure 6.12 Basic structure of PLL for timing recovery for both decision direction and blind timing recovery. There are five major blocks that measure, control, and correct for the timing error in the received signal y.

of these techniques and covers purely analog designs as well as hybrid analog and digital designs. Here we will focus on MATLAB implementations and algorithmic structural features.

6.3.1 Phase-Locked Loops

The timing correction we will be using is a feedback or closed-loop method based on PLL theory. The structure of this algorithm is provided in Figure 6.11 derived from [6, Chapter 7], which essentially locks when an error signal is driven to zero. There are many different designs for PLLs and implementation strategies, but here we will outline four basic components that we will interchange here for timing correction and in the following chapter on carrier recovery. This all-digital PLL-based algorithm shown here works by first measuring some offset, such as timing error or phase error, of the received sample in the error detector (ED), which we call the error signal e. The ED is designed based on the structure of the desired receive constellation/symbols or the nature of the sequence itself. Next, the loop filter helps govern the dynamics of the overall PLL. The loop filter can determine operational ranges, lock time, and dampness/responsiveness of the PLL. Next, we have the correction generator. The correction generator is responsible for generation of the correction signal for the input, which again will be fed back into the system. Finally is the corrector itself, which modifies the input signal based on input from the correction generator. Working together, these blocks should eventually minimize e over time and contually adapt to future changes from the environment or the system itself.

The correction generator, error detector, and corrector are specific to the purpose of the PLL structure, such as timing recovery or carrier recovery. However, the loop filter can be shared among the designs with modification to its numerical configuration. The loop filter in all PLL designs is the most challenging aspect, but provides the most control over the adaption of the system. Here we will use a proportional-plus-integrator (PI) filter as our loop filter, which maintains a simple transfer function:

$$F(s) = g_1 + \frac{g_2}{s}, \tag{6.4}$$

where g_1 and g_2 are selectable gains. PI filters produce second-order PLLs and only consist of a single pole in their transfer function; therefore, they are relatively easy to analyze. Since we are dealing with discrete time signals a z-domain representation is preferable:

$$F(z) = G_1 + \frac{G_2}{1 - z^{-1}}, \tag{6.5}$$

where $G_1 \neq g_1$ and $G_2 \neq g_2$.[1] The fractional portion of (6.5) can be represented nicely by a biquad filter.[2] For the calculation of the gain values (G_1, G_2) utilize the following equations based on a preferred damping factor ζ and loop bandwidth B_{Loop}:

$$\theta = \frac{B_{Loop}}{M(\zeta + 0.25/\zeta)} \qquad \Delta = 1 + 2\zeta\theta + \theta^2 \tag{6.6}$$

$$G_1 = \frac{4\zeta\theta/\Delta}{M} \qquad G_2 = \frac{4\theta^2/\Delta}{M} \tag{6.7}$$

where M is the samples per symbol associated with the input signal. Note that B_{Loop} is a normalized frequency and can range $B_{Loop} \in [0, 1]$. If you are interested in how these are derived, see [6, Appendix C]. For the selection of ζ:

$$\zeta = \begin{cases} < 1, \text{Underdamp} \\ = 1, \text{Critically Damped} \\ > 1, \text{Overdamped,} \end{cases} \tag{6.8}$$

which will determine the responsiveness and stability of the PLL.

6.3.2 Feedback Timing Correction

The timing synchronization PLL used in all three algorithms consists of four main blocks: interpolator, timing ED (TED), loop filter, and an interpolator controller. Their interaction and flow is presented in Figure 6.14. These operations first estimate an unknown offset error, scale the error proportionally, and apply an update for future symbols to be corrected. To provide necessary perspective on the timing error, let us considered the eye diagram presented in Figure 6.13. This eye diagram has been upsampled by twenty times so we can examine the transitions more closely, which we notice are smooth unlike Figure 6.6. In the optimal case, we chose to sample our input signal at the dark gray instances at the widest openings of the eye. However, there will be an unknown fractional delay τ that shifts this sampling period. This shifted sampling position is presented by the light gray selections. To help work around this issue, our receive signal is typically not decimated fully,

1. A simple way to translate between (6.4) and (6.5) is to utilize a bilinear transform.
2. See `dsp.BiquadFilter` for a simple realization of a biquad filter.

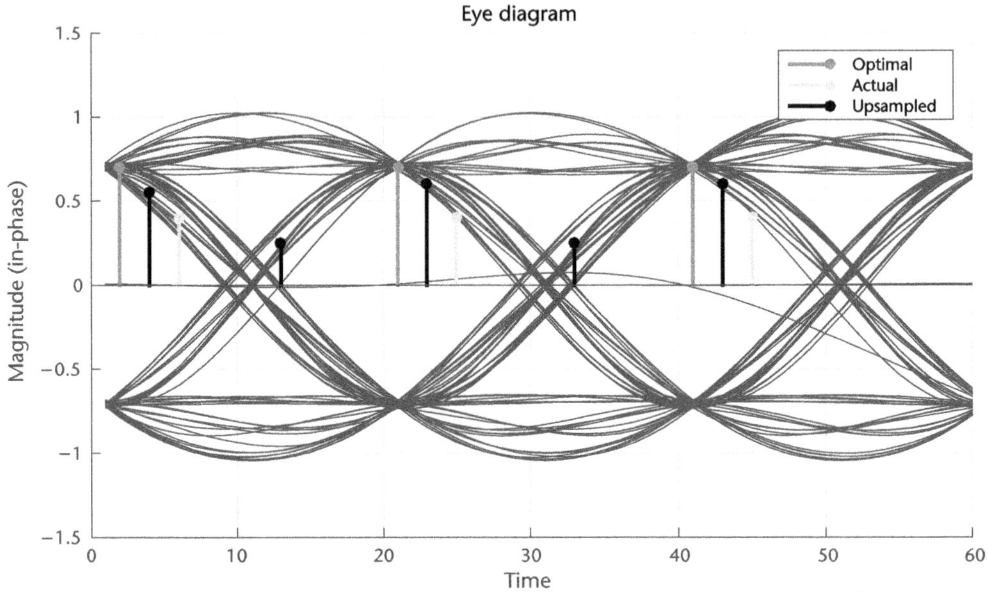

Figure 6.13 Eye diagram of received signal marking positions where received samples may exist. This figure is highly oversampled to show many positions, but a received sample could lie anywhere on the eye.

providing the receiver with multiple samples per symbol (this is not always the case). Therefore, if we are straddling the optimal sampling position instead as in the black markers, we can simply interpolate across these points to get our desired period. This interpolation has the effect of causing a fractional delay to our sampling, essentially shifting to a new position in our eye diagram. Since τ is unknown we must weight this interpolation correctly so we do not overshoot or undershoot the desired correction. This is similar to the idea presented at the end of Section 6.2. Finally, controlling the instances in time when an output is produced or sampled from the input data is the function of the interpolator control block, which will be at the symbol rate. This correction loop, when implemented properly, that will cause the eye diagram to open for input signals with clock timing missmatches. However, a constellation diagram may also be useful tool for evaluating timing correction as presented in Figures 6.4 and 6.9.

We will initially discuss the specifics of the blocks in Figure 6.12 through the perspective of the zero-crossing (ZC) method, since it is the most straightforward to understand. Then we will provide extensions to the alternative methods. ZC, as the name suggests, will produce an error signal $e(n)$ of zero when one of the sampling positions is at the zero intersection. ZC requires two samples per symbol or more, resulting in the other sampling position occurring at or near the optimal position. The TED for ZC [4] is evaluated as

$$\begin{aligned} e(n) =& Re(y((n-1/2)T_s+\tau))[sgn\{Re(y((n-1)T_s+\tau))\} - sgn\{Re(y(nT_s+\tau))\}] \\ &+ Im(y((n-1/2)T_s+\tau))[sgn\{Im(y((n-1)T_s+\tau))\} \\ &- sgn\{Im(y(nT_s+\tau))\}], \end{aligned} \tag{6.9}$$

where *Re* and *Im* extract the real and imaginary components of a sample, and *sgn* process the sign (-1 or 1) for the sample. In (6.9) it is important to note that these indexes are with respect to samples, not symbols, and to be specifc $y(nT_s + \tau)$ is simply the latest output of the interpolator filter. Looking at (6.9) it first provides a direction for the error with respect to the *sgn* operation, and the shift required to compensate is determined by the midpoints. The in-phase and quadrature portions operate independently, which is desirable.

Once the error is calculated it is passed to the loop filter, which we can entirely borrow from Section 6.3.1. The same principles apply here, with a near identical formulation for our equations we have provided in a slightly more compact form.

$$G_1 = \frac{-4\zeta\theta}{G_D N\Delta} \qquad G_2 = \frac{-4\theta^2}{G_D N\Delta} \tag{6.10}$$

Here B_{Loop} is the normalized loop bandwidth, ζ is our damping factor, N is our samples per symbol, and G_D is our detector gain. The new variable G_D provides an additional step size scaling to our correction. Again the loop filter's purpose is to maintain stability of the correction rate. This filter can be implemented with a simple linear equation:

$$y(t) = G_1 x(t) + G_2 \sum_{n=0} y(n), \tag{6.11}$$

or with a biquad filter.

The next block to consider is the Interpolation Controller, which is responsible to providing the necessary signaling to the interpolator. With respect to our original PLL structure in Figure 6.11 the interpolation controller takes the place of the correction generator. Since the interpolator is responsible for fractionally delaying the signal, this controller must provide this information and generally the starting interpolant sample. By starting interpolant sample we are referring to the sample on the left side of the straddle, as shown by the second black sampling position from the left in Figure 6.13. The interpolation controller implemented here will utilize a counter-based mechanism to effectively trigger at the appropriate symbol positions. At these trigger positions the interpolator is signaled and updated, as well as an output symbol is produced from the system.

The main idea behind a counter-based controller is to maintain a specific triggering gap between updates to the interpolator, with an update period on average equal to symbol rate N of the input stream. In Figure 6.14 a logical flowchart of the interpolation controller is provided to better understand the complex flow. If we consider the case when the timing is optimal and the output of the loop filter $g(n)$ is zero, we would want to produce a trigger every N samples. Therefore, it is logical that the weighting or decrement for the counter would be

$$d(n) = g(n) + \frac{1}{N}. \tag{6.12}$$

resulting in a maximum value of 1 under modulo-1 subtraction of the counter $c(n)$, where wraps of the modulus occur every N subtractions. This modulus counter update is defined as

$$c(n+1) = (c(n) - d(n)) \mod 1. \tag{6.13}$$

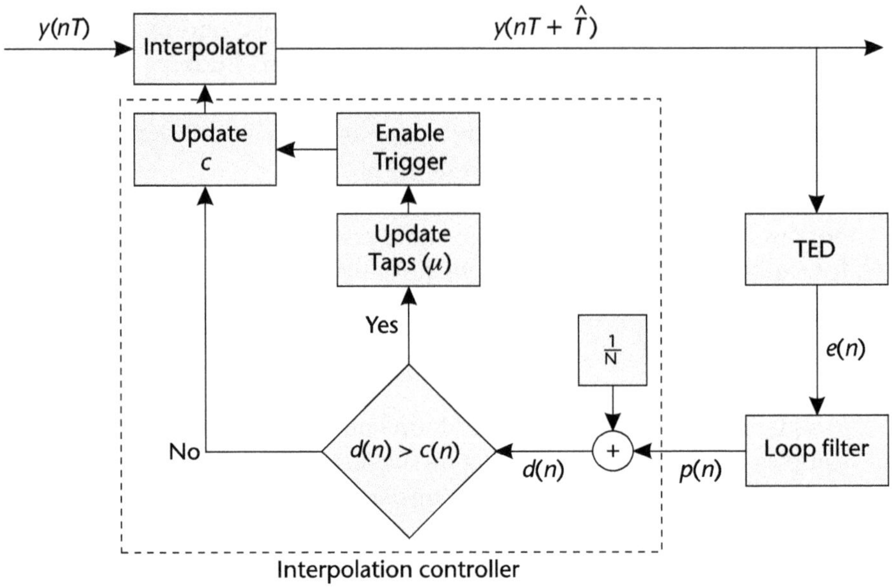

Figure 6.14 Timing recovery triggering logic used to maintain accurate interpolation of input signal.

We determine a trigger condition, which is checked before the counter is updated, based on when these modulus wraps occur. We can easily check for this condition before the update, such as

$$Trigger = \begin{cases} c(n) < d(n) & True \\ \text{Otherwise} & False \end{cases}. \tag{6.14}$$

This triggering signal is the method used to define the start of a new symbol; therefore, it can also be used to make sure we are estimating error over the correct samples. When the trigger occurs we will update $\mu(n)$ our estimated gap between the interpolant point and the optimal sampling position. This update is a function of the new counter step $d(n)$ and our current count $c(n)$:

$$\mu(k) = c(n)/d(n). \tag{6.15}$$

This μ will be passed to our interpolator to update the delay it applies.

We want to avoid performing timing estimates that span over multiple symbols, which would provide incorrect error signals and incorrect updates for our system. We can avoid this by adding conditions into the TED block. We provide additional structure to the TED block in Figure 6.15, along with additional logic to help identify how we can effectively utilize our trigger signals. Based on this TED structure, only when a trigger occurs the output error e can be nonzero. Looking downstream in Figure 6.14 from the TED, since we are using a PI loop filter only nonzero inputs can update the output, and as a result modify the period of the triggering associated d. When the system enters steady state, where the PLL has locked, the TED output can be nonzero every N samples.

The final piece of the timing recovery we have not yet discussed is the interpolator itself. With respect to our original PLL structure in Figure 6.11 the interpolator takes the place of the corrector. Interpolation here is simply a linear

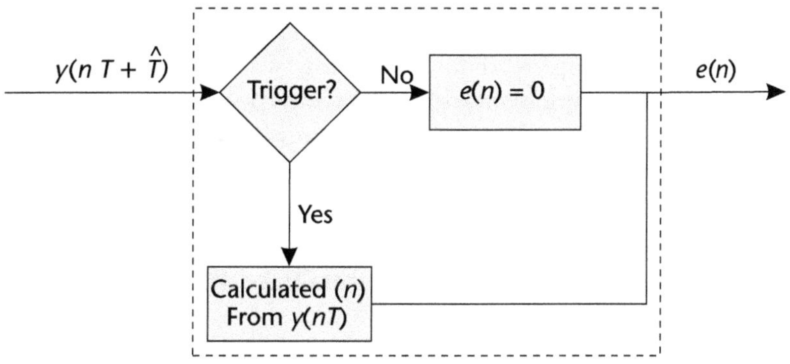

Figure 6.15 An internal view of the timing error detector to outline the error and triggering control signals relation to operations of other blocks in Figure 6.14.

combination of the current and past inputs y, which in essence can be thought of as a filter. However, to create a FIR filter with any arbitrary delay $\tau \in [0, ..., T_s]$ cannot be realized [7]. Realizations for ideal interpolation IIR filters do exist, but the computation of their taps are impractical in real systems [8]. Therefore, we will use an adaptive implementation of a FIR lowpass filter called a piecewise polynomial filter (PPF) [6]. The PPF can only provide estimations of offsets to a polynomial degree. Alternative implementations exists such as polyphase-filterbank designs, but depending on the required resolution the necessary phases become large. However, they can be straightforward to implement [9].

The PPF are useful since we can easily control the form of interpolations by determining the order of the filter, which at most is equivalent to the order of the polynomial used to estimate the underlying received signal. Here we will use a second order, or quadratic, interpolation requiring a four-tap filter. The general form of the interpolator's output is given by

$$y(kT_s + \mu(k)T_s) = \sum_{n=1}^{2} h(n)y((k-n)T_s), \tag{6.16}$$

where h_k are the filter coefficients at time instance k determined by [10]:

$$\begin{aligned} h =&[\alpha\mu(k)(\mu(k)-1), \\ &-\alpha\mu(k)^2 - (1-\alpha)\mu(k) + 1, \\ &-\alpha\mu(k)^2 + (1+\alpha)\mu(k), \\ &\alpha\mu(k)(\mu(k)-1)], \end{aligned} \tag{6.17}$$

where $\alpha = 0.5$. $\mu(k)$ is related to the fractional delay, which is provided by the interpolator control block, which relates the symbol period T_s to the estimated offset. Therefore, we can estimate the true delay τ as

$$\hat{\tau} \sim \mu(k)T_s. \tag{6.18}$$

Without any offset ($\mu = 0$), the interpolator acts as a two-sample delay or single-symbol delay for the ZC implementation. We can extend the PPF to utilize more samples creating cubic and greater interpolations, but their implementations become more complex. The underlying waveform should be considered when determining the implementation of the interpolator as well as the required degrees of freedom to accurately capture the required shape.

This design using four samples in a quadratic form can be considered irregular, since the degree of taps does not reach three. However, odd length realizations (using an odd number of samples) are not desirable since we are trying to find values in-between the provided samples. We also do not want a two-sample implementation due to the curvature of the eye in Figure 6.13.

In MATLAB we can realize this interpolator with a few lines of code that are dependent on the input data y and the last output of the interpolator controller μ provided in Code 6.2.

Code 6.2 Interpolator: **interpFilter.m**

```
% Define interpolator coefficients
alpha = 0.5;
InterpFilterCoeff = ...
    [ 0,        0,          1,        0;    % Constant
    -alpha, 1+alpha, -(1-alpha), -alpha; % Linear
    alpha,   -alpha,    -alpha,    alpha]; % Quadratic
% Filter input data
ySeq = [y(i); InterpFilterState]; % Update delay line
 % Produce filter output
filtOut = sum((InterpFilterCoeff * ySeq) .* [1; mu; mu^2]);
InterpFilterState = ySeq(1:3); % Save filter input data
```

From this output *filtOut* we can drive our TED using the ZC equation (6.9) to create error signals for the loop filter and interpolator controller. Based on Figure 6.14 we know that this TED calculation will be based on a triggered signal from the interpolator controller. Additional historical triggers are also checked which prevent driving the output of the timing loop faster than the symbol rate. This logic and TED measurement is captured in Code 6.3.

Additional logic is added to the TED from lines 13 to 22, which manage symbol stuffing. Symbol stuffing is basically forcing an additional trigger from the synchronizer routine. This is necessary when clock deviations force the interpolator to minimally straddle the symbol of interest. To compensate we must insert an additional output symbol. Note that at the output of the system, the sample rate will equal the symbol rate, essentially downsampling our signal when $N > 1$.

Following the TED is the loop filter, which has already been discussed in Section 6.3.1. Since the filter is quite simple it can be implemented in a straightforward way without filter objects. However, using a biquad filter object provides more compact code as shown Code 6.4.

Finally, we can evaluate the filtered error at the interpolator control block. In steady state this block should produce a trigger every N input samples. This trigger signal can be considered a valid output signal, which will concide with output data from the whole algorithm. In the coding context here, when *Trigger* is true at time n the output of the interpolation filter at input $n+1$ should be processed

Code 6.3 ZC TED: **zcTED.m**

```
% ZC-TED calculation occurs on a strobe
if Trigger && all(~TriggerHistory(2:end))
    % Calculate the midsample point for odd or even samples per symbol
    t1 = TEDBuffer(end/2 + 1 - rem(N,2));
    t2 = TEDBuffer(end/2 + 1);
    midSample = (t1+t2)/2;
    e = real(midSample)*(sign(real(TEDBuffer(1)))-sign(real(filtOut))) ...
        imag(midSample)*(sign(imag(TEDBuffer(1)))-sign(imag(filtOut)));
else
    e = 0;
end
% Update TED buffer to manage symbol stuffs
switch sum([TriggerHistory(2:end), Trigger])
    case 0
      % No update required
    case 1
      % Shift TED buffer regularly if ONE trigger across N samples
      TEDBuffer = [TEDBuffer(2:end), filtOut];
    otherwise % > 1
      % Stuff a missing sample if TWO triggers across N samples
      TEDBuffer = [TEDBuffer(3:end), 0, filtOut];
end
```

Code 6.4 Loop Filter: **loopFilter.m**

```
% Loop filter
loopFiltOut = LoopPreviousInput + LoopFilterState;
g = e*ProportionalGain + loopFiltOut; % Filter error signal
LoopFilterState = loopFiltOut;
LoopPreviousInput = e*IntegratorGain;
% Loop filter (alternative with filter objects)
lf = dsp.BiquadFilter('SOSMatrix',tf2sos([1 0],[1 -1])); % Create filter
g = lf(IntegratorGain*e) + ProportionalGain*e; % Filter error signal
```

downstream. The interpolation controller itself will utilize the filtered error signal g and will update the internal counter as data is processed in Code 6.5.

Code 6.5 Interpolator Control Logic: **interpControl.m**

```
% Interpolation Controller with modulo-1 counter
d = g + 1/N;
TriggerHistory = [TriggerHistory(2:end), Trigger];
Trigger = (Counter < d); % Check if a trigger condition
if Trigger % Update mu if a trigger
    mu = Counter / d;
end
Counter = mod(Counter - d, 1); % Update counter
```

The overall design of the timing synchronizer can be complex and implementations do operate at different relative rates. Therefore, we have provided Table 6.1 as a guide to a recommended implementation. These rates align with

Table 6.1 Operational Rates of Timing Recovery Blocks

Block	*Operational rate*
Interpolator	Sample rate
TED	Symbol rate
Loop filter	Symbol rate
Interpolator controller	Sample rate

the trigger implementation outlined in Figure 6.14. This system will result in one sample per symbol (N) when output samples of the interpolator are aligned with the triggers.

> **Q** Starting with script `TimingError`, which models a timing offset, implement ZC timing correction.

6.4 Alternative Error Detectors and System Requirements

Within the discussed PLL framework alternative TEDs can be used if the application or system arrangement is different. For example, the discussed method of ZC cannot operate under carrier phase or frequency offsets. Therefore, such a nonideality would require compensation first before application of ZC, which is not true for other methods. Besides carrier offsets, a requirement of the ZC method is an upsample factor N of at least two, which may not be possible for certain systems due to bandwidth and data rate constraints.

6.4.1 Gardner

The second TED we will considered is called Gardner [11], which is very similar to ZC. The error signal is determined by

$$\begin{aligned} e(n) = &Re(y((n-1/2)T_s+\tau))\big[Re(y((n-1)T_s+\tau)) - Re(y(nT_s+\tau))\big]+ \\ &Im(y((n-1/2)T_s+\tau))\big[Im(y((n-1)T_s+\tau)) - Im(y(nT_s+\tau))\big]. \end{aligned} \tag{6.19}$$

This method also requires two samples per symbol and differs only in the quantization of the error direction from ZC. One useful aspect of Gardner is that it does not require carrier phase correction and works especially well with BPSK and QPSK signals. However, since Gardner is not a decision-directed method, for best performance the excess bandwidth of the transmit filters should be $\beta \in (0.4, 1)$.

> **Q** Implement the Gardner TED inside your existing timing error detector. Introduce a small phase shift into the received signal of $\pi/4$. Compare ZC and Gardner in these cases.

6.4.2 Müller and Mueller

Next is the Müller and Mueller (MM) method named after Kurt Mueller and Markus Müller [12]. This can be considered the most efficient method since it does not require upsampling of the source data, operating at one sample per symbol. The error signal is determined by [6]

$$\begin{aligned} e(k) = & Re(y((k)T_s + \tau)) \times sgn\{Re(y((k-1)T_s + \tau))\} \\ & - Re(y((k-1)T_s + \tau)) \times sgn\{Re(y((k)T_s + \tau))\} \\ & + Im(y((k)T_s + \tau)) \times sgn\{Im(y((k-1)T_s + \tau))\} \\ & - Im(y((k-1)T_s + \tau)) \times sgn\{Im(y((k)T_s + \tau))\}. \end{aligned} \tag{6.20}$$

MM also operates best when the matched filtering used minimizes the excess bandwidth, meaning β is small. It is important to note when the excess bandwidth of the receiver or transmitter filters is high the decisions produced by the *sgn* operation can be invalid. Therefore, this trade-off must be considered during implementation. However, even though MM is technically most efficient performance can be questionable at $N = 1$ due to the lack of information available per symbol.

> Add phase and frequency offsets to the input signal and compare the performance of ZC, Gardner, and MM estimation methods. Do this for fixed fractional delays $\frac{T_s}{2}$, $\frac{T_s}{4}$, $\frac{T_s}{5}$ in the channel and plot the error output of the TEDs for Gardner and ZC.

6.5 Putting the Pieces Together

Throughout this chapter we have outlined the structure and logic behind a PLL-based timing recovery algorithm and the associated MATLAB code. In the remaining sections we will discuss putting the algorithmic components together and provide some intuition on what happens during evaluation. Here we will also address parameterization and the relation to system dynamics.

The system-level scripts have shown a constant theme throughout where data is modulated, transmit filtered, passed through a channel with timing offset, filtered again, then is timing recovered. Many rate changes can happen in this series of steps. To help understand these relations better we can map things out as in Figure 6.16, which takes into account these stages. Here the modulator produces symbols equal to the sample rate. Once passing through the transmit filter we acquire our upsampling factor N, which increases our samples per symbol to N. At the receiver we can perform decimation in the receive filter by a factor N_F where $N_F \leq N$. Finally, we will perform timing recovery across the remaining samples and remove the fractional offset τ, returning to the original rate of one sample per symbol. The rate pattern outlined in Figure 6.16 is identical to that of the first MATLAB script in Code 6.1. That script can be modified to produce a slightly dynamic timing offset, which we provide below:

From Code 6.6 we can evaluate the receive filtered signal with a variable offset over time. Figure 6.17(a) provides the direct output of `rxFilt` when `samplesPerSymbol` is equal to `decimation`, where we can observe the constellation of the signal collapsing into constellations over time similar to Figure 6.9. This is essentially when no timing recovery is being used. Next,

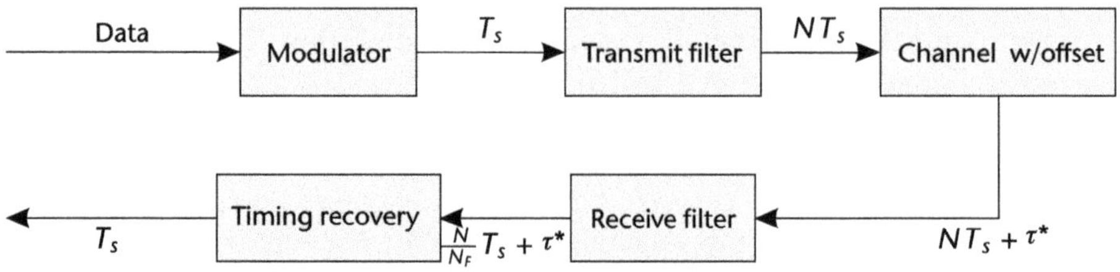

Figure 6.16 Relative rates of transmit and receive chains with respect to the sample rate at different stages. Here τ^* represents a timing shift not an increase in the data rate. This is a slight abuse of notation.

Code 6.6 Transmit Filter Data: **srrcFilterData.m**

```
% User tunable (samplesPerSymbol>=decimation)
samplesPerSymbol = 4; decimation = 2;
% Create a QPSK modulator System object and modulate data
qpskMod = comm.QPSKModulator('BitInput',true);
% Set up filters
rctFilt = comm.RaisedCosineTransmitFilter( ...
    'OutputSamplesPerSymbol', samplesPerSymbol);
rcrFilt = comm.RaisedCosineReceiveFilter( ...
    'InputSamplesPerSymbol',  samplesPerSymbol, ...
    'DecimationFactor',       decimation);
% Set up delay object
VFD = dsp.VariableFractionalDelay;
% Delay data with slowly changing delay
rxFilt = [];
for index = 1:1e3
    % Generate, modulate, and tx filter data
    data = randi([0 1],100,1);
    modFiltData = rctFilt(qpskMod(data));
    % Delay signal
    tau_hat = index/30;
    delayedsig = VFD(modFiltData, tau_hat);
    rxSig = awgn(delayedsig,25); % Add noise
    rxFilt = [rxFilt;rcrFilt(rxSig)]; % Rx filter
end
```

taking the lessons from this chapter and utilizing the implementation of timing recovery proposed, we can adapt to this changing fractional delay. Figure 6.17(b) demonstrates the recovery for the ZC technique when $\frac{N}{N_F} = 2$. Here we can observe clear division between the level for the real component of the signal, meaning our output constellation is collapsing to a correct QPSK signal. In Figure 6.17(c) we increase B_{Loop} from 0.001 to 0.01, which causes the system to react faster. However, for $B_{Loop} = 0.001$ once converged the residual symbols have less noise than for $B_{Loop} = 0.01$.

Q Regenerate Figure 6.17 and utilize alternative $\zeta = \{0.5, \sqrt{2}, 10, 20\}$. Comment on the dynamic of the recovery algorithm.

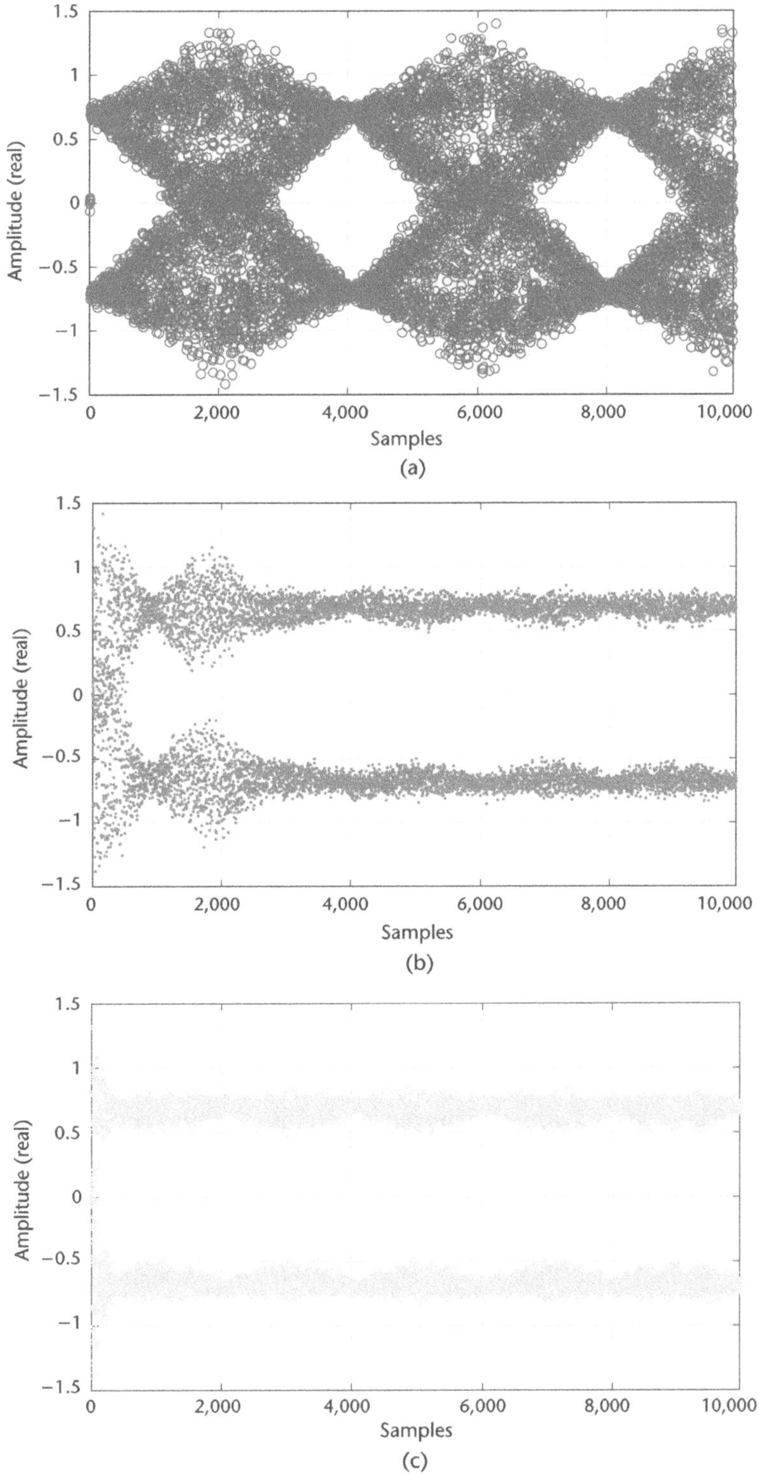

Figure 6.17 Comparison of a signal that requires timing recovery, and outputs of two parameterization of ZC timing recovery after application. (a) Receive signal without timing recovery, (b) receive signal with ZC timing recovery for parameterization $\{N, \zeta, B_{Loop}, G_D\} = \{2, 1, 0.001, 2.7\}$, and (c) receive signal with ZC timing recovery for parameterization $\{N, \zeta, B_{Loop}, G_D\} = \{2, 1, 0.01, 2.7\}$.

> **Q** Regenerate Figure 6.17, but utilize Pluto SDR in loopback as the channel. Tune the recovery algorithm (N, ζ, B_{Loop}, G_D) and measure the best conference you can achieve.

6.6 Chapter Summary

Timing recovery is a fundamental tool for successful transmission between nodes with independent oscillators. In this chapter, a model for timing offset was introduced mathematically, in simulation, and demonstrated with Pluto SDR. To combat this offset, a PLL-based timing recovery methodology was introduced that included several timing error detectors. This included an review and extension to matched filtering introduced in Chapter 6. MATLAB code was provided for the different components of the timing recovery algorithms, and a considerable amount of time was spent examining their configuration and interactions. Finally, once all the components were investigated, portions of the design's parameterization were explored. In subsequent chapters, the implementations developed here will be utilized to created a full receiver design which can recover signals transmitted between separate Pluto SDR devices.

References

[1] Proakis, J., and M. Salehi, *Digital Communications*, Fifth Edition, Boston: McGraw-Hill, 2007.

[2] Saleh, A. A. M., "Frequency-Independent and Frequency-Dependent Nonlinear Models of TWT Amplifiers," *IEEE Transactions on Communications*, Vol. 29, No. 11, November 1981, pp. 1715–1720.

[3] Boumaiza, S., T. Liu, and F. M. Ghannouchi, "On the Wireless Transmitters Linear and Nonlionear Distortions Detection and Pre-correction," in *2006 Canadian Conference on Electrical and Computer Engineering*, May 2006, pp. 1510–1513.

[4] Mengali, U., *Synchronization Techniques for Digital Receivers*, Applications of Communications Theory, New York: Springer, 2013.

[5] Oerder, M., and H. Meyr, "Digital Filter and Square Timing Recovery," *IEEE Transactions on Communications*, Vol. 36, No. 5, May 1988, pp. 605–612.

[6] Rice, M., *Digital Communications: A Discrete-Time Approach*, Third Edition, Pearson/Prentice Hall, 2009.

[7] Laakso, T. I., V. Valimaki, M. Karjalainen, and U. K. Laine, "Splitting the Unit Delay [FIR/All Pass Filters Design]," *IEEE Signal Processing Magazine*, Vol. 13, No. 1, January 1996, pp. 30–60.

[8] Thiran, J. P., "Recursive Digital Filters with Maximally Flat Group Delay," *IEEE Transactions on Circuit Theory*, Vol. 18, No. 6, November 1971, pp. 659–664.

[9] Rice, M., and F. Harris, "Polyphase Filterbanks for Symbol Timing Synchronization in Sampled Data Receivers," in *MILCOM 2002, Proceedings*, Vol. 2, October 2002, pp. 982–986.

[10] Erup, L., F. M. Gardner, and R. A. Harris, "Interpolation in Digital Modems. ii. Implementation and Performance," *IEEE Transactions on Communications*, Vol. 41, No. 6, June 1993, pp. 998–1008.

[11] Gardner, F., "A BPSK/QPSK Timing-Error Detector for Sampled Receivers," *IEEE Transactions on Communications*, Vol. 34, No. 5, May 1986, pp. 423–429.

[12] Mueller, K., and M. Muller, "Timing Recovery in Digital Synchronous Data Receivers," *IEEE Transactions on Communications*, Vol. 24, No. 5, May 1976, pp. 516–531.

CHAPTER 7

Carrier Synchronization

This chapter will introduce the concept of carrier frequency offset between transmitting and receiving nodes. Specifically, a simplified error model will be discussed along with two recovery methods that can operate jointly or independently based on their implementation. Carrier recovery complements timing recovery, which was implemented in the previous Chapter 6, and is necessary for maintaining wireless links between radios with independent oscillators.

Throughout this chapter we will assume that timing mismatches between the transmitting and receiving radios have already been corrected. However, this is not a requirement in all cases, specifically in the initial implementation provided here, but will become a necessary condition for optimal performance of the final implementation provided. For the sake of simplicity we will also ignore timing effects in our simulations except when discussing Pluto SDR itself, since obviously timing correction cannot be overlooked in that case. With regard to our full receiver diagram outline in Figure 7.1, we are now considering the carrier recovery and CFO blocks.

7.1 Carrier Offsets

The receiving and transmitting nodes are generally two distinct and spatially separate units. Therefore, relative frequency offsets will exist between their LOs due to natural effects such as impurities, electrical noise, and temperature differences, among others. Since these differences can also be relatively dynamic the LOs will drift with respect to one another. These offsets can contain random phase noise, frequency offset, frequency drift, and initial phase mismatches. However, for simplicity we will only model this offset as a fixed value. This is a reasonable assumption at the time scale of RF communications.

When considering commercial oscillators, the frequency offset is provided in parts per million (PPM), which we can translate into a maximum carrier offset for a given frequency. In the case of the Pluto SDR the internal LO is rated at 25 PPM [1] (2 PPM when calibrated) and we can use (7.1) to relate maximum carrier offset Δf to our operating carrier frequency f_c.

$$f_{o,max} = \frac{f_c \times PPM}{10^6} \tag{7.1}$$

The determination of $f_{o,max}$ is important because it provides our carrier recovery design criteria. There is no point wasting resources on a capability to correct for a frequencies beyond our operational range. However, scanning techniques can be used in such cases but are beyond the scope of this book.

Figure 7.1 Receiver block diagram.

Mathematically we can model a corrupted source signal at baseband $s(k)$ with a carrier frequency offset of f_o (or ω_o) as

$$r(k) = s(k)e^{j(2\pi f_o kT+\theta)} + n(k) = s(k)e^{j(\omega_o kT+\theta)} + n(k). \tag{7.2}$$

where $n(k)$ is a zero-mean Gaussian random process, T is the symbol period, θ is the carrier phase, and ω_o the angular frequency.

In the literature, carrier recovery is sometimes defined as carrier phase recovery or carrier frequency recovery. These generally all have the same goal of providing a stable constellation at the output of the synchronizer. However, it is important to understand the relation of frequency and phase, which will make these naming conventions clear. An angular frequency ω, or equivalently in frequency $2\pi f$, is purely a measure of a changing phase θ over time:

$$\omega = \frac{d\theta}{dt} = 2\pi f. \tag{7.3}$$

Hence, recovering the phase of the signal is essentially recovering that signal's frequency. Through this relation is the common method for estimating frequency of a signal since it cannot be measured directly unlike phase. We can demonstrate this technique with a simple MATLAB script shown in Code 7.1. There we generate a simple continuous wave (CW) tone at a given frequency, measure the instantaneous phase of the signal, and then take the difference of those measurements as our frequency estimate. The instantaneous phase θ of any complex signal $x(k)$ can be measured as

$$\theta = tan^{-1}\left(\frac{Im(x(k))}{Re(x(k))}\right), \tag{7.4}$$

where Re and Im capture the real and imaginary components of the signal respectively. In Code 7.1 we also provide a complex sinusoid generation through a Hilbert transform with the function `hilbert` from our real signal. Hilbert transforms are very useful for generating analytic or complex representations of real signals. If you wish to learn more about Hilbert transforms, Oppenheim [2] is a suggested reading based in signal processing theory.

In Figure 7.2 we provide the outputs from Code 7.1. In Figure 7.2(a) it first can be observed that the Hilbert transform's output is equal to the CW tone generated from our sine (imag) and cosine (real) signal. In Figure 7.2(b) we can clearly see that the estimation technique based on phase difference correctly estimates the frequency of the signal in question. In this script we also utilized the function `unwrap` to prevent our phase estimates from becoming bounded between $\pm\pi$. This estimation is a straightforward application of the relation from (7.3). Alternatively, it can be useful to examine a frequency offset, but usually only large offsets, in the frequency domain itself. This is useful since time domain signals alone, especially when containing modulated data and noise, can be difficult to interpret for such an offset. In Figure 7.3, PSDs of an original and offset signal are shown, which

Code 7.1 `freqEstimate.m`

```
% Sinusoid parameters
fs = 1e3; fc = 30; N = 1e3;
t = 0:1/fs:(N-1)/fs;
% Create CW Tone
r = cos(2*pi*fc*t); i = sin(2*pi*fc*t);
% Alternatively we can use a hilbert transform from our real signal
y = hilbert(r);
% Estimate frequency from phase
phaseEstHib = unwrap(angle(y))*fs/(2*pi); freqEstHib = diff(phaseEstHib);
phaseEstCW = unwrap(atan2(i,r))*fs/(2*pi); freqEstCW = diff(phaseEstCW);
tDiff = t(1:end-1);
```

> **Q** From the MATLAB Code 7.1 examine the frequency range of this estimation technique with respect to the sampling rate f_s and the frequency of the tone f_c. (Ignore the output of the Hilbert transform for this exercise.) What is roughly the maximum frequency that can be correctly estimated and what happens when the frequency offset exceeds this point?

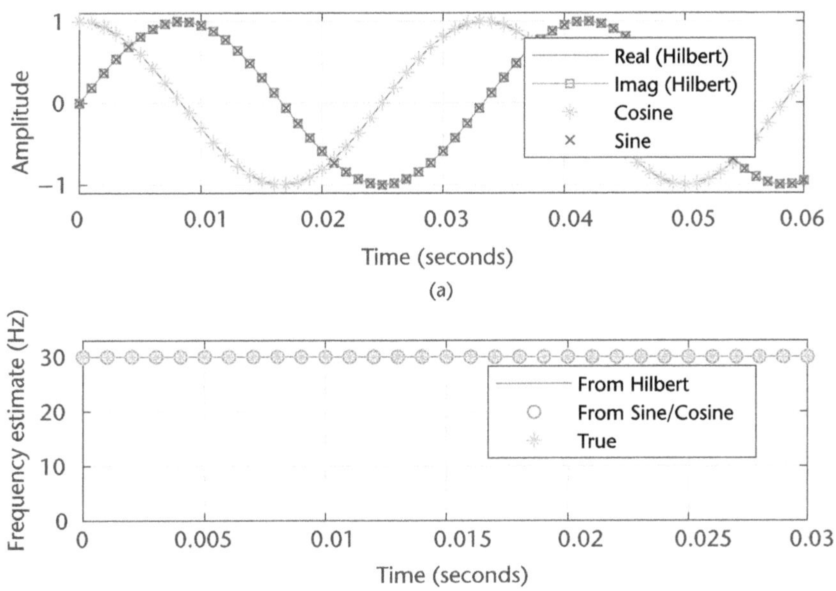

Figure 7.2 Outputs of MATLAB scripts for a simple frequency estimation technique compared with the true offset. (a) CW tones generated from sine/cosine and Hilbert transform, and (b) frequency estimates of CW tones.

clearly demonstrates this perspective. Here the signal maintains a 10-kHz offset with respect to the original signal, which is well within the 25-PPM specification of communicating Pluto SDR above 200 MHz.

Moving complex signals in frequency is a simple application of (7.2), which was how Figure 7.3(b) was generated. The example MATLAB script in Code 7.2

demonstrated how to shift a complex signal using an exponential function. Alternatively, sinusoids can be used directly if desired. In the script provided it is important to upsample or oversample the signal first, as performed by the SRRC filter in Code 7.2. This makes the frequency shift obvious since the main signal energy is limited to a fraction of the bandwidth.

Code 7.2 `freqShiftFFT.m`

```
% General system details
fs = 1e6; samplesPerSymbol = 1; frameSize = 2^8;
modulationOrder = 2; filterOversample = 4; filterSymbolSpan = 8;
% Impairments
frequencyOffsetHz = 1e5;
% Generate symbols
data = randi([0 samplesPerSymbol], frameSize, 1);
mod = comm.BPSKModulator(); modulatedData = mod(data);
% Add TX Filter
TxFlt = comm.RaisedCosineTransmitFilter('OutputSamplesPerSymbol',...
    filterOversample, 'FilterSpanInSymbols', filterSymbolSpan);
filteredData = TxFlt(modulatedData);
% Shift signal in frequency
t = 0:1/fs:(frameSize*filterOversample-1)/fs;
freqShift = exp(1i.*2*pi*frequencyOffsetHz*t.');
offsetData = filteredData.*freqShift;
```

7.2 Frequency Offset Compensation

There are many different ways to design a wireless receiver, using many different recovery techniques and arrangement of algorithms. In this section we will consider

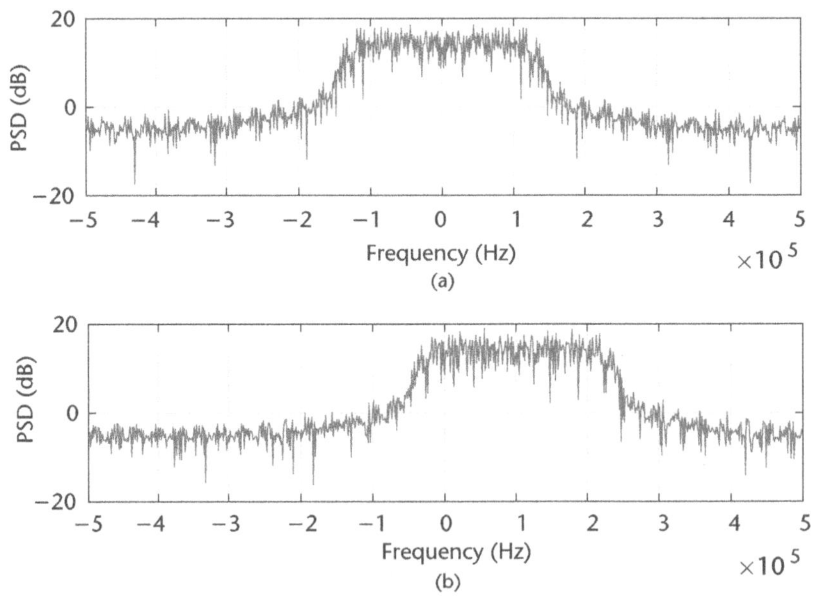

Figure 7.3 Comparison of frequency domain signals with and without frequency offsets. (a) PSD of BPSK signal without frequency offset, and (b) PSD of BPSK signal with 10-kHz offset.

> **Q** Change `filterOversample` in Code 7.2 above and observe the spectrum. Explain what you observe. Next with the original script increase the frequency offset in units of $0.1F_s$, where F_s is the sample rate, from $0.1F_s$ to $1.0F_s$. Explain the observed effect.

frequency offset first and then proceed to manage the remaining synchronization tasks. As discussed in Section 10.3, the oscillator of Pluto SDR is rated at 25 PPM. Transmitting signals in an unlicensed band, such as 2.4 GHz, can produce a maximum offset of 120 kHz between the radios. Since this is quite a large range we will develop a two-stage frequency compensation technique separated into coarse and fine frequency correction. This design is favorable, since it can reduce convergence or locking time for estimation of the relative carrier.

7.2.1 Coarse Frequency Correction

There are two primary categories of coarse frequency correction in the literature: data-aided (DA) and blind correction. DA techniques utilize correlation type structures that use knowledge of the received signal, usually in the form of a preamble, to estimate the carrier offset f_o. Although DA methods can provide accurate estimates, their performance is generally limited by the length of the preambles [3], and as the preamble length is increased this decreases system throughput.

Alternatively, blind or nondata-aided (NDA) methods can operate over the entire duration of the signal. Therefore, it can be argued in a realistic system NDA can outperform DA algorithms. These coarse techniques are typically implemented in an open-loop methodology, for ease of use. Here we will both outline and implement a NDA FFT-based technique for coarse compensation. The concept applied here is straightforward, and based on our initial inspection provided in Figure 7.3, we can provide a rough estimate on the symbols offsets. However, directly taking the peak from the FFT will not be very accurate, especially if the signal is not symmetrical in frequency. To compensate for this fact, we will remove the modulation components of the signal itself by raising the signal to its modulation order M. From our model in (7.2), ignoring noise, we can observe the following:

$$r^M(k) = s^M(k)e^{j(2\pi f_o kT+\theta)M}. \tag{7.5}$$

This will shift the offset to M times its original location and make $s(t)$ purely real or purely complex. Therefore, the $s^M(t)$ term can be ignored and only the remaining exponential or tone will remain. To estimate the position of this tone we will take the FFT of $r^M(t)$ and relate the bin with the most energy to the location of this tone. Figure 7.4 is an example frequency plot of $r^M(t)$ for a BPSK signal generated from the MATLAB Code 7.2 offset by 10 kHz. The peak is clearly visible at twice this frequency as expected. Formally this frequency estimation can be written in a single equation as [4]

$$\hat{f}_o = \frac{1}{2TK}\arg\Big|\sum_{k=0}^{K-1} r^M(k)e^{-j2\pi kT/K}\Big| \tag{7.6}$$

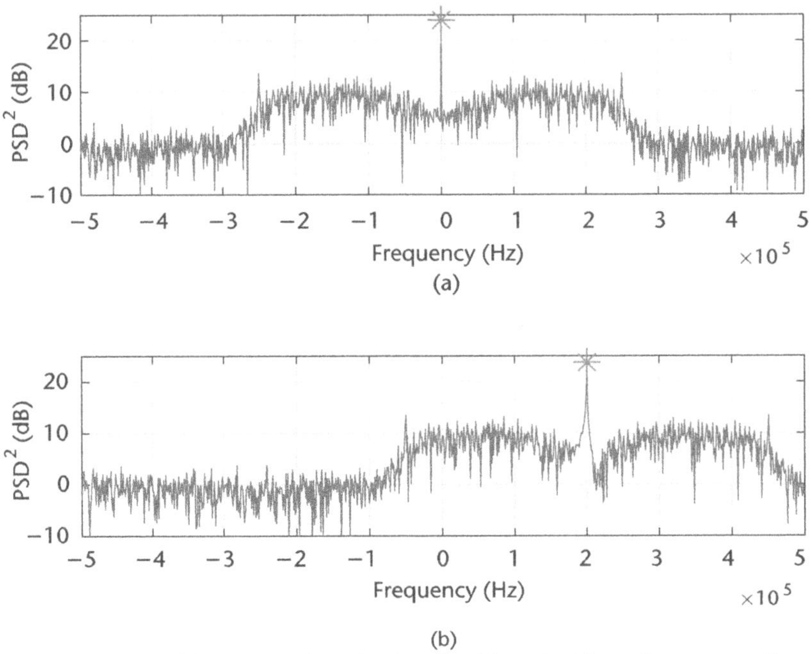

Figure 7.4 Comparison of frequency domain signals with and without frequency offsets. (a) *PSD* of squared BPSK signal without frequency offset, and (b) *PSD* of squared BPSK signal with 10-kHz offset.

where K is the *FFT* length. The estimation in (7.6) is defined as coarse since the resulting $\hat{f}_o$ can only be one of K values produced by the FFT. However, we can extend this accuracy by interpolating across a fixed set of FFT bins over multiple estimates if desired. The frequency resolution of each FFT bin for the signal is simply

$$f_r = \frac{1}{MTK}. \tag{7.7}$$

Therefore, we can increase the performance of our estimator by increasing the FFT size or by decreasing the sample rate of the system. However, do not reduce the sample below the bandwidth of your signal of interest.

> **Q** What is the limitation of this method? (What happens when the M becomes larger?) Finally, add AWGN to the receive signal at different SNR value and examine when the peak become difficult to determine. Provide a plot of peak estimated MSE versus SNR for this analysis.

Implementing this method in MATLAB is straightforward and for efficiency K should alway be the base two number for efficiency of the FFT. In Code 7.3 we produce an estimate for each K samples of data, and compensate for the arrangement of frequencies from the `fft` function. When using this technique we should also consider other aspects of the system or impacts this operation can have. From the perspective of downstream algorithms, they will observe a frequency

Code 7.3 `fftFreqEst.m`

```
%% Estimation of error
fftOrder = 2^10; k = 1;
frequencyRange = linspace(-sampleRateHz/2,sampleRateHz/2,fftOrder);
% Precalculate constants
offsetEstimates = zeros(floor(length(noisyData)/fftOrder),1);
indexToHz = sampleRateHz/(modulationOrder*fftOrder);
for est=1:length(offsetEstimates)
    % Increment indexes
    timeIndex = (k:k+fftOrder-1).';
    k = k + fftOrder;
    % Remove modulation effects
    sigNoMod = offsetData(timeIndex).^modulationOrder;
    % Take FFT and ABS
    freqHist = abs(fft(sigNoMod));
    % Determine most likely offset
    [~,maxInd] = max(freqHist);
    offsetInd = maxInd - 1;
    if maxInd>=fftOrder/2 % Compensate for spectrum shift
       offsetInd = offsetInd - fftOrder;
    end
    % Convert to Hz from normalized frequency index
    offsetEstimates(est) = offsetInd * indexToHz;
end
```

correction every K samples. Ideally $\hat{f}_o$ remains constant, but this is unlikely if the offset is close to an FFT bin boundary. Resulting is frequency jumps in the signal $\pm f_r$ from previous signals. Unfortunately these abrupt changes can disrupt feedback algorithm downstream, which are ill-equipped to deal to sudden shift in frequency or phase of the signal they are estimating/correcting. To combat this we have two main strategies. First, the estimates can be averaged over time with a filter, smoothing out the changes over time. The second option would be to only apply this correction at the start of a frame. Since the offset should be relatively stationary across a reasonably sized frame, a single measurement should be accurate over that duration of time. This correction is also considered coarse since it can only be accurate to within f_r, which only enforces this type of correction interval.

With that said a weakness of this FFT-based technique is that it requires a significant amount of data for a reasonable estimate. This technique will also produce unpure tones when oversampled at the transmitter with transmit filters. However, other techniques such as from Luise [5] are designed for burst-type applications where less data relative to the FFT method above is required. Unfortunately, the Luise method is a biased estimator unlike the FFT method.

7.2.2 Fine Frequency Correction

After coarse frequency correction (CFC) there will still be offset based on the configured resolution chosen f_r. Fine frequency correction (FFC), also called carrier phase correction, should produce a stable constellation for eventual demodulation. Essentially this will drive the remaining frequency offset of the received signal to zero. We can describe this correction as producing a stable constellation due to how fine frequency offset effects are typically examined with a constellation diagram. If a

> **Q** Using loopback with Pluto SDR and Code 7.3, measure the frequency estimate's mean squared error as a function of the difference between the center frequencies ($f_{o,max}$) of the transmitter and receiver. Use BSPK signal here and examine f_Δ from 0 – 100 kHz at 1-MHz baseband sampling rate.
> Repeat this, but fix the transmitter to a gain of −30 and take estimates with the receiver in manual gain mode at 10, 30, and 50.

discrete digitally modulated signal exhibits frequency offset, this will cause rotation over time as examined in a constellation diagram. In Figure 7.5 we demonstrate this effect where each number relates a sample's relative occurrence in time, which provides this perspective of rotation. The signal itself is BPSK, causing it to jump across the origin with different source symbols. If a positive frequency offset is applied the rotation will be counterclockwise and clockwise with a negative offset. The rate of the rotation is equal to the frequency offset, which is where our notion of ω (angular frequency) comes from, as previously defined in (7.3).

This offset can also be observed with Pluto SDR in a similar way. In Figure 7.6 we transmitted a BPSK signal in loopback with 1-kHz difference between transmit and receive LOs. We observe a similar rotation as in Figure 7.5 in Figure 7.6(b). In order to correctly visualize this effect we needed to perform timing correction, which was borrowed from Chapter 6. Without timing correction the signal is difficult to interpret from the constellation plot as observed in Figure 7.6(a). Since timing correction was performed in this case before the frequency was corrected, this required use of the Gardner technique as detailed in Section 6.4.1. Unlike CFC,

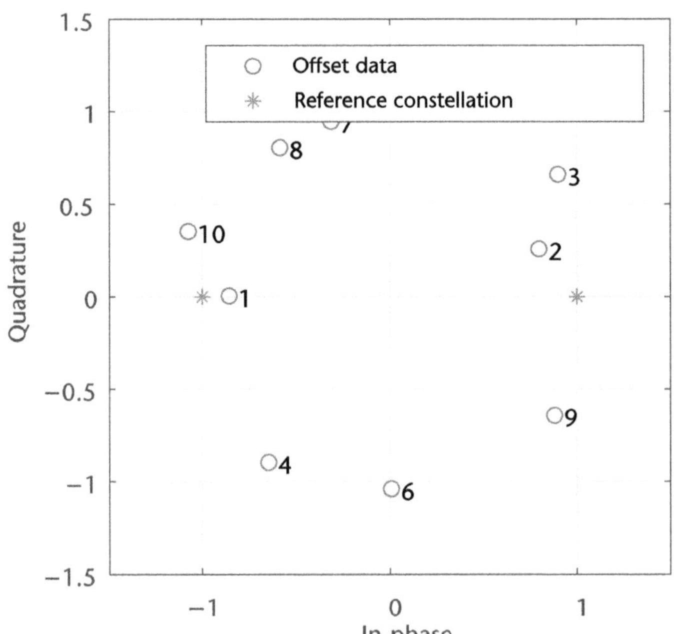

Figure 7.5 Rotating constellation of BPSK source signal with frequency offset.

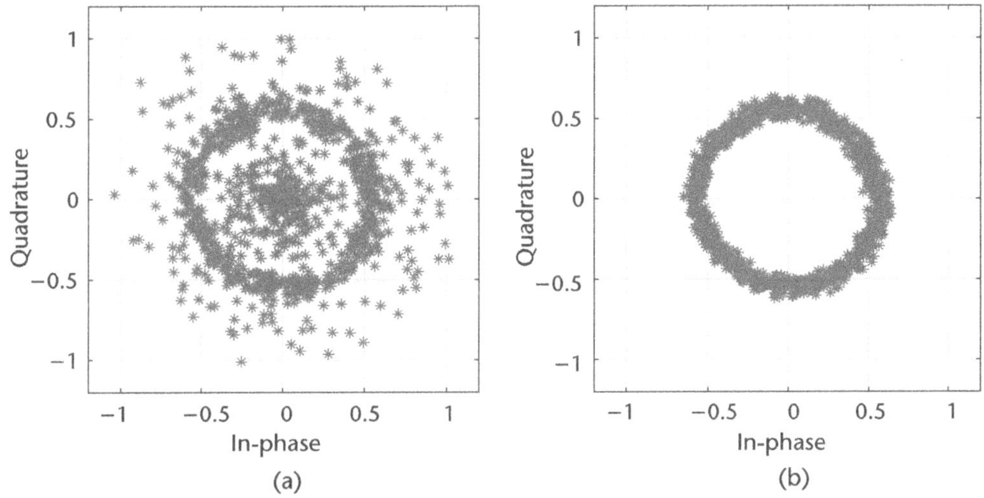

Figure 7.6 BPSK signal transmit through Pluto SDR in loopback with 1-kHz offset at 1 MHz. (a) BPSK signal before timing correction, and (b) BPSK signal after timing correction.

which uses a feedforward technique, for FFC we will utilize a feedback or closed-loop method based PLL theory as examined in Chapter 4. The structure of this algorithm is provided in Figure 6.11 derived from [6, Chapter 7], which relates back our original outline in Figure 6.11.

This all-digital PLL-based algorithm works by first measuring the phase offset of a received sample in the phase error detector (PED), which we call the error signal $e(n)$. The PED is designed based on the structure of the desired receive constellation/symbols. Next, the loop filter helps govern the dynamics of the overall PLL. The loop filter can determine operational frequency (sometimes called pull-in range), lock time, and responsiveness of the PLL, as well as smoothing out the error signal. Finally, we have the direct digital synthesizer (DDS), whose name is a remnant of analog PLL designs with voltage-controlled oscillators (VCOs). The DDS is responsible for generation of the correction signal for the input, which again will be fed back into the system. In the case of the FFC design, this PLL should eventually produce an output signal with desired offset equal to zero.

Starting with the PED, the goal of this block is simply to measure the phase or radial offset of the input complex data from a desired reference constellation. By reference and by extension $e(n)$, we are actually referring to the distance from the constellation bases. In the case of QAM, PSK, and PAM these will always be the real and imaginary axes. However, you may need to extend this perspective with regard to FSK or other modulation schemes. The primary reasoning behind this idea is that it will remove the scaling aspect in a specific dimension, and instead consider the ratio of energy or amplitude in a specific basis. To better understand this concept let us consider QPSK, which has the following PED equation:

$$e(n) = sign(Re(y(n))) \times Im(y(n)) - sign(Im(y(n))) \times Re(y(n)). \tag{7.8}$$

In (7.8) $e(n)$ is essentially measuring the difference between the real and imaginary portions of $y(n)$, and will only be zero when $Re(y(n)) = Im(y(n))$. You will notice that this will force the output constellation only to a specific orientation, but not a specific norm value. However, if $y(n)$ requires a different orientation this can be accomplished after passing through the synchronizer with a simple multiply with the desired phase shift of ϕ_{POST} as

$$y_{SHIFT}(n) = y(n)e^{j*\phi_{POST}}. \tag{7.9}$$

Note that $y_{SHIFT}(n)$ should not be fed into the PED.

In the case of other modulation schemes the PED error estimation will change based on the desired signal arrangement. BPSK or PAM for example will have the following error estimation:

$$e(n) = sign(Re(y(n))) \times Im(y(n)). \tag{7.10}$$

This PED error function again has the same goal of providing the error signal only for the orientation of $y(n)$. For (7.10) $e(n)$ will only be zero when $y(n)$ is purely real.

The reasoning behind (7.8) and (7.10) is straightforward. On the other hand, the loop filter in all PLL designs is the most challenging aspect, but it provides the most control over the adaptation of the system. Again here we will use a PI filter as our loop filter, which was detailed in Section 6.3.1. The last piece to this FFC synchronizer is the DDS, which is just an integrator. Since the loop filter produces a control signal, which is equivalent to the frequency of the input signal, it becomes necessary to extract the phase of this signal instead. The transfer functions used for the integrator here are

$$D(s) = G_3\frac{1}{s} \quad \rightarrow \quad D(z) = G_3\frac{z^{-1}}{1 - z^{-1}}. \tag{7.11}$$

Note that we have added an additional delay of a single sample in the discrete domain, and since we are producing a correction signal $G_3 = -1$. Again this integrator can be implemented with a biquad filter.

In this arrangement of the PLL shown in Figure 7.7, the system should produce an output $y(n)$, which has minimal phase and frequency offsets. Going around the loop again in Figure 7.7, the PED will first produce an error equal to the phase offset associated with the observed corrected[1] symbol $y(n)$, then the loop filter will relate this observed error and weight it against all previous errors. Finally, the DDS will convert the weighted error/control signal $f(n)$ to a phase $\phi(n)$, which we use to correct the next input sample $x(n + 1)$. In the case of frequency offsets, ϕ will continuously change since is it a phase value, not a frequency value. However, if the input signal is too dynamic or the gains of the filters are not set appropriately, the PLL will not be able to keep up with the changing phase (frequency) of x.

For the calculation of the gain values (G_1, G_2) of the loop filter, utilize the following equations based on a preferred damping factor ζ and loop bandwidth B_{Loop}:

$$\theta = \frac{B_{Loop}}{M(\zeta + 0.25/\zeta)} \qquad \Delta = 1 + 2\zeta\theta + \theta^2 \tag{7.12}$$

1. We define this as a corrected symbol since it has passed through the rotator and we will not apply additional phase shifts to this sample. This is also the output of the PLL.

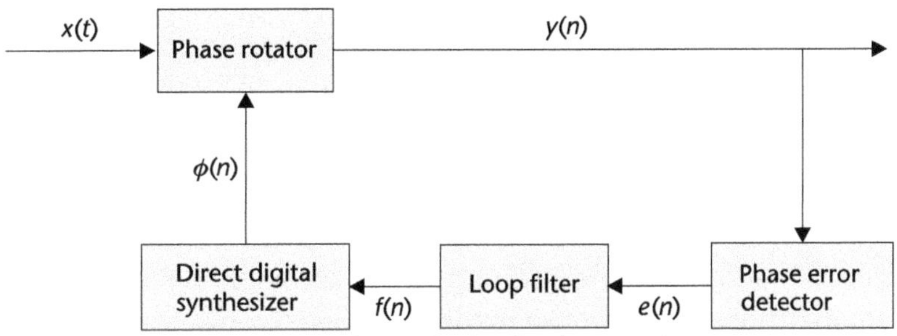

Figure 7.7 FFC structure based on PLL design for feedback corrections.

$$G_1 = \frac{4\zeta\theta/\Delta}{MK} \qquad G_2 = \frac{(4/M)\theta^2/\Delta}{MK} \tag{7.13}$$

where M is the number of sample per symbol and K is the detector gain. For QPSK and rectangular QAM $K = 2$, but for PAM and PSK $K = 1$. Note that B_{Loop} is a normalized frequency. If you are interested in how these are derived, consult [6, Appendix C] for a full detailed analysis. For the selection of ζ refer back to Section 6.3.1, which has the same definition here. The selection of B_{Loop} should be related to the maximum estimated normalized frequency locking range $\Delta_{f,lock}$ range desired:

$$\Delta_{f,pull} \sim 2\pi\sqrt{2}\zeta B_{Loop}. \tag{7.14}$$

Note that this value is an estimate based off a linearized model of the PLL. Therefore inconsistencies may exist in the simulated versions. However, this PLL design should perform well even under strong noise conditions when configured correctly. Unlike the CFC correction this FFC will generally not have the same operational range. In your designs, it may be useful to start with a damping factor of $\zeta = 1$ and a loop bandwidth of $B_{Loop} = 0.01$. From experience, using an overdamped system here is preferable since it directly increases the pull-in range. However, it will take the loop longer to converge.

> **Q** Starting from Code 7.4 implement a carrier recovery algorithm for BPSK. Tune this implementation for a normalized frequency offset of 0.001 and 0.004. Evaluate these implementations over a range of SNR for the MSE of their frequency estimates.

We now have all the necessary pieces to implement the FFC synchronizer, for which we provide a full reference implementation in Code 7.4. However, it is important to discuss some of the design considerations. First, we have stated the output of the FFC synchronizer can have a target of a specific orientation of the output constellation, which is solely determined by the PED. However, the synchronizer may not always be able to achieve this target constellation orientation, meaning the constellation may appear slightly rotated or appear at multiples of the expected position. This will result from signals with larger carrier offsets than the FFC was configured to handle or most notably when the system is configured in an underdamped way. Alternatively, if the received signal has poor

SNR this will also degrade the effective pull-in-range of the synchronize of cause a nondesirable lock position. This is illustrated in Code 7.4, where two different ζ configurations are used. The system is also driven close to the estimated maximum offset for the configuration. In general these estimates will be conservative and will require empirical testing for a specific modulation scheme, SNR, and loop filter configuration. However, in this case if we examine the converged signals in Figure 7.8 we notice an interesting set of outcomes. In Figure 7.8(b) the constellation actually converges to a false minimum. This is a result of the dynamics of the PLL, which is in an underdamped state. Forcing the system to be more rigid will provide the correct result as in Figure 7.8(a). However, if ζ is too large the synchronize will not converge or can take a very long time to do so.

> **Q** Introduce timing offset into the model for Code 7.4. For the recovery process take your implementation from Chapter 4 for timing recovery and place this into the system. Evaluate these implementations over a range of SNR for the MSE of their frequency estimates.

When implementing and testing your own system it can be useful to actually measure the frequency estimation of the synchronizer itself. Since we know that the output of the DDS ϕ is the instantaneous phase correction needed for the next symbol, we can simply apply (7.3) with similar computations as in Code 7.1. From the angular frequency estimates we can translate this to a more tangible frequency estimate in hertz as in (7.3). From inspecting the derivative of `Phase` (ϕ) for Code 7.4 we can examine the convergence of the estimate for an offset of 20 Hz with $f_s = 1000$ Hz. In Figure 7.9 we plot f_{est} where there is an obvious convergence around the correct value. However, since the signal contains noise and there is inherent noise to the PLL, the estimate will not be static. This is useful in a real system since the offsets between transmitter and receiver LOs will always be dynamic with respect to one another.

7.2.3 Performance Analysis

To evaluate the synchronization performance a number of variables can be considered. These include but are not limited to lock time, effective pull-in range, and converged error vector magnitude (EVM). These metrics should be balanced in a way that meets the needs for a specific design since they will clash with one another. For example, it can be simple to design a system with a fast lock time, but it will probably have limited pull-in range. This is a direct relation to (7.14) and a secondary measurement from [6, Appendix C], which defines the normalized frequency lock delay:

$$t_{\Delta,Max} \sim \frac{32\zeta^2}{B_{Loop}}. \tag{7.15}$$

We can demonstrate this trade-off between ζ and B_{Loop} if we focus on the error signal directly from the PED. Modifying the code from 7.4 by fixing the normalized carrier offset to 0.01, $\zeta = 1.3$, and selecting two different values for B_{Loop} we

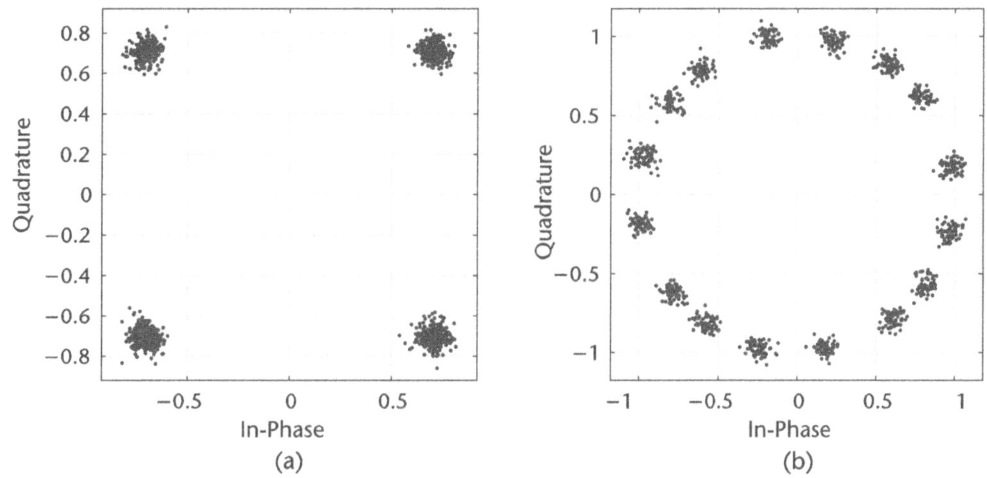

Figure 7.8 Converged QPSK signals after carrier recovery with different damping factors, both (a) overdamped ($\zeta = 1.3$), and (b) underdamped ($\zeta = 0.9$).

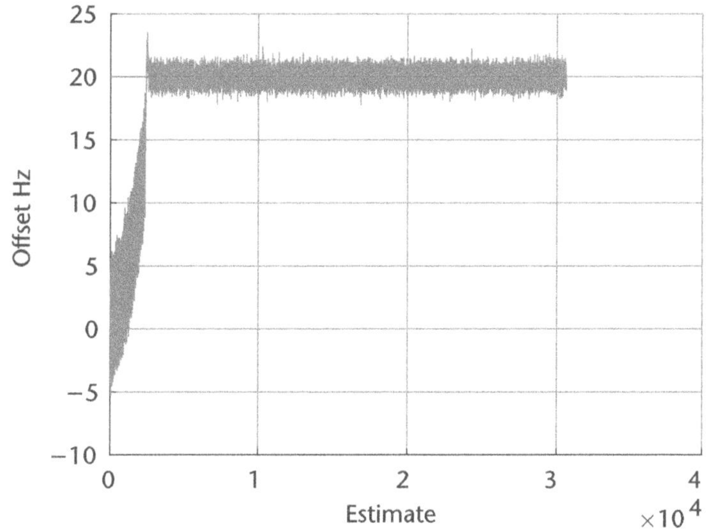

Figure 7.9 Estimations over time and eventual convergence of implemented FFC for 20-Hz offset.

> **Q** Based off your existing simulation solutions that recover signals with both timing and carrier offset, introduce Pluto SDR as the channel mechanism. It may be useful to start with Code 6.1 and 7.4. Evaluate your implementation in loopback with increasing frequency difference between transmit and receive LOs.

can observe $e(n)$ in Figure 7.10. In both configurations of B_{Loop} the normalized offset is less than $\Delta_{f,pull}$. In the case for $B_{Loop} = 0.24$, the system converges to a solution within a few tens of samples, while the $B_{Loop} = 0.03$ case is an order of magnitude slower. However, the variance of the converged error signal σ_e^2 is three times smaller for the case when $B_{Loop} = 0.03$. This error will appear as phase noise on $y(n)$, which will affect the demodulation correctness of the signal.

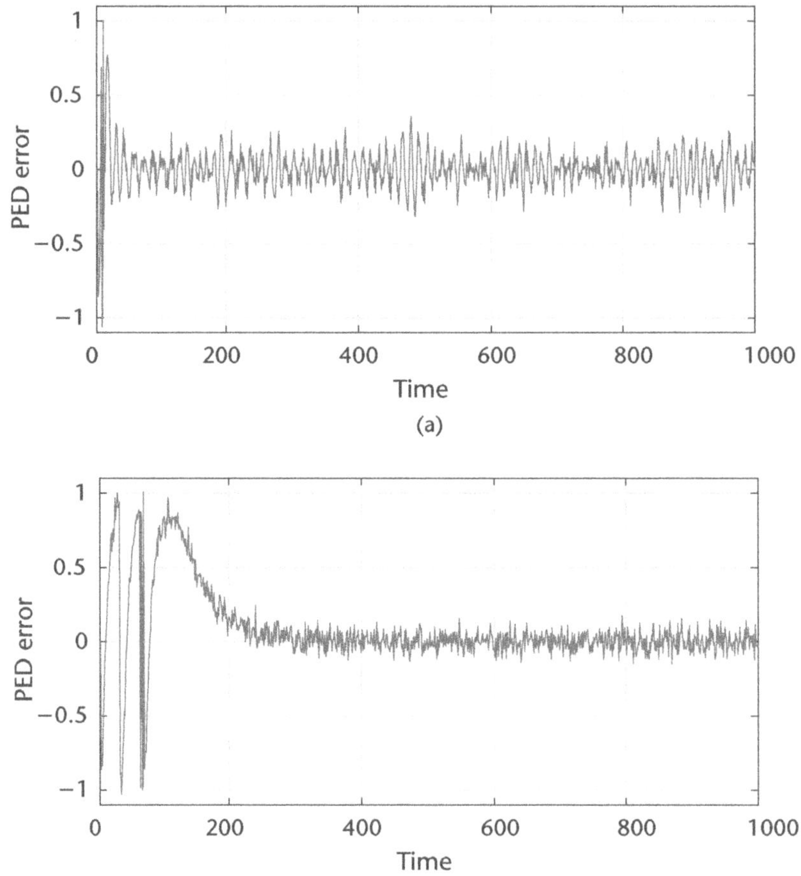

(b)

Figure 7.10 Error signal from QPSK PED for different loop bandwidth for time. (a) $B_{Loop} = 0.24$ with $\sigma_e^2 = 0.0103$ after convergence, and (b) $B_{Loop} = 0.03$ with $\sigma_e^2 = 0.0031$ after convergence.

7.2.4 Error Vector Magnitude Measurements

Evaluating the EVM for $y(n)$ will provide a measure of this phase noise in the recovered signal. EVM is a very useful measurement to understand the algorithmic performance in the system. EVM measures the residual error of the constellation with respect to a reference position. To calculate EVM in percent RMS we can use the following equation:

$$EVM_{RMS} = 100 \times \sqrt{\sum_{k=0}^{N-1} e_{const}(k) \sum_{k=0}^{N-1} (Re(\bar{y}(k))^2 + Im(\bar{y}(k))^2)}, \tag{7.16}$$

where

$$e_{const}(k) = (Re(y(k)) - Re(\bar{y}(k)))^2 + (Im(y(k)) - Im(\bar{y}(k)))^2 \tag{7.17}$$

and $\bar{y}(k)$ is the reference symbol for $y(k)$. EVM is a measure on the dispersiveness of the received signal. Therefore, the lower the EVM values the better. In some situations it can be useful to calculate EVM in decibels, which can be converted

Code 7.4 badlock.m

```
%% General system details
sampleRateHz = 1e6; samplesPerSymbol = 1; frameSize = 2^10;
numFrames = 10; nSamples = numFrames*frameSize;
DampingFactors = [0.9,1.3]; NormalizedLoopBandwidth = 0.09;
%% Generate symbols
order = 4; data = pskmod(randi([0 order-1], nSamples, 1),order,0); % QPSK
%% Configure LF and PI
LoopFilter = dsp.IIRFilter('Structure', 'Direct form II transposed', ...
    'Numerator', [1 0], 'Denominator', [1 -1]);
Integrator = dsp.IIRFilter('Structure', 'Direct form II transposed', ...
    'Numerator', [0 1], 'Denominator', [1 -1]);
for DampingFactor = DampingFactors
    %% Calculate range estimates
    NormalizedPullInRange = min(1, 2*pi*sqrt(2)*DampingFactor*...
        NormalizedLoopBandwidth);
    MaxFrequencyLockDelay = (4*NormalizedPullInRange^2)/...
        (NormalizedLoopBandwidth)^3;
    MaxPhaseLockDelay = 1.3/(NormalizedLoopBandwidth);
    %% Impairments
    frequencyOffsetHz = sampleRateHz*(NormalizedPullInRange);
    snr = 25; noisyData = awgn(data,snr);% Add noise
    % Add frequency offset to baseband signal
    freqShift=exp(1i.*2*pi*frequencyOffsetHz./sampleRateHz*(1:nSamples)).';
    offsetData = noisyData.*freqShift;
    %% Calculate coefficients for FFC
    PhaseRecoveryLoopBandwidth = NormalizedLoopBandwidth*samplesPerSymbol;
    PhaseRecoveryGain = samplesPerSymbol;
    PhaseErrorDetectorGain = log2(order); DigitalSynthesizerGain = -1;
    theta = PhaseRecoveryLoopBandwidth/...
        ((DampingFactor + 0.25/DampingFactor)*samplesPerSymbol);
    delta = 1 + 2*DampingFactor*theta + theta*theta;
    % G1
    ProportionalGain = (4*DampingFactor*theta/delta)/...
        (PhaseErrorDetectorGain*PhaseRecoveryGain);
    % G3
    IntegratorGain = (4/samplesPerSymbol*theta*theta/delta)/...
        (PhaseErrorDetectorGain*PhaseRecoveryGain);
    %% Correct carrier offset
    output = zeros(size(offsetData));
    Phase = 0; previousSample = complex(0);
    LoopFilter.release();Integrator.release();
    for k = 1:length(offsetData)-1
        % Complex phase shift
        output(k) = offsetData(k+1)*exp(1i*Phase);
        % PED
        phErr = sign(real(previousSample)).*imag(previousSample)...
            - sign(imag(previousSample)).*real(previousSample);
        % Loop Filter
        loopFiltOut = step(LoopFilter,phErr*IntegratorGain);
        % Direct Digital Synthesizer
        DDSOut = step(Integrator,phErr*ProportionalGain + loopFiltOut);
        Phase =  DigitalSynthesizerGain * DDSOut;
        previousSample = output(k);
    end
    scatterplot(output(end-1024:end-10));title('');
end
```

from (7.16) as

$$EVM_{dB} = 20\,log_{10}\Big(\frac{EVM_{RMS}}{100}\Big). \tag{7.18}$$

Calculating EVM in decibels is very common in OFDM standards due to high-order constellations that can be transmitting, which require a significant EVM margin to recover. For convience, the Communications System Toolbox include a system object called `comm.EVM` to provide these calculations for us.

> **Q** Starting with Code 7.4, evaluate the EVM of converged signals with regard to ζ and B_{Loop}. Select values of ζ in underdamped, overdamped, and critically damped configurations.

7.3 Phase Ambiguity

The last topic to consider for carrier synchronization is phase ambiguity. Phase ambiguity arises from the fact that the FFC synchronizer outlined here is blind to the true orientation of the transmitted signal. For a given symmetrical modulation scheme there can be a number of convergent orientations, which can be related to the modulation order. For example, PAM will have two possible orientations, QPSK and rectangular QAM will have four, while MPSK with have M possible orientations. However, there are a number of solutions to compensate for this problem, which includes code words, use of an equalizer with training data, and differential encoding. There are different use cases for each implementation.

7.3.1 Code Words

The use of code words is a common practice for resolution of phase ambiguity, which relies on a known sequence in the received data. This is typically just the preamble itself, which will exist in each frame and is known at the receiver. This strategy can be used before or after demodulation if desired. If performed post demodulation, the output bits must be remapped onto their true positions. This process is best explained through an example. Consider the source words w and associated QPSK symbols s:

$$w = [1,\ 0,\ 3] \qquad s = [(-1, 1i)\,(1, 1i)\,(-1, -1i)]. \tag{7.19}$$

The possible received symbols would be

$$\begin{aligned} s_1 &= [(-1, 1i)\,(1, 1i)\,(-1, -1i)] \\ s_2 &= [(-1, -1i)\,(-1, 1i)\,(1, -1i)] \\ s_3 &= [(1, -1i)\,(-1, -1i)\,(1, 1i)] \\ s_4 &= [(1, 1i)\,(1, -1i)\,(-1, 1i)]. \end{aligned} \tag{7.20}$$

Demodulating each code word symbol and comparing with the expected result would provide the necessary mapping to correctly demodulate the remaining data symbols. For an implementation it would be useful to demodulate all the preamble symbols and take the most common orientation mapping, since relying on a single symbol can be error-prone.

Alternatively, the phase offset θ_p from the correct orientation can be measured directly where p is the received preamble symbols, p_r is the reference or true preamble symbols, and the correction required is simply

$$\theta_p = tan^{-1}\left(\sum_n \frac{Im(p(n)^* \times p_r(n))}{Re(p(n)^* \times p_r(n))}\right), \tag{7.21}$$

assuming $p(n)$ has a desirable orientation. Then the remaining signal y would be corrected as

$$y_c = y\, e^{-j\theta_p}. \tag{7.22}$$

7.3.2 Differential Encoding

The second option to deal with phase ambiquity is to differentially encode the source bits themselves. The goal here is to make the true data dependent on the difference between successive bits, not on the received bits themselves. To encode the source data we apply the following at the transmitter:

$$b_t(n) = b_t(n-1) \oplus b(n), \tag{7.23}$$

where b_t are the transmitted encoded bits, b are the uncoded bits, and $\oplus$ is a modulo two addition. To decode the signal we basically perform (7.23) in reverse as

$$b(n) = b_t(n) \oplus b_t(n-1). \tag{7.24}$$

Usually in this scheme the first bit is ignored since it is only based on itself, not the difference between two consecutive bits. Engineers may point to this as wasteful, but this reduces any complex mathematics associated with measuring offsets with symbols and only requires bit-level operations. This can also reduces the bit error rate of a received signal due to propagation of bit errors.

7.3.3 Equalizers

The third popular option is to rely on an equalizer to correct this ambiguity for the system. Using training data the equalizer can learn and correct for this phase shift, which in essence is just a complex multiplication. Equalizers will be discussed in detail in Chapter 9. However, this is a small task for an equalizer implementation if channel correct or synchronization are not performed by the equalizer as well.

7.4 Chapter Summary

In this chapter we have discussed and provided a model of carrier offset and how it relates to receiver operations. From this model we have provided two schemes for compensating for carrier offset including coarse and fine recovery algorithms. However, other implementations do exist that can jointly perform timing and carrier recovery [6] if desired. We have examined how these algorithms can be used at the system level, as well as how individual performance can be evaluated. These include characterization of their parameterization as well as metric on the recovered data.

In summary, carrier offset compensation is a necessary synchronization technique when transmitting data between two disjoint nodes with independent LOs.

References

[1] Analog Devices, Inc., ADALM-PLUTO SDR Active Learning Module, http://www.analog.com/media/en/news-marketing-collateral/product-highlight/ADALM-PLUTO-Product-Highlight.pdf.

[2] Oppenheim, A.V., and R.W. Schafer, *Discrete-Time Signal Processing*, Prentice Hall, 1989.

[3] Morelli, M., and U. Mengali, "Feedforward Frequency Estimation for PSK: A Tutorial Review," *European Transactions on Telecommunications*, Vol. 9, No. 2, 1998, pp. 103–116.

[4] Wang, Y., K. Shi, and E. Serpedin,"Non-Data-Aided Feedforward Carrier Frequency Offset Estimators for QAM Constellations: A Nonlinear Least-Squares Approach," *EURASIP Journal on Advances in Signal Processing*, (2004) 2004: 856139, https://doi.org/10.1155/S1110865704403175.

[5] Luise, M. and R. Reggiannini, "Carrier Frequency Recovery in All-Digital Modems for Burst-Mode Transmissions," *IEEE Transactions on Communications*, Vol. 43, No. 2, 1995, pp. 1169–1178.

[6] Rice, M., *Digital Communications: A Discrete-Time Approach*, Third Edition, Pearson/Prentice Hall, 2009.

CHAPTER 8

Frame Synchronization and Channel Coding

In this chapter we will cover the topics of frame synchronization and channel coding. First, frame synchronization will be discussed to complete our full reception of data from the transmitter. As in Chapter 7, which required timing recovery to realize, for frame synchronization to be accomplished, it requires that the signal has been timing and frequency corrected. However, once frame synchronization has been completed we can fully decode data over our wireless link. Once this has been accomplished, we can move on toward channel coding, where we will discuss popular coding techniques and some of the implementation details from the perspective of a system integrator.

With regard to our receiver outline in Figure 8.1, this chapter will address the second-to-last block, Frame Sync, which is highlighted.

8.1 O Frame, Where Art Thou?

In previous chapters we have discussed frequency correction, timing compensation, and matched filtering. The final aspect of synchronization is frame synchronization. At this point it is assumed that the available samples represent single symbols and are corrected for timing, frequency, and phase offsets. However, since in a realistic system the start of a frame will still be unknown, we need to perform an additional correction or estimation. We demonstrate this issue visually in Figure 8.2, which contains a sample synchronized frame with an unknown offset of p samples. Mathematically, this is simply an unknown delay in our signal y:

$$u[n] = y[n - p], \tag{8.1}$$

where $p \in \mathbb{Z}$. Once we have an estimate $\hat{p}$ we can extract data from the desired frame, demodulated to bits, and perform any additional channel decoding or source decode originally applied to the signal. There are various way to accomplish this estimation but the implemented outline in this chapter is based around cross-correlation.

Depending on the receiver structure and waveform it may be possible to perform frame synchronization after demodulation, where we mark the start of a frame with a specific sequence of bits. However, this cannot be used if symbols are required downstream for an equalizer or if the preamble contains configuration parameters for downstream modulation. This is the case in IEEE 802.11 [1], where the preamble can have a different modulation than the payload. Alternatively, if the system is packet-based and does not continuously transmit data it can be difficult

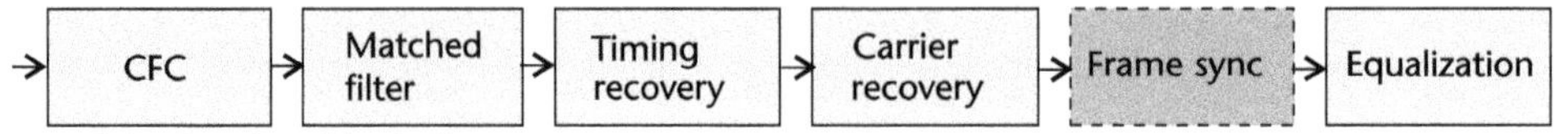

Figure 8.1 Receiver block diagram.

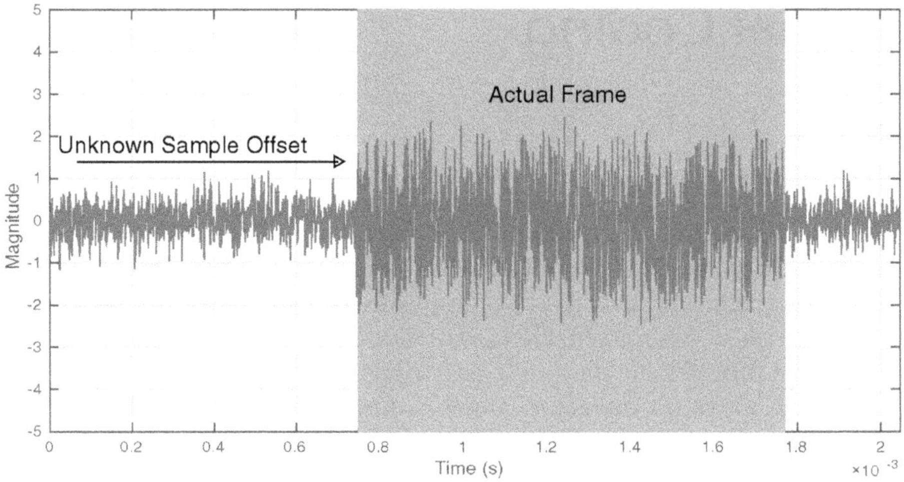

Figure 8.2 Example received frame in AWGN with an unknown sample offset.

to distinguish noise from actual received signal. However, if only bits are consult the relative gain of the signal is removed, which is information that is useful when determining if signal is present.

8.2 Frame Synchronization

The common method of determining the start of a given frame is with the use of markers, even in wired networking. However, in the case of wireless signals, this problem becomes more difficult, as made visible in Figure 8.2, which actually uses a marker. Due to the high degree of noise content in the signal, specifically designed preamble sequences are appended to frames before modulation. Such sequences are typically known exactly at the receiver and have certain qualities that make frame estimation accurate. In Figure 8.3 we outline a typical frame ordering containing preamble, header, and payload data. Header and payloads are unknown are the receiver, but will maintain some structure so they can be decoded correctly.

Before we discuss the typical sequences utilized we will introduce a technique for estimation of the start of a known sequence starting at an unknown sample in time. Let us consider a set of N different binary sequences b_n, where $n \in [1, ..., N]$, each of length L. Given an additional binary sequence d, we want to determine how similar d is to the existing N sequences. The use of a cross correlation would provide us the appropriate estimate, which we perform as

$$C_{d,b}(k) = \sum_m d^*(m) b_n(m + k), \tag{8.2}$$

which is identical to a convolution without a time reversal on the second term. When $d = b_n$ for a given n, $C_{d,b}$ will be maximized compared with the other $n - 1$

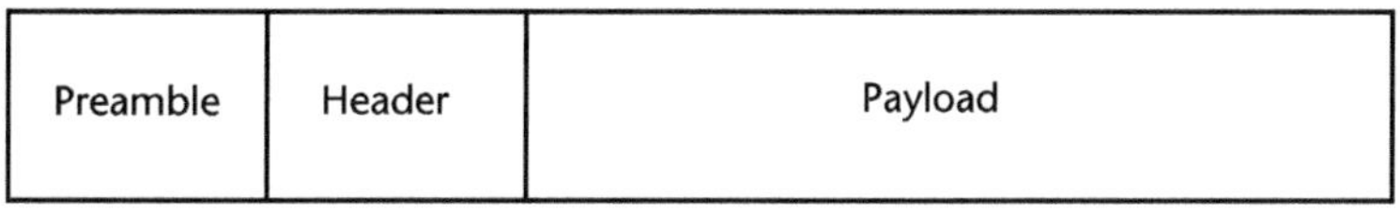

Figure 8.3 Common frame structure of wireless packet with preamble followed by header and payload data.

sequences, and produce a peak at L^{th} index at least. We can use this concept to help build our frame start estimator, which as discussed will contain a known sequence called the preamble.

Common sequences utilized in preambles for narrowband communications are Barker codes [2]. Barker codes are utilized since they have unique autocorrelation properties that have minimal or ideal off-peak correlation. Specifically, such codes or sequences $a(i)$ have autocorrelation functions defined as

$$c(k) = \sum_{i=1}^{N-k} a(i)a(i+k), \tag{8.3}$$

such that

$$|c(v)| \leq 1, \quad 1 \leq v < N. \tag{8.4}$$

However, only nine sequences are known $N \in [1, 2, 3, 4, 5, 7, 11, 13]$, provided in Table 8.1. We provide a visualization of these autocorrelations in Figure 8.5 for a select set of lengths. As the sequence becomes longer the central peak becomes more pronounced. For communication systems we typically append multiple such codes together to produce longer sequences for better performance, as well as for other identifications.

Using these codes we have implemented a small example to show how a Barker sequence $a(k)$ can be used to locate sequences in a larger set of data $r(k)$, which we have provided in Code 8.1. In this example we insert a Barker code within a larger random sequence at an unknown position p, similar to our original error model in Section 8.1, shown as Figure 8.4(a). A cross correlation is performed using MATLAB's `xcorr` function, provided in Figure 8.4(b). The cross correlation will be of length $2L_r - 1$, where L_r is the length of r. Since `xcorr` will pad zeros to a so its length is equal to L_r [3], this will result in at least $L_r - L_a$ zeros to appear in the correlation where L_a is the original length of a. From Figure 8.5, we know that the peak will appear at L_a samples from the start of the sequence. Taking this into account we can directly determine at what offset position of our desired sequence:

$$\hat{p} = \underset{k}{\text{argmax}}\ C_{ra}(k) - L_r, \tag{8.5}$$

which is what we observe from our estimates in Figure 8.4(a).

The `xcorr` function is a useful tool in MATLAB and will actually utilize the `fft` function for large sequences for performance. Since we know from Chapter 2 that convolution is just a multiplication in the frequency domain, and from above the relation of correlation and convolution, this strategy is obvious for `xcorr`. However, the process actually inflates the data processed since the sequences must be of equal length for correlation. We can observe inflation from the zeros in

Table 8.1 Barker Codes from `comm.BarkerCode`

N	*Code*
2	−1, +1
3	−1, −1, +1
4	−1, −1, +1, −1
5	−1, −1, −1, +1, −1
7	−1, −1, −1, +1, +1, −1, +1
11	−1, −1, −1, +1, +1, +1, −1, +1, +1, −1, +1
13	−1, −1, −1, −1, −1, +1, +1, −1, −1, +1, −1, +1, −1

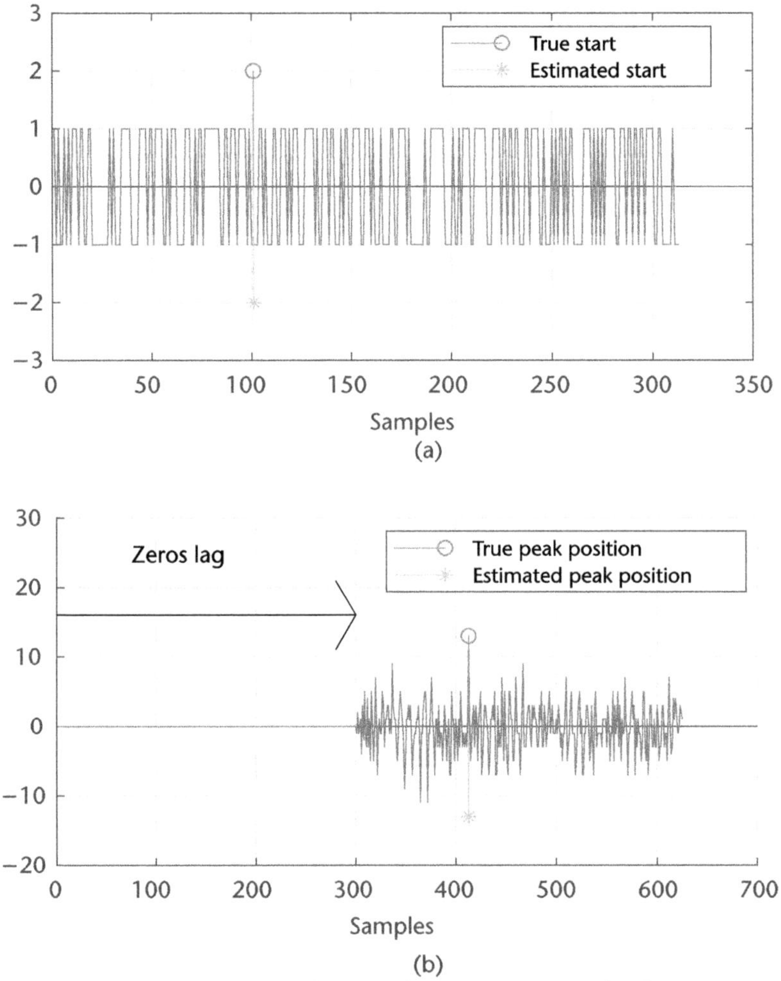

Figure 8.4 Example of using cross correlation to find a sequence with a larger sequence of data. (a) Random bit sequence with Barker code embedded at delay p, and (b) crosscorrelation of Barker code with random sequence containing code.

Figure 8.4(b). A more efficient implementation would be to utilize a filter. The output y of an FIR filter with taps b_i can be written as

$$y[n] = \sum_{i=0}^{N} b_i u[n-i], \tag{8.6}$$

where u is our received signal that contains the sequence of interest. Equation (8.6) is almost identical to (8.2) except for a time reversal. Therefore, to use an FIR filter

Code 8.1 Barker Sequence Example: **barkerBits13.m**

```
% Show Barker Autocorrelations search example
sequenceLength = 13;
hBCode = comm.BarkerCode('Length',7,'SamplesPerFrame', sequenceLength);
seq = hBCode(); gapLen = 100; gapLenEnd = 200;
gen = @(Len) 2*randi([0 1],Len,1)-1;
y = [gen(gapLen); seq; gen(gapLenEnd)];
corr = xcorr(y,seq);
L = length(corr);
[v,i] = max(corr);
% Estimation of peak position
% The correlation sequence should be 2*L-1, where L is the length of the
% longest of the two sequences
%
% The first N-M will be zeros, where N is the length of the long sequence
% and N is the length of the shorter sequence
%
% The peak itself will occur at zero lag, or when they are directly
```

as a cross correlator we could simply replace b_i with the sequence of interest, but in reverse order. This implementation would not require padding of the sequence of interest d, and can be efficiently implemented in hardware.

> **Q** Based on Code 8.1, reimplement the sequence search with an FIR filter.

8.2.1 Signal Detection

Now that we have a method for estimating the start of a frame, let us consider a slightly simpler problem. Can we determine that a frame exists in the correlation? This can be useful if wish to handle data in smaller pieces rather than working with complete frames, or possibly a mechanism for determining if a channel is occupied. When we consider signal detection, we typically define this feature as a power sensitivity or the minimum received power at the receiver to be detected. However, this sensitivity will be based on some source waveform and cannot be generalized in most cases. Therefore, such a value should never be given on its own, unless given with respect to some standard transmission. Even when considering formal methods of detection theory, such as Neyman-Pearson or even Bayesian, you must have some knowledge or reference to the source signal [4]. The receiver sensitivity requirement for IEEE 802.11ac specifically is defined as the minimum received signal power to maintain a packet error rate of 10%, for a give modulation and coding scheme [1].

In a formal mathematic context, this process devolves into a simple binary hypothesis test:

$$\begin{aligned} \mathcal{H}_0 &: \text{no signals}, \\ \mathcal{H}_1 &: \text{signals exist}, \end{aligned} \tag{8.7}$$

where $\mathcal{H}_0$ is usually referred to as a null hypothesis and $\mathcal{H}_1$ is usually called an alternative hypothesis.

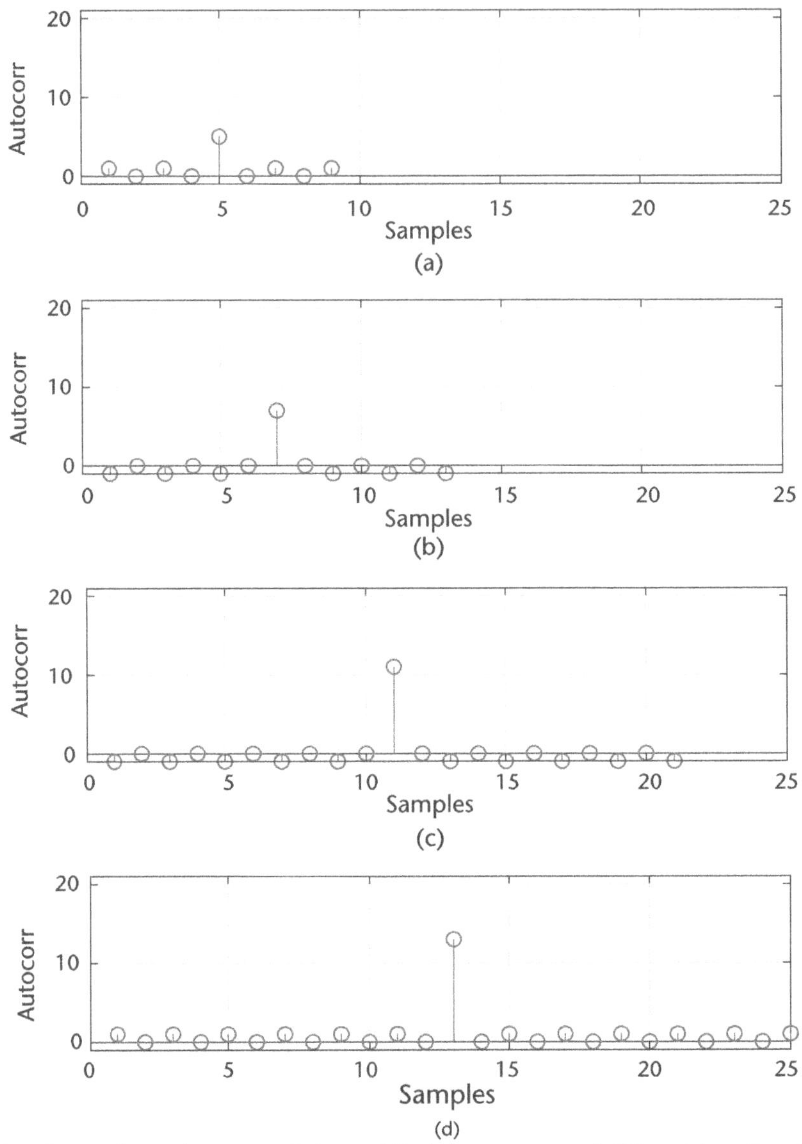

Figure 8.5 Comparison of autocorrelations of Barker sequences of different lengths. (a) $N = 5$, (b) $N = 7$, (c) $N = 11$, and (d) $N = 13$.

For a null hypothesis, since there are no primary signals present, the received signal is just the noise in the RF environment. On the other hand, for the alternative hypothesis, the received signal would be the superposition of the noise and the primary signals. Thus, the two hypotheses in (8.14) can be represented as

$$\begin{aligned} \mathcal{H}_0 &: r[n] = n[n], \\ \mathcal{H}_1 &: r[n] = x[n] + n[n], \end{aligned} \tag{8.8}$$

where $r[n]$ is the received signal, $n[k]$ is the noise in the RF environment, and $x[n]$ is the signal we are trying to detect. Based on the observation r, we need to decide among two possible statistical situations describing the observation, which can be expressed as

$$\delta(x) = \begin{cases} 1 & x \in \Gamma_1, \\ 0 & x \in \Gamma_1^c. \end{cases} \tag{8.9}$$

When the observation x falls inside the region Γ_1, we will choose $\mathcal{H}_1$. However, if the observation falls outside the region Γ_1, we will choose $\mathcal{H}_0$. Therefore, (8.9) is known as *decision rule*, which is a function that maps an observation to an appropriate hypothesis [5]. In the context of packet detection, thresholding is actually the implementation of a decision rule.

Regardless of the precise signal model or detector used, sensing errors are inevitable due to additive noise, limited observations, and the inherent randomness of the observed data [6]. In testing $\mathcal{H}_0$ versus $\mathcal{H}_1$ in (8.14), there are two types of errors that can be made; namely, $\mathcal{H}_0$ can be falsely rejected or $\mathcal{H}_1$ can be falsely rejected [5]. In the first hypothesis, there are actually no signals in the channel, but the testing detects an occupied channel, so this type of error is called a *false alarm* or *Type I error*. In the second hypothesis, there actually exist signals in the channel, but the testing detects only a vacant channel. Thus, we refer to this type of error as a *missed detection* or *Type II error*. Consequently, a false alarm may lead to a potentially poor data recovery, while a missed detection ignores an entire frame of data requiring retransmission [6].

Given these two types of errors, the performance of a detector can be characterized by two parameters; namely, the *probability of false alarm* (P_F), and the *probability of missed detection* (P_M) [7], which correspond to Type I and Type II errors, respectively, and thus can be defined as

$$P_F = P\{\text{Decide } \mathcal{H}_1 | \mathcal{H}_0\}, \tag{8.10}$$

and

$$P_M = P\{\text{Decide } \mathcal{H}_0 | \mathcal{H}_1\}. \tag{8.11}$$

Note that based on P_M, another frequently used parameter is the *probability of detection*, which can be derived as follows:

$$P_D = 1 - P_M = P\{\text{Decide } \mathcal{H}_1 | \mathcal{H}_1\}, \tag{8.12}$$

which characterizes the detector's ability to identify the primary signals in the channel, so P_D is usually referred to as the power of the detector.

As for detectors, we would like their probability of false alarm to be as low as possible, and at the same time, their probability of detection as high as possible. However, in a real-world situation, this is not achievable, because these two parameters are constraining each other. To show their relationship, a plot called *receiver operating characteristic* is usually employed [8], as shown in Figure 8.6, where its x-axis is the probability of false alarm and its y-axis is the probability of detection. From this plot, we observe that as P_D increases, the P_F is also increasing. Such an optimal point that reaches the highest P_D and the lowest P_F does not exist. Therefore, the detection problem is also a trade-off, which depends on how the Type I and Type II errors should be balanced.

When we consider the implementation consequences of detecting a signal, our design become more complicated than for example from Code 8.1. In the most basic sense detection becomes a thresholding problem for our correlator. Therefore,

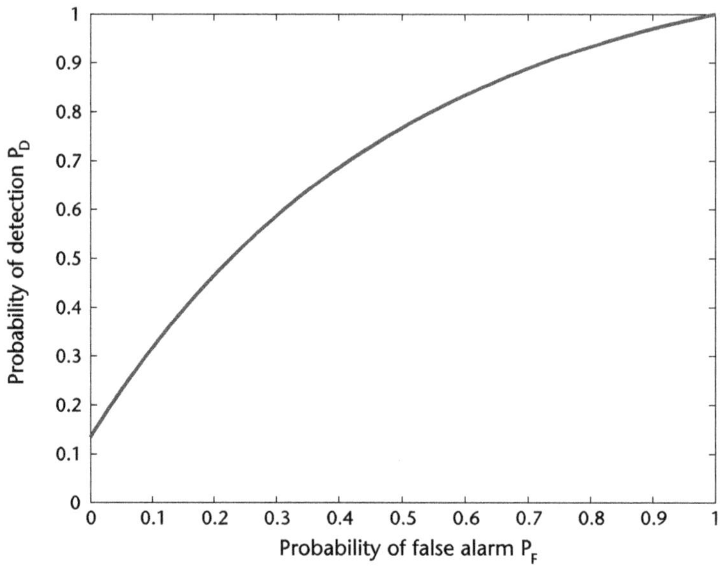

Figure 8.6 A typical receiver operating characteristic, where the x-axis is the probability of false alarm (P_F), and the y-axis is the probability of detection (P_D).

the objective becomes determining a reference or criteria for validating a peak, which can be radically different over time depending on channel noise and the automatic gain control of the Pluto SDR. However, even in simulations appropriate thresholding becomes nontrivial, which we can demonstrate with Figure 8.7(a) and 8.7(b). In these figures the peak appears larger relative to the rest of the correlation in the case where no frame exists in the receive signal compared to the condition when a frame exists. Therefore, for an implementation that performs well it should handle such conditions and operate regardless of the input scaling.

A common technique to aid with this thresholding process is to self-normalize the received signal. If we look back at Figure 8.7, we will notice that the magnitude can be quite different, which makes thresholding even more difficult. If we self-normalize the signal we can force it into a range closely between $\in [0, 1]$. A simple way to accomplish this operation is to scale our cross-correlation metric $C_{y,x}$ by the mean energy of the input signal x. To do this in an efficient way we can again utilize filtering to accomplish this task by implementing a moving average filter. Mathematically, this moving averaging would be modeled as another sum:

$$u_{ma}[n] = \sum_{i=0}^{N} u[n-i], \tag{8.13}$$

where N is the length of the preamble or sequence of interest. A useful aspect of (8.6) and (8.13) is that these are simple FIR filter implementations, making them simple to implement in hardware. In fact (8.13) requires no multiplication like the CIC filter discussed in Section 2.6.4. This aspect is import since in many systems this frame synchronize may also be used as packet detection mechanism at the front of the receiver, requiring it to run at the fastest rate of the input data without decimation. Combining our correlator from (8.6) and scaler from (8.13) we can

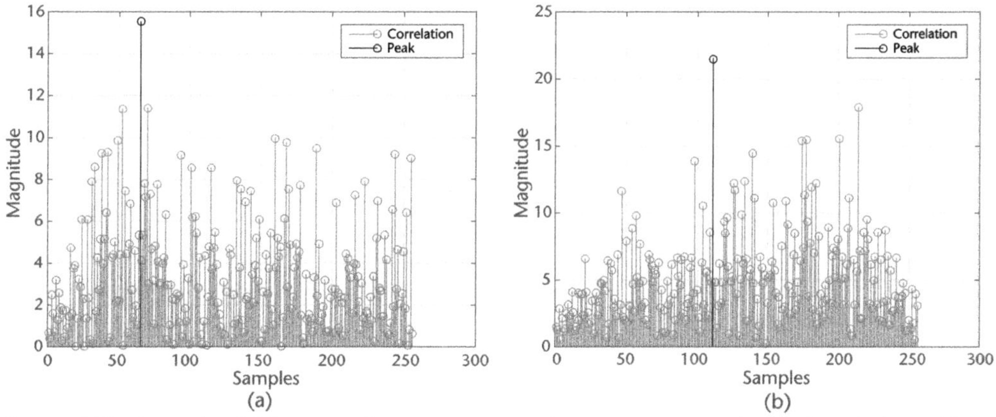

Figure 8.7 Example of false peaks in a cross-correlation sequence search. (a) Correlation without signal present, and (b) Correlation with signal present.

write our detector as

$$\mathcal{H}_0 : \frac{y[n]}{u_{ma}[n]} < T \text{ no signals,}$$
$$\mathcal{H}_1 : \frac{y[n]}{u_{ma}[n]} \geq T \text{ signals exist,} \tag{8.14}$$

where T is our threshold value.

In MATLAB code `lst:findSignalStartTemplate` we have provided a template that nicely compensates for the transmit filter delay in the system, providing the true delay of a given packet at the receiver.

> **Q** From the code provided in 8.2, implement a preamble start estimator using `xcorr` and the `filter` function. Evaluation the estimation accuracy over SNRs $\in [0, 12]$ dB in single dB steps.

8.2.2 Alternative Sequences

Besides Barker sequences, there are other sequences that have similar properties of minimal cross correlation except at specific instances. Two popular options are Zadoff-Chu sequences and Golay complementary sequences, which are currently both part of existing wireless standards.

Zadoff-Chu sequences, named after authors Solomon Zadoff and David Chu [9], are used for LTE synchronization and channel sounding operations. They are useful since they have a constant amplitude, zero circular autocorrelation, and very low correlation between different sequences. This properly of limited correlation between themselves is useful in a multiaccess environment where many users can transmit signals. Mathematically, the sequence numbers are generated as

$$s_n = exp\left(-j\frac{\pi\, k\, n\,(n+1+2q)}{L} \right), \tag{8.15}$$

where L is the sequence length, n the sequence index, q and integer, and k, which is coprime with L. Unlike Barker sequences, which are purely integers 1 or -1, Zadoff-Chu sequences are complex valued.

Code 8.2 Loopback Pluto Example: **`findSignalStartTemplate.m`**

```
%% General system details
sampleRateHz = 1e6; samplesPerSymbol = 8; numFrames = 1e2;
modulationOrder = 2;  filterSymbolSpan = 4;
barkerLength = 26; % Must be even
%% Impairments
snr = 15;
%% Generate symbols and Preamble
bits = randi([0 3], modulationOrder*1e3,1);
hBCode = comm.BarkerCode('Length',7,'SamplesPerFrame', barkerLength/2);
barker = hBCode()>0; frame=[barker;barker;bits];frameSize = length(frame);
% Modulate
modD = comm.DBPSKModulator(); bMod = clone(modD);
modulatedData = modD(frame);
%% Add TX/RX Filters
TxFlt = comm.RaisedCosineTransmitFilter(...
    'OutputSamplesPerSymbol', samplesPerSymbol,...
    'FilterSpanInSymbols', filterSymbolSpan);
RxFlt = comm.RaisedCosineReceiveFilter(...
    'InputSamplesPerSymbol', samplesPerSymbol,...
    'FilterSpanInSymbols', filterSymbolSpan,...
    'DecimationFactor', samplesPerSymbol);
RxFltRef = clone(RxFlt);
%% Setup visualization object(s)
hts1 = dsp.TimeScope('SampleRate', sampleRateHz,'TimeSpan', ...
    frameSize*2/sampleRateHz);
hAP = dsp.ArrayPlot;hAP.YLimits = [-3 35];
%% Demodulator
demod = comm.DBPSKDemodulator;
%% Model of error
BER = zeros(numFrames,1);PER = zeros(numFrames,1);
for k=1:numFrames
    % Insert random delay and append zeros
    delay = randi([0 frameSize-1-TxFlt.FilterSpanInSymbols]);
    delayedSignal = [zeros(delay,1); modulatedData;...
        zeros(frameSize-delay,1)];
    % Filter signal
    filteredTXDataDelayed = TxFlt(delayedSignal);
    % Pass through channel
    noisyData = awgn(filteredTXDataDelayed,snr,'measured')
    % Filter signal
    filteredData = RxFlt(noisyData);
    % Visualize Correlation
    hts1(filteredData);pause(0.1);
    % Remove offset and filter delay
    frameStart = delay + RxFlt.FilterSpanInSymbols + 1;
    frameHatNoPreamble = filteredData(frameStart:frameStart+frameSize-1);
    % Demodulate and check
    dataHat = demod(frameHatNoPreamble);
    demod.release(); % Reset reference
    BER(k) = mean(dataHat-frame);PER(k) = BER(k)>0;
end
% Result
fprintf('PER %2.2fn',mean(PER));
```

The second sequence of interest are Golay complementary sequences, which are currently used in IEEE 802.11ad. Again they are used for channel estimation and synchronization within the preamble of IEEE 802.11ad packets. Golay complementary sequences are sequences of bipolar symbols with minimal autocorrelation properties. Therefore, they have a very similar to concept to Barker codes. However, as the name suggests these sequences come in complementary pairs that are typically denoted as Ga_n and Gb_n, where n is the sequence length. IEEE 802.11ad uses pairs Ga_{32}, Ga_{64}, and Gb_{64}. Using these sequences with BPSK is exceptional since performing the autocorrelations under even severe phase rotation is high. Another important aspect with Golay or specifically Ga and Gb sequence pairs is that their autocorrelation can be performed in parallel in hardware. This is very useful for a standard like 802.11ad, which is targeting transfer rates of 7 Gbits/s [10]. Building on this concept of minimal autocorrelation pairs and parallel processing of sequences, the preamble in IEEE 802.11ad can be used to provide signaling information to the receiver just based on its autocorrelation properties. This means that depending on the packet type a correlator bank can be used to identify that specific structure, conditioning the processing receiver to a specific decoder path for that type of packet.

8.3 Putting the Pieces Together

At this point we have all the necessary pieces to build a wireless receiver that can handle carrier offsets, timing mismatches, and random packet delay. With all these components in our tool belt, now it is a good time to talk about the system as a whole and determine the arrangement of components based on system requirements. This discussion on algorithm arrangements will be based on what we have learned from the previous chapters.

Starting at the front of the receiver we need to first accomplish two goals: carrier offset removal and timing synchronization with the received signal. In the system proposed so far we have first implemented timing recovery in the receive chain, but this requires usage of a TED, which is insensitive to phase rotation. Based on the options provided in Chapter 4, this would require a technique such as Gardner or a polyphase type implementation from Harris [11]. It is possible to utilize the FFC implementation described in Chapter 7 before timing recovery, but there can be residual phase noise left in the system. The receiver would be arranged similar to Figure 8.8. However, it is definitely useful to place a CFO before all processing, even before matched filtering, to reduce the work of other recovery algorithm in the receive chain. With that said, inserting CFO after a matched filter can make the estimates more accurate from CFO since the received signal will be SNR maximized. You must consider the trade-off in all these situations.

In Figure 8.9 we have outlined a possible receiver flow that contains the relative sample rates R_n between the recovery stages. The blocks with dashed outlines, the

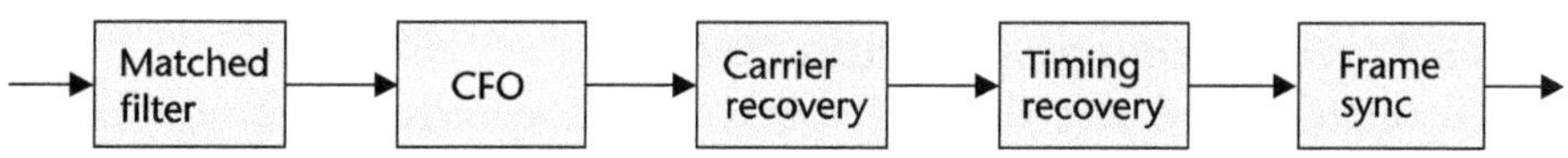

Figure 8.8 Example receiver processing flow to recover transmitted frames where frequency recovery is considered first.

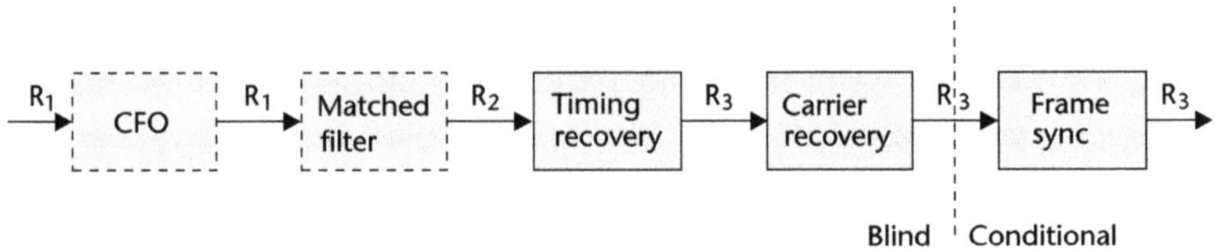

Figure 8.9 Complete receiver processing flow to recover transmitted frames. The relative sample rates are defined by R_n.

matched filter and CFO, can be optional if a RRC filter is used at the transmitter or the carrier offset is not severe. We specifically place the CFO before matched filtering since the matched filter can reduce the bandwidth CFO can utilize. With regard to the rates $R_m \geq R_k$ where $m < k$, meaning that the system will never upsample downstream or run at faster rates. Overall, downsampling will only occur in two possible stages: matched filtering and timing recovery. It is not required to do so in either stage but they can have specific benefits. For example, decimating at the matched filter stage will provide useful averaging, making symbols easier to distinguish. From a hardware perspective decimating reduces the sample rate and the constrains on downstream processing, lowering the bounds on clock rates for intensive operations in our downstream loops. When considering our timing recovery algorithms we already know from Chapter 6 that we can have specific requirements on R_n for the TED utilized. Providing the timing recovery loop with more samples per symbol can provide better performance as well.

In Figure 8.9 between the carrier recovery and frame sync block we have partitioned the system into two domains, blind and conditional. This is to define which processing blocks are dependent on the actual data transmitted and those blocks which are essentially blind to this fact. We can utilize this aspect of the design to introduce training data into our system that can be used to help the system converge to lock states in the recovery loops before actual data needs to be recovered. In this type of configuration we could prepend random bits that would be modulated and filtered as actual data of the transmitter to be sent. This extra or training data could just be prepended to the start of each frame or continuously transmitted in between data frames to keep the receiver locked. This would remove convergence delays in our system. In hardware we could simply connect a linear-feedback shift register (LFSR) to our modulator, which is a convenient mechanism of generating random bits. In MATLAB this is represented by the `comm.PNSequence` System object. Once converged it may not be necessary to continuously transmit training data, which would increase the throughput of the system.

When implementing a system with Pluto SDR or even in simulation it can be helpful to introduce training information into your system. In this configuration our frame sync block from Figure 8.9 would act as a gateway to downstream processing, only allowing data to pass through once a preamble or marker was detected.

8.3.1 Full Recovery with Pluto SDR

Throughout the last three chapters we have introduced templates of code and provided guidance on how to implement scenarios with Pluto SDR. However, when considering full frame recovery and actually demodulation of data we need to reenforce some implementation details. Since the receiver is a complex system requiring many processing components, real-time operation should not be an initial goal. Once your receiver algorithms are working you can extend them to work in real-time if desired. Therefore, for your implementations you should focus on coding templates Code 5.3, 5.4, and 5.5. Performing processing in between calls to Pluto SDR similar to Code 5.6 will likely result in overflows and missed data, making it difficult to recover full frames of data.

A second important tool to utilize is the `transmitRepeat` method of Pluto SDR as in code example 5.7. This will deterministically transmit data continuously. With regard to this data at the receiver, since the delay will be random from the transmitter, you should always collect at least $2L$ samples where L is the length of the desired frame. Setting Pluto SDR's `SamplesPerFrame` property to $2L$ will guarantee at least one full packet received when the transmitter is in `transmitRepeat` mode. This was unnecessary in Chapters 4 and 7 since we could lose data to an overflow and this would have little impact on our tests. However, this is paramount when checking for full frames. Here is a simple example to follow in Code 8.3. In this example we actually step the receive several times to remove possible stale data in its IIO buffers.

Code 8.3 Capture Repeated Frame: **captureExample.m**

```
% Transmit frame repeatedly
tx = sdrtx('Pluto');
tx = sdrtx('Pluto','SamplesPerFrame',length(frame)*2);
tx.transmitRepeat(frame);
for k=1:4,rx();end; % Remove stale data from buffers
rxBuffer = rx();
```

> Using the template from Code 8.3 and the synchronization blocks developed in Chapters 4, 7, and in this chapter, begin to estimate the start of packets. First, utilizing `transmitRepeat` collect $L \times N$ samples of data where $N = 1, 2, 3, 4, 5$. Your packet detector should be able to locate at least $L \times (N - 1)$ of data for each iteration of N. Evaluate the performance of your system for this growing number of packets. Collect 1,000 packets and determine your probability of detection (try to collect 10 packets at a time and repeat this process).

A useful tool when evaluating complete recovered frames is a cyclic redundancy check (CRC) that provides a binary value if a packet contains errors or not. Conveniently, MATLAB's Communication Systems Toolbox contains a system object with this functionality called `comm.CRCGenerator` and its mirrored detector `comm.CRCDetector`. They can be called in the following way in Code 8.4 where we utilize the polynomial $z^3 + 1$ in both objects.

A CRC works by appending a sequence of length L_c, called a checksum, to the end of the our data. L_c will be equal to the order of the polynomial, which is

Code 8.4 Loopback Pluto Example: **crcExample.m**

```
x = logical([1 0 1 1 0 1 0 1 1 1 0 1]');
crcGen = comm.CRCGenerator('z^3 + 1');
crcDet = comm.CRCDetector('z^3 + 1');
codeword = crcGen(x);
codewordWithError = codeword; codewordWithError(1) = ~codewordWithError(1);
[tx, err]    = crcDet(codeword);
[tx1, err1] = crcDet(codewordWithError);
```

three in the case of Code 8.4. This polynomial determines how bits are combined (XORed) together to produce the checksum. The larger the value L_c the lower the probability of Type I or Type II errors, as outlined in Section 8.2.1. However, this is also dependent on the length of the data related to the checksum. In practice, the data related to a checksum size will be orders of magnitude greater than L_c. With regard to our frame synchronization testing we can utilize CRCs in transmitted frames to easily check if we have recovered all our transmitted bits.

> **Q** Again, using the template from Code 8.3, and the synchronization blocks developed in Chapters 4, 7, and in this chapter, begin to estimate the start of packets. This appends CRC to each frame before transmission. At the receiver demodulate the recovered symbols, check the CRC values for 1,000 packets. Repeat this process but calculate the bit error rate for each frame recovered. Skip lost frames.

8.4 Channel Coding

Now that we can successfully recover data across the wireless link, we can discuss techniques of making this process more robust. Channel coding is an obvious option and is ubiquitous in any digital communications standard.

8.4.1 Repetition Coding

One of key building blocks of any communication system is the forward error correction (FEC), where redundant data is added to the transmitted stream to make it more robust to channel errors. There are many types of FEC techniques, such as the *repetition coding* approach, where each transmitted bit is repeated multiple times. In this section, we will explore together one technique for combating the introduction of errors to data transmissions by implementing a simple repetition coder (repetition factor $R = 4$). So what does it mean by a repetition coder with repetition factor $R = 4$? A simple definition is that if a "0" symbol is to be transmitted, this "0" symbol will be repeated four times by the repetition coder, such that the output would be "0000."

Let us first start by double-clicking on the repetition coder block, which will result in a MATLAB function block editor opening, in which we can write customized MATLAB code. As mentioned previously, setting break points is a great way for understanding and debugging M-files. For more information about break points and how they can be used to debug and evaluate the code, please refer to Appendix B.

> ⓘ The *repmat* function in MATLAB can be used to realize a simplistic repetition coding scheme. For example, to repeat a vector *u* for 4 times, the following expression can obtain this result:
> ```
> y=repmat(u,4,1);
> ```

> **Q** What are the trade-offs to consider when choosing between a high or a low repetition factor?

8.4.2 Interleaving

A repetition code is one of several useful tools for a communication systems engineer in order to enhance a robust data transmission. However, it is sometimes not enough, since it does not address the issue when a large quantity of data is corrupted in contiguous blocks. For instance, if a transmitter sends the data stream "101101," a repetition coder with a repetition factor of 4 will yield

$$111100001111111100001111,$$

where each input bit is repeated four times. While this encoding scheme may appear robust to error, it is still possible during a data transmission that a significant noise burst occurs over many consecutive bits, corrupting numerous binary digits in the transmission, and yields the following outcome:

$$111100 - - - - - - - - -1100001111,$$

where some of the original data is completely irretrievable.

> **Q** Why is it that even with repetition coding, our data transmission can still be severely affected? What could be done to make it even more robust?

Interleaving is an approach where binary data is reordered such that the correlation existing between the individual bits within a specific sequence is significantly reduced. Since errors usually occur across a consecutive series of bits, interleaving a bit sequence prior to transmission and deinterleaving the intercepted sequence at the receiver allows for the dispersion of bit errors across the entire sequence, thus minimizing its impact on the transmitted message. A simple interleaver will mix up the repeated bits to make the redundancy in the data even more robust to error. It reorders the duplicated bits among each other to ensure that at least one redundant copy of each will arrive even if a series of bits are lost. For example, if we use an interleaving step of 4, it means we reorder the vector by index [1, 5, 9, ..., 2, 6, 10, ...]. As a result, running "111100001111111100001111" through such an interleaver will yield the following output:

$$101101101101101101101101.$$

The interleaving step can be any of the factoring numbers of the data length. However, different mixing algorithms will change the effectiveness of the interleaver.

> The *reshape* function in MATLAB can be used to realize the interleaving.

Once we have implemented the interleaver, let us combine the repetition coder and the interleaver into a single FEC subsystem. Although the simple interleaving technique introduced above is sufficient for our implementation, there are various other forms of interleaving, that we will investigate in the next two sections.

8.4.2.1 Block Interleaving

The first approach to interleaving is to employ a block interleaver, as shown in Figure 8.10. Block interleaving is one method for reordering a bit sequence, where $N \times M$ bits fill an N column by M row matrix on a column basis, and then each resulting row is concatenated with each other in serial and outputted from the interleave block. At the transmitter side, the block interleaver is loaded column by column with N codewords, each of length M bits. These N codewords are then transmitted row by row until the interleaver is emptied. Then the interleaver is loaded again and the cycle repeats. The main drawback of block interleavers is the delay introduced with each column-by-column fill of the interleaver [12].

8.4.2.2 Convolutional Interleaving

Another approach to interleaving is to employ a convolutional interleaver [13], as shown in Figure 8.11. At the transmitter, the bit sequence is shifted into a bank of N registers, each possessing an increasing amount of buffer memory. The bits in the bank of registers are then recombined via a commutator and transmitted across the channel. At the receiver, the reverse process is performed in order to recover the original sequence. Compared with block interleavers, convolutional interleavers reduce memory requirements by about one-half [14]. However, the delay problem associated with the initial fill still exists.

8.4.3 Encoding

Besides interleaving multiple copies of data, we can instead encode the data into alternative sequences that introduce redundancy. A unique property of many encoding schemes is the ability to introduce redundancy without increases in data size without integer order. For example, in the case of repetitive coding that duplicates every bit with $R = 2$, this number is usually inverted in FEC discussions as a rate of $\frac{1}{2}$, a convolutional encoding scheme can introduce rates closer to 1. This makes them more efficient and provides more effective throughput. In this

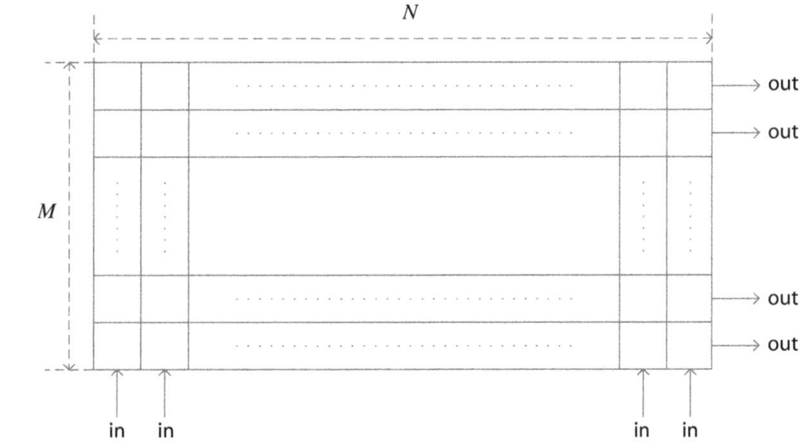

Figure 8.10 Schematic of a block interleaver.

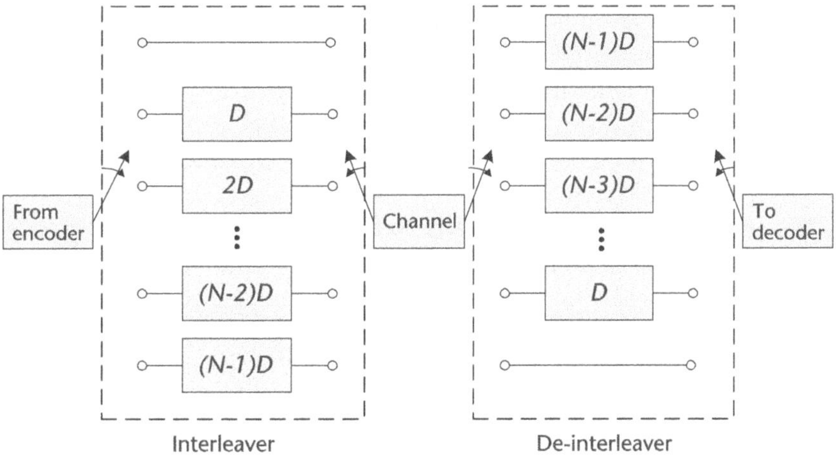

Figure 8.11 Schematic of a convolutional interleaver. (From [13].)

section we will discuss several common channel encoding schemes, with some basic information on how they function. In general, channel encoding is a mathematically complex area in information theory. Instead of diving into the theoretical designs of these scheme, we will compare their relative performance as well as some implementation details and drawbacks.

Similar to interleavers, encoders can typically be categorized into two basic types: block encoders and convolutional type encoders. Block encoders work on specific predefined groups or blocks of bits. Alternatively, convolutional encoders work on streams of data of indeterminate size but can be made to work on blocks of data if necessary.

The first coding scheme we will discuss is Reed-Solomon (RS) codes, which are linear-block-code developed in the 1960s. RS codes work by inserting symbols into a given frame or block of data, which are then used to correct symbol errors that occur. If we define M as the length of a given frame, sometimes called the message length, and define E as the encoded frame then we can correct up to $\lfloor \frac{E-M}{2} \rfloor$ symbols. RS are interesting since they can even provide information on how many errors were found and corrected as part of their implementation. However, RS encoders can be specific on what inputs they can process, which is why we have so far only considered symbols not bits. The symbols you can encode with a RS can be integers between $[0, 2^N - 1]$, where N is the exponent of our finite Galois field $GF(2^N)$. A Galois field is a field, a set which certain mathematical operations are defined, which has a finite number of objects. Due to this definition there will be some restrictions on E and N, but they are beyond the scope of this book. This set is how RS takes advantage of during decoding, which will reduce solutions spaces based on received data and selection of M, E and N.

With regard to implementation, it can be argued that RS codes are useful in bursty error situations where a swath of symbols close to each other are corrupted. When considering transmitted bits, each symbol will represent B bits, and since RS operate on symbols it can correct B bits of data in a group. Therefore, a bursty error corrupting $B + 1$ bits in a group can corrupt at most 2 symbols.

A similar code to RS is Bose Chaudhuri Hocquenghem (BCH) codes, which also relies on the concept of Galois fields. BCH codes are better at correcting errors that do not occur in groups, unlike RS. To reduce this probability of grouped data it can be useful to shuffle or scramble bits before and after transmission to reduce error locality, which is better for BCH when errors are sparse. However, this is a poor thing to do with RS to some extent. BCH codes can also correct more errors for the same amount of parity bits, but in general BCH codes require more computational power to decode than RS.

The final popular block code to consider are low-density parity check (LDPC) codes, which have even begun to replace the dominant Turbo codes, which we will consider next. LDPC codes have been around since the 1960s, but due to their complexity have only been considered for hardware implementation in the last decade. Developed by Robert Gallager, LDPC codes can approach the theoretical Shannon limit [15] for certain redundancy rates unlike RS and BCH. However, the computation required to use LDPC is considerable higher. Nonetheless, they exist in some modes of 802.11n and DVB-S2 standards.

When utilizing LDPC the implementor must select a parity matrix which the encoder and decoder will utilize, and the characteristics of this matrix will determine performance of the code. Strong performing codes will typically come in large block lengths like 648, 1296, and 1944 IEEE 802.11n/ac. This means that, you need to encode a significant amount of bits compared to the other codes to utilize LPDC efficiently in many cases.

Besides block codes, an alternative or stream-based coding implementation is convolutional codes. These codes convolutionally encode data, meaning redundancy is introduced by the succession of information passed through the encoder/decoder, essentially creating dependency on consecutive symbols or bits. A convolutional encoder is best understood by an example. Let us consider an encoding scheme with $R = 2$ with a recursive encoder, with three registers. Figure 8.12 provides a possible structure for such an encoder, which outputs two bits for every bit pushed into the system. You will notice that the current output is at least dependent on the last three inputs, similar to the memory of an FIR or IIR filter.

Figure 8.12 can be interpreted as two equations, provided in (8.16).

$$\begin{aligned} y_{n,1} &= (x_n + x_{n-2} + x_{n-3}) + x_{n-1} + x_{n-3} \\ y_{n,2} &= x_{n-2} + x_{n-3} \end{aligned} \tag{8.16}$$

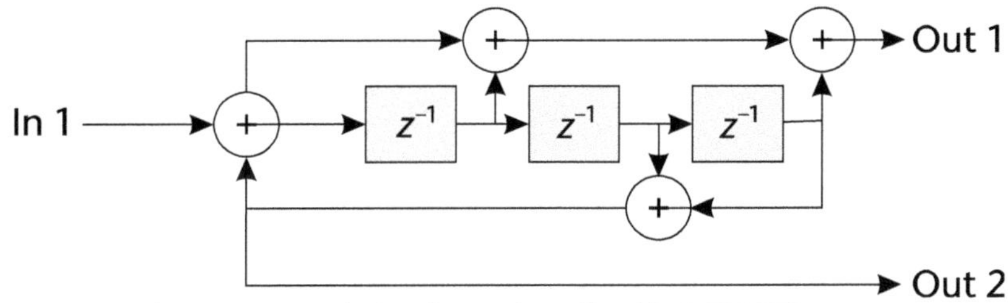

Figure 8.12 Example $R = 2$ convolutional encoder utilized in 3GPP LTE.

The decoder itself will utilize this dependency in the data to help remove errors that can occur. The most popular algorithm to accomplish this task is called the Viterbi algorithm [15], sometimes called a trellis algorithm or decoder. The concept of the Viterbi/trellis decoder is to trace back through previous decisions made and utilize them to best determine the most likely current bit or sample. This is what naturally leads to Figure 8.13 and hence the name trellis. In Figure 8.13, the left-most position represents the most recently receiver symbols or bits. The lines connecting the dots represent possible previous symbols, where the thick gray line represents the more probable symbols based on previous decisions made. The depth of this trellis is called the traceback length, and the deeper this trace becomes the better the recovery of bits will be. However, this process tends to plateau at a traceback around 34 symbols, and the deeper the traceback the increased time required to decode a specific symbol. This traceback limitation can be seen in Figure 8.14, where we examine a variety of traceback lengths across EbN0 for a 16-QAM signal.

In the early 1990s turbo codes were introduced, which are part of the convolutional code family [16]. Turbo codes have been heavily utilized by both third and fourth generation cellular standards as their primary FEC scheme. Like LDPC, turbo code can operate near the Shannon limit for performance but are less computationally intensive than LDPC with less correction performance. Turbo inherently utilizes the Viterbi algorithm internally for decoding with some additional interleaving, as well as using a set of decoders, and performs likelihood estimation between them. This is an extreme simplification of how turbo decoders actually work, and analysis of their operation is a very difficult topic area. However, they are a very powerful coding technique as long as you have the resources on your hardware to implement the decoder at the necessary speeds.

When utilizing FEC one should always consider the trade-offs with regard to computational complexity, performance required, and coding overhead allowed for the link. This is important since heavy coding may not be required in a clear transmission channel where additional throughput could be gained at a better coding rate closer to one. Therefore, modern standards like LTE and IEEE 802.11, will utilize adaptive modulation and coding schemes (MCSs), which reduce coding redundancy and increase modulation order, providing much higher throughput across a link. IEEE 802.11 itself has 32 MCS states or indexes, for which we have provided the first four entries in Table 8.2, for perspective on how code rates and modulation are used to trade off redundancy and data rate.

MATLAB itself provides all of the coding techniques we have described so far. However, some of the advanced codes are more complex to utilize, especially in

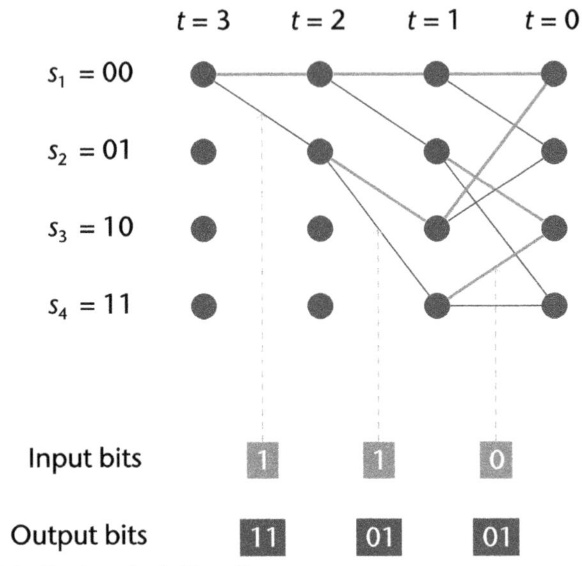

Figure 8.13 Viterbi/trellis decoder lattice diagram.

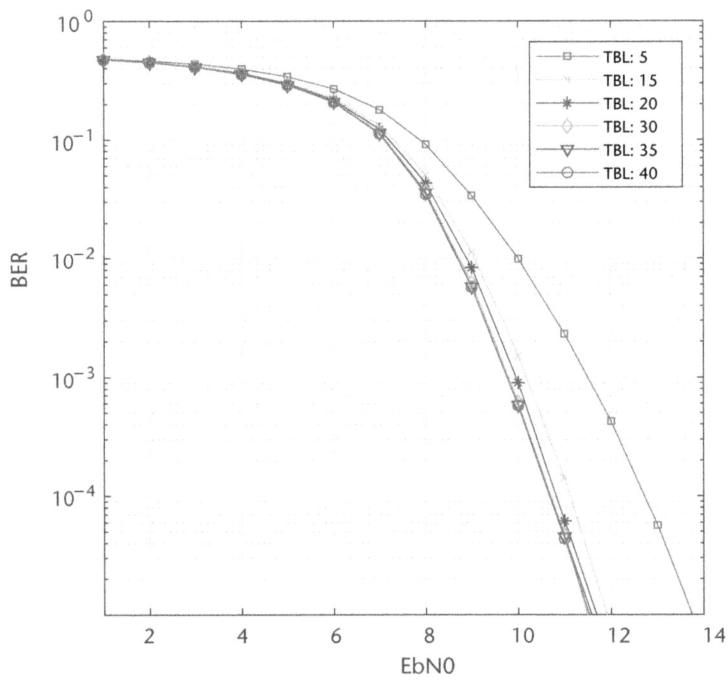

Figure 8.14 BER results of Viterbi decoder for 16-QAM with increasing traceback length.

different modes. For example, due to how turbo decoders work they require channel estimates to correctly determine the noise variance of the channel. Therefore, in a give receiver design this information must be provided for effective decoding. LDPC, as we have discussed before, requires parity matrices in their design. By default MATLAB provides the parity matrix for DVB, which requires 32,000 bits per block, which is rather unreasonable for many applications. Designing a smaller matrix can be complex, and this is rather outside the scope of MATLAB's provided

Table 8.2 Shortended Modulation and Coding Schemes List for IEEE 802.11*

MCS Index	*Streams*	*Modulation*	*R*	*Data Rate (Mbits/s)*
0	1	BPSK	2	6.5
1	1	QPSK	2	13
2	1	16-QAM	4/3	19.5
3	1	16-QAM	2	26

* From [1]

tools. Nonetheless, RS, BCH, and general Viterbi decoding are extremely easy to utilize out of the box and are simple to parameterize.

8.4.4 BER Calculator

After examing several techniques for protecting the reliability of a data transmission subject to interference and noise, we now need an approach to measure how well these techniques perform quantitatively. Referring back to Chapter 4, we saw that BER is a commonly used metric for the evaluation and comparison of digital communication systems. One straightforward way of calculating the BER is to count the number of received bits of information and then determine which ones are received in error. In other words, the ratio of bit errors to the total number of bits received can provide us with an approximate BER calculation. Note that the more bits that are received, the more accurate this link level metric becomes.

8.5 Chapter Summary

This chapter examined the concept of frame synchronization through correlation techniques and covered some common channel coding techniques through redundancy insertion. Given the last piece of the receiver with regard to synchronization provided here, a full receiver can be implemented for frame recovery. Possible arrangement for the receiver algorithms as discussed throughout the book so far have been examined, focusing on requirements from the design. Once full synchronization was covered, we moved on to techniques for making our links more robust through encoding implementations, including a discussion on their drawbacks and advantages with regard to implementation and use.

References

[1] IEEE Standard for Information Technology–Telecommunications and Information Exchange between Systems Local and Metropolitan Area Networks–Specific Requirements–Part 11: Wireless LAN Medium Access Control (MAC) and Physical Layer (PHY) Specifications–Amendment 4: Enhancementsfor Very High Throughput for Operation in Bands below 6 GHz, IEEE Std 802.11ac-2013, (Amendment to IEEE Std 802.11-2012, as amended by IEEE Std 802.11ae-2012, IEEE Std 802.11aa-2012, and IEEE Std 802.11ad-2012), December 2013, pp. 1–425.

[2] Barke, R. H., "Group Synchronizing of Binary Digital Sequences," in *Communication Theory*, London: Butterworth, 1953, pp. 273–287.

[3] The Math Works Inc., xcorr [online], 2017 https://www.mathworks.com/help/signal/ref/xcorr.html.

[4] Kay, S. M., *Fundamentals of Statistical Signal Processing*, Volume II: Detection Theory. Upper Saddle River, NJ: Prentice Hall, 1998.

[5] Poor, H. V., *An Introduction to Signal Detection and Estimation*, New York: Springer, 2010.

[6] Zhao, Q., and A. Swami, "Spectrum Sensing and Identification" in *Cognitive Radio Communications and Networks: Principles and Practice*, Burlington, MA: Academic Press, 2009.

[7] Kay, S. M., "Statistical Decision Theory I," in *Fundamentals of Statistical Signal Processing*, Volume II: Detection Theory, Upper Saddle River, NJ: Prentice Hall, 1998.

[8] Shanmugan, K. S., and A. M. Breipohl, "Signal Detection," in *Random Signals: Detection, Estimation and Data Analysis*, Wiley, 1988.

[9] Finger, A., *Pseudo Random Signal Processing: Theory and Application*, Hoboken, NJ: Wiley, 2013.

[10] IEEE Standard for Information Technology–Telecommunications and Information Exchange between Systems–Local and Metropolitan Area Networks–Specific Requirements-Part 11: Wireless LAN Medium Access Control (MAC) and Physical Layer (PHY) Specifications Amendment 3: Enhancements for Very High Throughput in the 60 GHz Band, IEEE Std 802.11ad-2012 (amendment to IEEEStd 802.11-2012, as amended by IEEE Std 802.11ae-2012 and IEEE Std 802.11aa-2012), December 2012,pp. 1-628.

[11] Harris, F. J., and M. Rice, "Multirate Digital Filters for Symbol Timing Synchronization in Software Defined Radios," *IEEE Journal on Select Areas in Communications*, Vol. 19, October 2001, pp. 2346–2357.

[12] Jacobsmeyer, J. M., Introduction to Error-Control Coding, www.pericle.com/papers/Error_Control_Tutorial.pdf.

[13] Forney, G. D., "Burst-Correcting Codes for the Classic Bursty Channel, in *IEEE Transactions on Communications, COM-19*, 1971, pp. 772–781, 1971.

[14] Sklar, B., *Digital Communications Fundamentals and Applications*, Upper Saddle River, NJ: Prentice Hall, 1988.

[15] Anderson, J., and S. Mohan, *Source and Channel Coding: An Algorithmic Approach*, New York: Springer, 1991.

[16] Berrou, C., Error-Correction Coding Method with at Least Two Systematic Convolutional Codingsin Parallel, Corresponding Iterative Decoding Method, Decoding Module and Decoder, US Patent No. 5,446,747, 1995.

CHAPTER 9

Channel Estimation and Equalization

This chapter will introduce the concepts of channel estimation and channel equalization. A simplified error model will be discussed along with several practical strategies for added equalizers to a communication system. Specifically, we will include designs for both decision direction- and training-based equalization. Equalizers are a useful utility since they can be used to handle numerous sources of distortion present in the environment as well as from the mismatches between nodes themselves.

With regard to our receiver outline in Figure 9.1, this chapter will address the last block, equalization, which is highlighted.

9.1 You Shall Not Multipath!

In the previous chapters, we focused on the synchronization between the transmitter and the receiving nodes. By combining the previous chapters, frame recovery now becomes possible and we have reached the threshold of successfully decoding frames. However, under certain scenarios these implementations will not be enough. Assuming that we have adequate SNR of the received signal, the remaining challenge in the environment is the multipath and other additive interferers. Multipath, which is introduced by the dispersive nature of the channel, is the effect of scaled reflections of a transmitted signal traveling along different paths to reach the receiver. Since these reflections travel along different paths they will observe different attenuations and different delays from the perspective of the receiver. Naturally, these impairments can be considered echoes of the transmitted signal and can be easily modeled by a FIR filter.

The time between the first received signal and the final received echo is defined as the *delay spread* of the channel [1]. In Figures 9.2 and 9.3, we present a physical representation of multipath and the resulting time domain representation of the line-of-sight (LOS) signal and two scatterers. This is a *ray-tracing* representation of multipath, but in reality multipath is a continuum of reflections due to the signal radiation characteristics. Ray-tracing is a discretization of that continuum and is commonly used to model multipath since it is far simplier to understand and mathematically model.

When the delay spread has a long duration with respect to the signal bandwidth, multipath effects can cause significant distortion to the received signal and results in intersymbol interference (ISI) as discussed in Chapter 6. The delay spread is a function of the environment, and we must take this into consideration when designing a system. For example, for outdoor environments where the multipath distances are large, this will produce a large delay spread. Mathematically, we can

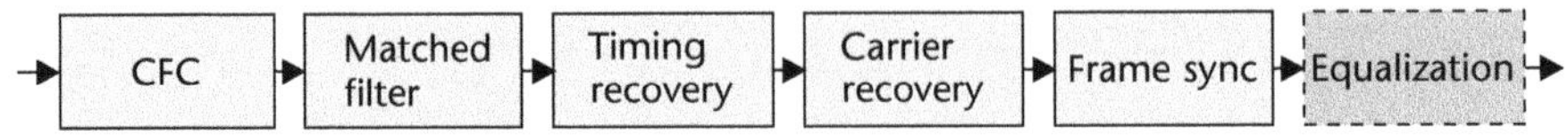

Figure 9.1 Receiver block diagram.

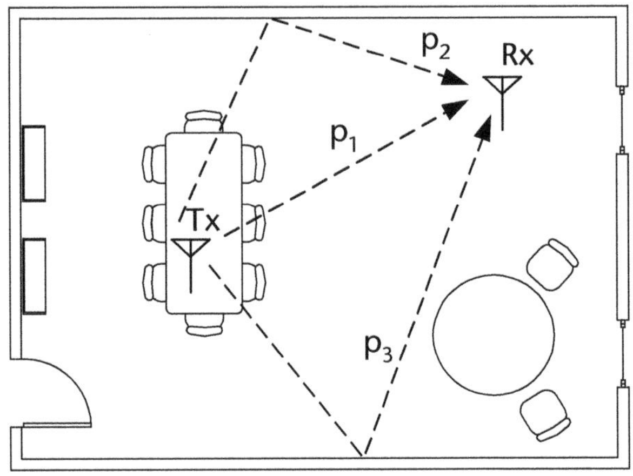

Figure 9.2 Example of multipath in an indoor environment.

relate the distance D that a scatter must travel to the sample delay t_s associated with that distance as

$$t_s = \frac{B \times D}{c}, \tag{9.1}$$

where c is the speed of light. A Wi-Fi signal at 20 MHz would have to travel approximately 15 extra meters to cause a single sample of delay. Therefore, Wi-Fi, which is commonly used indoors, will have a small delay spread. However, since the channel distances are short there can be a large number of high-powered scatterers. Due to path loss, the signal power of the interferers and delay spread are usually inversely related.

Mathematically, we can model a received multipath signal r as in impulse train at random time offsets Δ_n with associated gains α_n of the transmitted signal x as

$$r(t) = \mu(t) + \sum_{n=1}^{N} \alpha_n x(t - \Delta_n), \tag{9.2}$$

where there are $N-1$ scatters and μ is additional interferers or noise. As long as μ is uncorrelated with x and periodic or autoregressive, we can effectively filter it from the signal space [2]. In the case when we do have multipath, this will be experienced at the received as ISI. We can demonstrate such a condition in Figure 9.4 where we view a QPSK signal r. In Figure 9.4, we observe the effects of symbol smearing over time, which we view as symbol drift in the constellation diagram. Based on (9.2), we can simply model multipath effects using an FIR filter.

9.2 Channel Estimation

Before we consider correcting effects of a given channel, we can alternatively consider the estimation of an unknown channel. Channel estimation, as opposed to

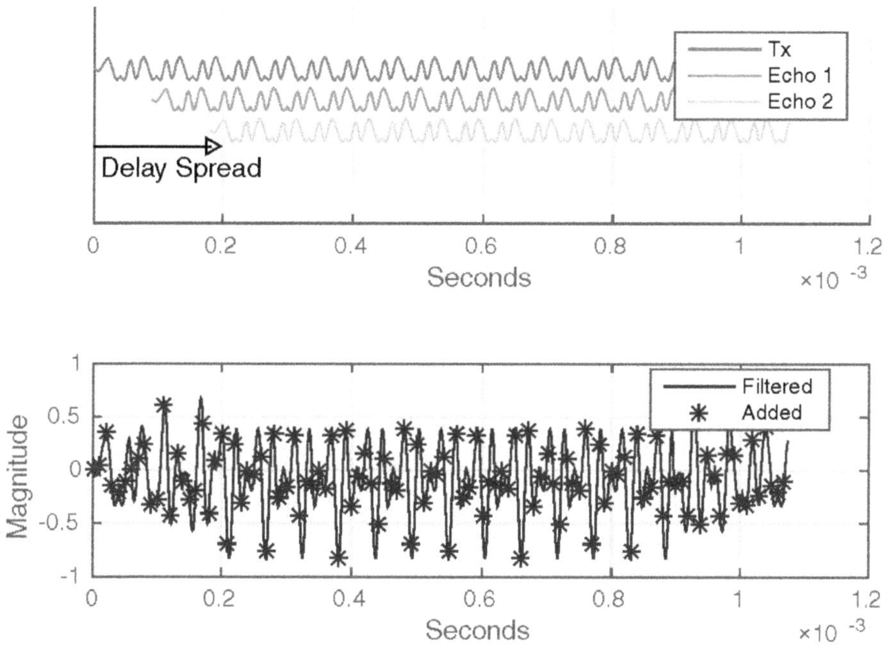

Figure 9.3 Physical characteristics of a multipath propagation environment.

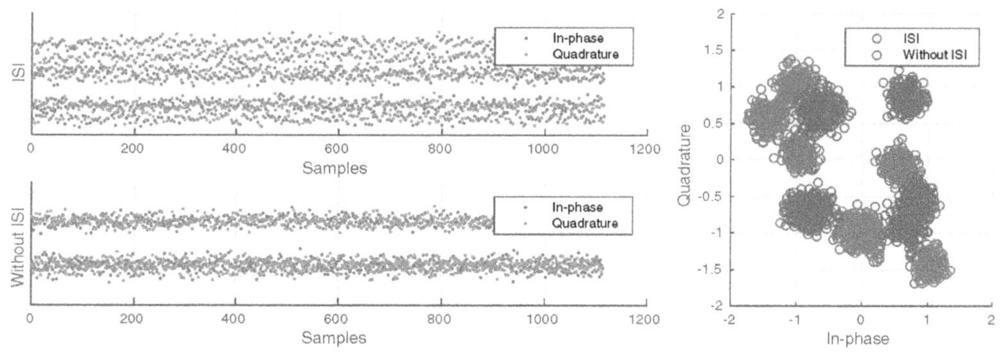

Figure 9.4 Effects of symbol smearing over time.

channel equalization, is a desirable place to start since we have a known solution in the simulation to test against, unlike equalization, which may not produce a unique solution depending on the channel and noise conditions. To perform channel estimation, we will utilize the least mean squares (LMS) algorithm developed by Widrow and Hoff [3]. LMS is a gradient descent or Newton method type algorithm, and can be considered the standard adaptive filter algorithm for signal processing. The LMS algorithm utilizes known information or symbols in the transmitted sequences in order to estimate the corruption of the receive data, which we model here as a FIR filter. We provide a model in Figure 9.5 of a common diagram for channel estimation, which we will use to help derive our system implementation. Alternative adaptive filter algorithms do exist, such as the recursive least squares (RLS) algorithm, which can outperform LMS in many situations. However, RLS can have stability concerns when designed improperly and is more computationally

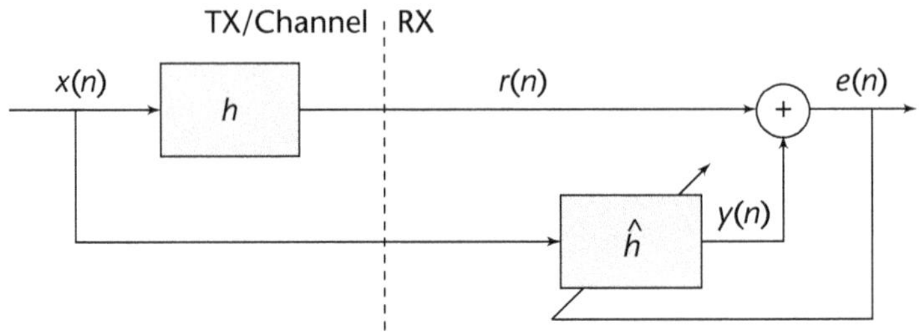

Figure 9.5 Adaptive FIR estimation of FIR channel h using training data.

complex due to a matrix inversion requirement. RLS is beyond the scope of this chapter, but Haykin [4] is a good reference.

We illustrate the pieces of our channel estimation system in Figure 9.5, where we have our unknown static channel $\mathbf{h}$, which affects our training data $x(t)$. The goal of the system shown in Figure 9.5 is to match $\mathbf{h}$ and $\hat{\mathbf{h}}$, which will drive our error down to zero. In order to achieve this goal, the adaptive algorithm will require an error signal $e(n)$ and the original signal $x(t)$. We will utilize LMS to estimate an unknown channel filter $\mathbf{h} \in^{\{L\times 1\}}$, whose estimate is defined as $\hat{\mathbf{h}} \in^{\{M\times 1\}}$ where $M \geq L$. For a transmitted signal $x(t)$ passing through the channel filter $\mathbf{h}$, $\mathbf{h}$ can be estimated using the following recursive algorithm:

$$y(n) = \hat{\mathbf{h}}^H(n)\mathbf{x}(n) \tag{9.3}$$

$$e(n) = r(n) - y(n) \tag{9.4}$$

$$\hat{\mathbf{h}}(n+1) = \hat{\mathbf{h}}(n) + \mu\, \mathbf{x}(n)e^*(n) \tag{9.5}$$

where

$$\mathbf{x}(n) = [x(n), x(n-1), ..., x(n-M-1)]^T. \tag{9.6}$$

and μ is the governing stepsize, which provides control over convergence rate and stability. Selection of the stepsize is implementation-specific but should be in the range $0 < \mu < \frac{2}{\lambda_{max}}$, where λ_{max} is the maximum eigenvalue of the autocorrelation of the true channel $\mathbf{R} = \mathrm{E}[\mathbf{h}\mathbf{h}^H]$. Alternatively, since $\mathbf{h}$ is unknown, a looser and safer bound is $M\sigma^2$ where σ^2 is the variance of the $r(n)$. For more information on LMS stability analysis, the interested reader should consult Chapter 9 of Haykin [4].

The channel length L is an unknown for an individual environment, but through measurements and the physical nature of the channel environment, estimates can be made for reasonable lengths. This relates to the previous discussion on delay spread, which is an alternative description of the channel length. Studies are commonly performed by standards committees to provide guidance for receiver designers.

Equations (9.3) to (9.6) can be implemented in a simple recursion as in lines 8-17 in Code 9.1 requiring $2L+1$ multiplications and $L+1$ additions. If we examine the mean-squared error (MSE) of $\hat{\mathbf{h}}$ over time for a $L = M = 2$ across a contour of the solution space in Figure 9.6, we can observe the descent to the true solution LMS will take. Figure 9.6 demonstrates this descent for four random starting positions for the estimate $\hat{\mathbf{h}}$. The use of contour plots is a common analysis tool in adaptive filters theory to model behavior of estimation evolution. However, when $M > 2$ we

Code 9.1 LMS Channel Estimation: **chanEst.m**

```
h = [0.5; 1; -0.6]; % Channel to estimate
mu = 0.01; % Stepsize
trainingSamples = 1000;
x = sign(randn(trainingSamples,1)); % Generate BPSK data
r = filter(h,1,x); % Apply channel
L = length(h); h_hat = zeros(L,1);
%% Estimate channel
for n = L:trainingSamples
    % Select part of training input
    in = x(n:-1:n-L+1);
    % Apply channel estimate to training data
    y = h_hat'*in;
    % Compute error
    e = r(n)-y;
    % Update taps
    h_hat = h_hat + mu*conj(e)*in;
end
```

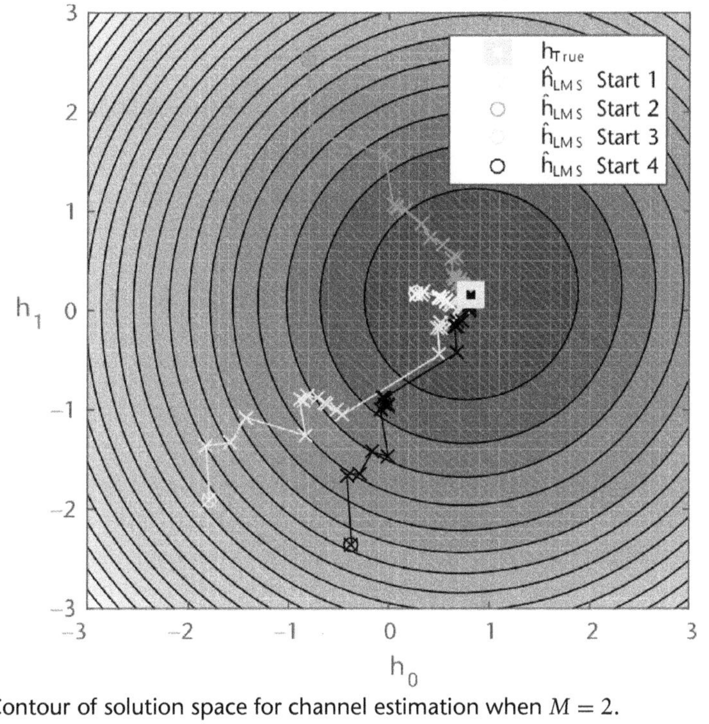

Figure 9.6 Contour of solution space for channel estimation when $M = 2$.

will use a MSE plot similar to Figure 9.7 since we cannot easily visualize all error dimensions on a contour.

We can extend Code 9.1 further and actually visualize the shape of the channel estimates as well to have better understanding of the accuracy of our designed system. Examining the response, especially for rather aggressive channels, can be very useful when determining parameterization of an equalizer design. In Code 9.2, we provide an example on how to visualize the channel responses and their

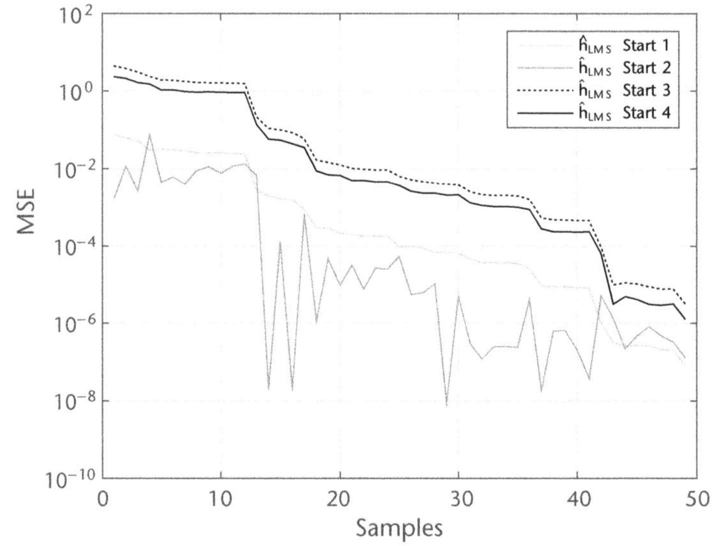

Figure 9.7 MSE of channel estimates over received samples from different starting locations.

estimates. The function `freqz` is utilized here, which will calculate the digital filter response using the `fft` function.

Code 9.2 Plot Responses: **`chanEst.m`**

```
%% Plot responses
Fs = 1e6; N = 64;
htrue=freqz(h,1,N,Fs,'whole');
[hb,we]=freqz(h_hat,1,N,Fs,'whole');
semilogy(we,abs(hb),'b')
hold on;semilogy(we,abs(htrue),'bo');hold off
grid on;xlabel('Frequency (Hz)');ylabel('Magnitude');
legend('Channel Est','Channel Actual','Location','Best');
```

The response and estimate provided in Figure 9.8 yields very accurate results. However, as noise is introduced into the system and the channel becomes more complex, this estimate will become worse. However, assuming the LMS algorithm has converged the error should only be limited based on our chosen stepsize μ and the noise of the channel.

9.3 Equalizers

Unlike channel estimation, an equalizer tries to undo the effects of the channel and remove interference if possible. Similar to channel estimation, some knowledge about the source data is necessary to train the equalizer at the receiver in order to reduce the channel effects. Typically, this information is part of the preamble sequence or header information of the frame since for logical operation we will always transmit some unknown data, which we call the payload. We will discuss several adaptive equalizer implementations here but many permutations and alternatives do exist in the literature [5].

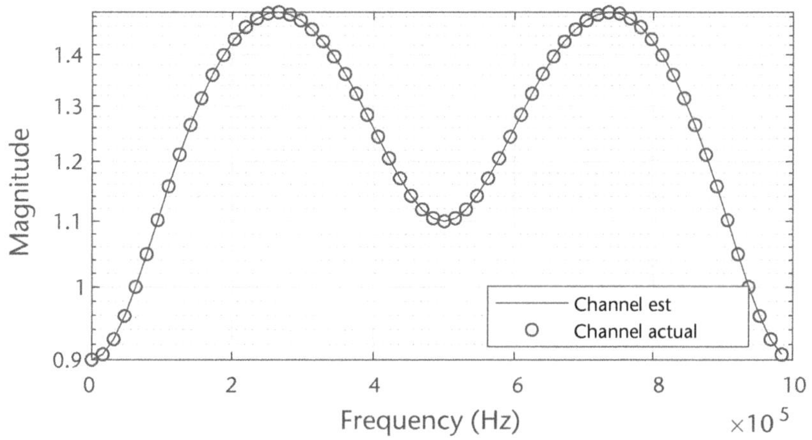

Figure 9.8 Example channel and associated estimate using the LMS algorithm.

The equalizer approaches the filter evolution problem from a different prespective than discussed in Section 9.2. Instead of evolving the filter to match the channel **h**, a filter **f** is evolved to compensate for the channel itself and reproduce the training data at the output of the filter accurately. This arrangement is provided in Figure 9.9, which places the equalizer filter or forward filter in the received data path. The known transmitted data is then used to adapt this filter. To solve for a filter $\mathbf{f} \in^{\{K\times 1\}}$ using LMS that will equalize the effects of a given channel **h**, we need to modify the recursion provided in (9.3) to (9.6) as

$$\hat{x}(n) = \mathbf{f}^H(n)\mathbf{r}(n) \tag{9.7}$$

$$e(n) = x(n) - \hat{x}(n) \tag{9.8}$$

$$\mathbf{f}(n+1) = \mathbf{f}(n) + \mu\, e^*(n)\mathbf{r}(n) \tag{9.9}$$

where:

$$\mathbf{r}(n) = [r(n), r(n-1), ..., r(n-K-1)]^T. \tag{9.10}$$

In this case, we are driving the received samples $r(n)$ to match the transmitted data $x(n)$. If this algorithm converges, the resulting output of the combined filters should just delay the signal:

$$x(n) * \mathbf{h} * \mathbf{f} = \hat{x}(n-\delta), \tag{9.11}$$

where $*$ represents the convolution operation and δ the group delay of the cascaded channel filter and equalizer. Therefore, the combined response of the equalizer and channel should be flat. The updated code is provided in Code 9.3.

The delay δ offsets the training sequence further into the filter, which is useful if the earliest filter taps do not produce the most gain. It can be argued this more efficiently utilizes the equalizer's additional tap advantage of the channel length. Since this equalizer must be a causal, $\delta > 0$ or else we would require knowledge of future samples (noncausal). Also, this delay will affect the output delay, which we must compensate for as well. To correctly evaluate the bit errors in the signal, use Code 9.4.

When we have exact knowledge of the source symbols, LMS is an effective algorithm to reach the Wiener solution for a given channel [4]. However, in many

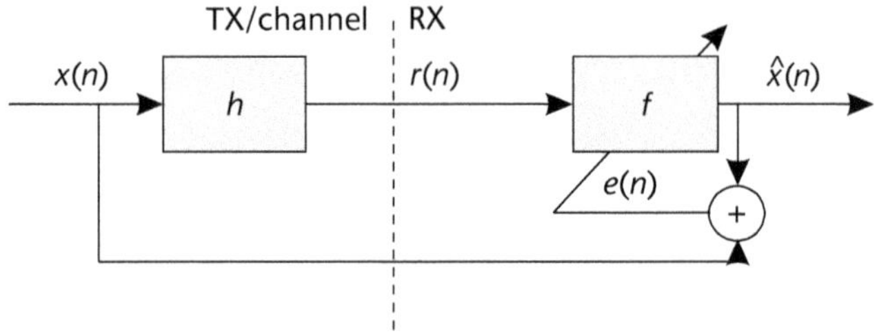

Figure 9.9 Adaptive FIR equalization of FIR channel h using known training data.

Code 9.3 Channel Equalization: **chanEQ.m**

```
h = [0.2; 1; 0.1; 0.02]; % Channel taps
mu = 0.001; % Stepsize
trainingSamples = 1e4;
x = sign(randn(trainingSamples,1)); % Generate BPSK data
r = filter(h,1,x); % Apply channel
K = length(h)+1; f = zeros(K,1);
delta = 3; % Reference tap
%% Equalize channel
index = 1;
[e,x_hat]=deal(zeros(trainingSamples-K+1,1));
for n = K:trainingSamples
    % Select part of training input
    in = r(n:-1:n-K+1);
    % Apply channel estimate to training data
    x_hat(index) = f'*in;
    % Compute error
    e = x(n-delta)-x_hat(index);
    % Update taps
    f = f + mu*conj(e)*in;
    index = index + 1;
end
```

Code 9.4 Determine Bit Errors: **chanEQ.m**

```
% Slice
x_bar = sign(x_hat);
% Calculate bit errors
eqDelay = K-delta;
fprintf('BER %2.4f n',mean(x(eqDelay:end-eqDelay-1) ~= x_bar));
```

cases it will not be possible to use training information and instead will require blind equalization techniques. *Blind equalization* is required when the received signal contains no known preamble or when equalizer updates want to be updated during the payload portions of frames. Such an implementation is necessary when channels have long filter lengths or require more data than provided in the preamble. Alternatively, for highly dynamic channels where the coherence of the channel is shorter than a frame, it may be necessary to update the equalizer to maintain a desired BER. Here we will consider a decision-directed (DD) LMS equalizer, although other implementations exist, such as the *constant modulus* (CM) equalizer, sometimes called the *dispersion-minimization* (DM) algorithm [6].

The DD equalizer operates by making hard decisions on received data to estimate the most likely symbol and updates the equalizer based on this estimate. The updated equalizer structure is provided in Figure 9.10, which inserts a decision block after the output of the equalizer. Therefore, the receiver has no knowledge of $x(n)$ from this design except for the modulation scheme. This results in an error term in (9.12) to be updated in the case of DD to

$$e(n) = \hat{x}(n) - \bar{x}(n) \tag{9.12}$$

where $\bar{x}(n)$ is the maximum likelihood estimate of the equalized received signal $x(n)$. For the case of BPSK and QPSK we have the following estimators:

$$\bar{x}(n) = \begin{cases} sign(y(n)) & \text{if BPSK} \\ sign(real(x(n))) + i \times sign(imag(x(n))) & \text{if QPSK.} \end{cases} \tag{9.13}$$

DD equalization can be effective when $x(n)$ and $\hat{x}(n)$ are relatively close, but when the difference is large DD equalizer can perform poorly. Alternatively, from a perspective provided in Chapter 6, the eye of the eye diagram plot must be open to some degree initially for the DD equalization to effectively invert the channel effectively or in a reasonable amount of symbols. We provide an received signal with ISI in Figure 9.11(a), the resultant signal after LMS equalization in Figure 9.11(b), and the resultant signal after DD equalization in Figure 9.11(c). The LMS implementation convergences within a few hundred samples, while the DD equalizer take roughly 2,000 samples to open the eye with still a significant amount of noise. LMS has an obvious advantage, but requires knowledge of the exact source symbols unlike the DD method. Note that both equalizers utilize the same μ, with the initialization of the equalizers being equal to

$$\mathbf{f}_{LMS} = \mathbf{f}_{DD} = [1, 0, ..., 0] \in^{\{K\times 1\}}, \tag{9.14}$$

$\mathbf{f}_{DD}$ cannot be initialized to all zeros.

9.3.1 Nonlinear Equalizers

So far we have only considered equalizers with forward filters, also known as linear equalizers. However, it is also possible to implement filters with both feedback and feedforward filters, which we call here *decision feedback equalizers* (DFEs). The feedback filter in the DFE will produce outputs that are subtracted from the

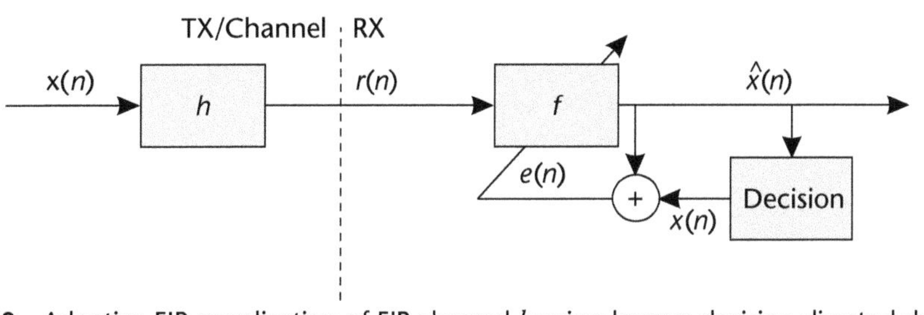

Figure 9.10 Adaptive FIR equalization of FIR channel h using known decision directed data.

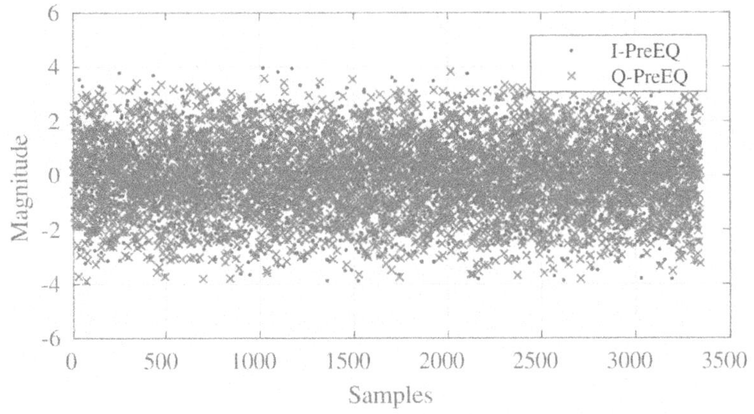

(a)

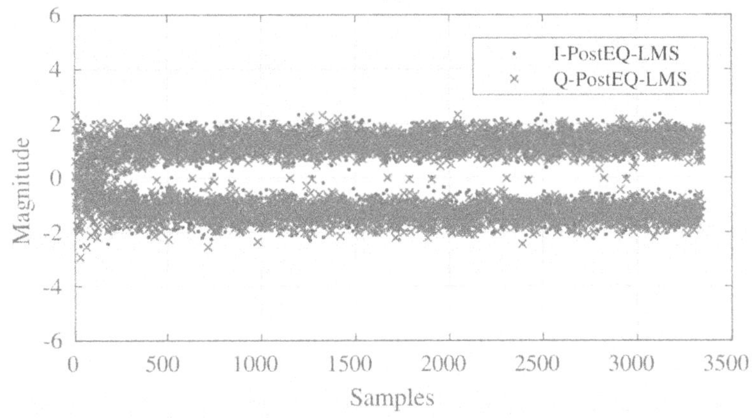

(b)

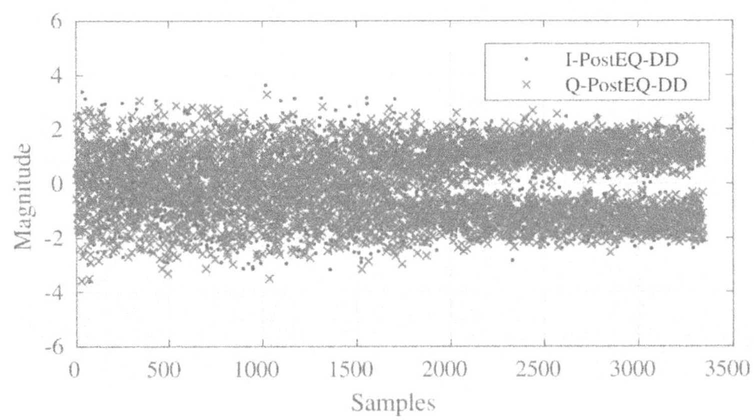

(c)

Figure 9.11 Example of using cross correlation to find a sequence with a larger sequence of data. (a) QPSK signal with ISI before equalization, (b) QPSK signal with ISI after LMS equalization, and (c) QPSK signal with ISI after DD equalization.

output of the feedforward filter. This is outlined in Figure 9.12, which shows both filters and their update paths. Since the feedback filter can only estimate the post-cursors symbols, it needs to be used in combination with a feedforward filter. After convergence, the feedback filter contains an estimate of the impulse response of the

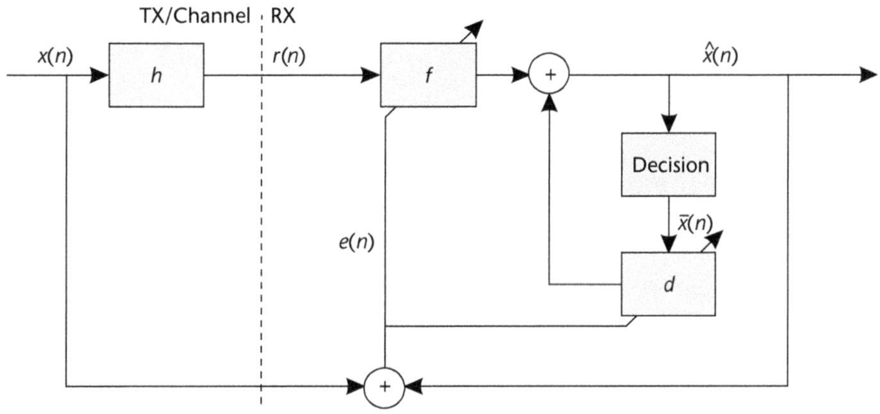

Figure 9.12 Adaptive FIR equalization of FIR channel h using decision feedback equalizer.

channel convolved with of the feedforward filter. Since the feedback filter utilizes this composite output, the DFE can compensate for severe amplitude distortion without increasing the noise in the highly distorted channel.

We update our previous equalizer equations with the addition of a new filter $\mathbf{d} \in^{\{P\times 1\}}$. To solve for a filter $\mathbf{d}$ using LMS that will equalize the effects of a given channel $\mathbf{h}$, we need to modify the recursion provided in (9.3) to (9.6) as

$$\hat{x}(n) = \mathbf{f}^H(n)\mathbf{r}(n) - \mathbf{d}^H(n)\bar{\mathbf{x}}(n) \tag{9.15}$$

$$e(n) = \bar{x}(n) - \hat{x}(n) \tag{9.16}$$

$$\mathbf{f}(n+1) = \mathbf{f}(n) + \mu\, e^*(n)\mathbf{r}(n) \tag{9.17}$$

$$\mathbf{d}(n+1) = \mathbf{d}(n) + \mu\, e^*(n)\bar{\mathbf{x}}(n) \tag{9.18}$$

where

$$\mathbf{r}(n) = [r(n), r(n-1), ..., r(n-K-1)]^T, \tag{9.19}$$

and

$$\bar{\mathbf{x}}(n) = [\bar{x}(n), \bar{x}(n-1), ..., \bar{x}(n-P-1)]^T. \tag{9.20}$$

Here $\bar{x}(n)$ will either be the DD version of $x(t)$ or $x(t)$ itself when training data is used. Again, we can update our MATLAB code to utilize this new feedback filter, which is provided in Code 9.5.

Note that the DFE implementation can be more difficult to tune since it contain two filters. The DFE implementation can also utilize different stepsizes per filter update as well, which may make tuning more managable.

9.4 Receiver Realization

For an actual implementation of an equalizer into our receiver structure, we have several design strategies that we can utilize depending on the system requirements. A reasonable design perspective to consider is the required amount of training data needed to equalize a given channel environment. There are three aspects here: the channel length, the convergence of the equalizer, and the dynamics of the channel. Training data is typically confined to the preamble sequence and the chosen length will be chosen on the maximum value of L, which also determines the value of M.

Code 9.5 Determine Bit Errors: **chanEQDFE.m**

```
h = [0.2; 1; 0.1; 0.02]; % Channel taps
mu = 0.001; % Stepsize
trainingSamples = 1e4;
x = sign(randn(trainingSamples,1)); % Generate BPSK data
r = filter(h,1,x); % Apply channel
K = length(h)+2; f = zeros(K,1);
P = length(h)-1; d = zeros(P,1); x_bar_vec = zeros(P,1);
delta = 4; % Reference tap
%% Equalize channel
index = 1;
[e,x_hat]=deal(zeros(trainingSamples-K+1,1));
for n = K:trainingSamples
    % Select part of training input
    in = r(n:-1:n-K+1);
    % Apply channel estimate to training data
    x_hat(index) = f'*in - d'*x_bar_vec;
    % Compute error
    e = x(n-delta)-x_hat(index);
    % Update taps
    f = f + mu*conj(e)*in;
    d = d - mu*conj(e)*x_bar_vec;
    % Update feedback filter
    x_bar_vec = [x(n-delta);x_bar_vec(1:end-1)];
    index = index + 1;
end
```

When the channel is dynamic, meaning that it can change over small periods of time when compared to the frame length, it becomes necessary to utilize multiple equalizers. For example, we can utilize a LMS equalizer to initially open the eye with the preamble sequence, then across the data we can utilize a DD equalizer to maintain the open eye (see Figure 9.13). If a secondary equalizer was not used, then the frame length would need to be shorted to meet the same BER requirements. However, the preamble must be long enough to allow for the appropriate adaption of the equalizer, which will be dependent on the values of M and L. For a larger equalizer, more data is required for it to converge, thus leading to longer preambles.

We can update the source code in Code 9.3 for a case of BPSK to utilize both LMS and DD with a few simple modifications. In Code 9.6, we simply add a condition to switch between the different modes.

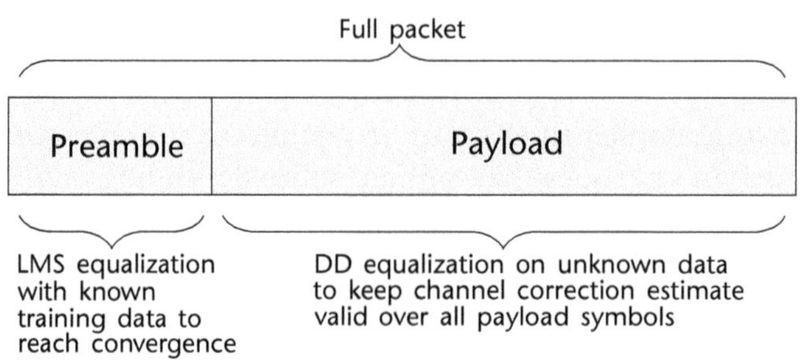

Figure 9.13 Example packet structure and application of different equalizer algorithms.

Code 9.6 Equalizer with DD and LMS Modes: **chanEQLMSDD.m**

```
for n = K+1:FrameLength
    % Apply equalizer to received data
    x_hat = f'*r(n:-1:n-K+1);
    % Estimate error
    if n<(PreambleLength-delta)
      e = x(i-delta) - x_hat;
    else
      e = sign(y) - x_hat;
    end
    % Update equalizer
    f = f + mu*conj(e)*r(n:-1:n-K+1);
end
```

Equalizers themselves are very useful utilities in the system beyond just compensating for multipath in a system. They can also be used to compensate for frequency offset, timing mismatches, and even frame synchronization errors. However, the equalizer must be configured in a way to compensate for such conditions. For example, when dealing with carrier offset, which can only be relatively small compared with conditions in Chapter 7, μ should be increased to deal with this condition. Since the equalizer is itself just a filter, compsenating for frequency offset simply forms the equalizer's taps in a complex gain equivalent to the instantaneous phase associated with the carrier. However, this must be updated continuously to compensate for such a condition, forcing a dual DD and training data implementation.

When considering timing compensation it can be useful to implement a fractional equalizer, which consumes multiple samples on the front feedforward filter. This makes the equalizer multirate, but also similar to the timing compensation designs in Chapter 6. In this configuration the equalizer will be able to weigh or interpolate across symbols to accurately correct for delay in the received signal. In addition to fractional timing offsets, by utilizing our delay variable δ we can compensate for sample errors during frame synchronization. As long as the desired sample, or start of frame sample, is within the feedforward filter we can compensate for an incorrect initial estimate. For a feedforward filter of length L and $\delta = \lfloor \frac{L}{2} \rfloor$, we should be able compensate for a sample offset of $\pm\delta$.

9.5 Chapter Summary

In this chapter we introduced the concept of channel estimation and equalization. We have built several equalizer designs around the gradient descent algorithm LMS, which is the dominant cost function in the field. We provided different strategies for implementing equalizers when training data is available and when blind operation is required. Finally, we provided a discussion on receiver implementation strategies that combine multiple equalizer designs as well as different use cases for nonidealities from the environment. For the interested reader, brief mathematical derivations for the basic linear equalizer design, zero-forcing equalizer, and decision feedback equalizer are available in Appendix C.

References

[1] Goldsmith, A., *Wireless Communications*, Cambridge, UK: Cambridge University Press, 2005.

[2] Johnson, C. R., Jr., W. A. Sethares, and A. G. Klein, *Software Receiver Design: Build Your Own Digital Communication System in Five Easy Steps*, Cambridge, UK: Cambridge University Press, 2011.

[3] Widrow, B., and M. E. Hoff, *Neurocomputing: Foundations of Research, Adaptive Switching Circuits*, Cambridge, MA: MIT Press, 1988, pp. 123–134.

[4] Haykin, S., and S. S. Haykin, *Adaptive Filter Theory*, Pearson Education, 2014.

[5] Farhang-Boroujeny, B., *Adaptive Filters: Theory and Applications*, Wiley, 1999.

[6] Johnson, R., P. Schniter, T. J. Endres, J. D. Behm, D. R. Brown, and R. A. Casas, Blind Equalization Using the Constant Modulus Criterion: A Review, *Proceedings of the IEEE*, Vol. 86, No. 10, 1998, pp. 1927–1950.

CHAPTER 10

Orthogonal Frequency Division Multiplexing

Until now, we have studied several single-carrier modulation schemes where the input binary bits are modulated by a carrier signal with a center frequency f_c. However, there are other approaches where data can be communicated across a channel, including a technique referred to as *multicarrier modulation*. Instead of having one center frequency, multicarrier modulation (MCM) multiplexes serial input data into several parallel streams and transmits them over independent subcarriers. These subcarriers can be individually modulated and manipulated, allowing for their optimization with respect to the channel. The ability of MCM to tailor its transmission parameters across the different subcarriers is especially useful when combating frequency selective fading channel environments. Consequently, most high data rate communication systems, including all digital subscriber line (xDSL) modems [1–3], as well as most commercial communication standards, including Wi-Fi [4, 5], Wi-MAX [6, 7] and LTE [8, 9], use some form of MCM at the core of their implementations. In this chapter, we will explore one of the most popular forms of MCM: orthogonal frequency division multiplexing (OFDM).

10.1 Rationale for MCM: Dispersive Channel Environments

From our high school physics courses, we learned about the fundamentals of wave propagation and how they combine constructively and destructively (e.g., the ripple tank experiment). The exact same principles hold in high-speed data transmission. For example, in a wireless communication system, the transmitter emanates radiation in all directions (unless the antenna is directional, in which case the energy is focused at a particular azimuth). In an open environment, like a barren farm field, the energy would continue to propagate until some of it reaches the receiver antenna. As for the rest of the energy, it continues on until it dissipates.

In an indoor environment, as depicted in Figure 10.1(c), the situation is different. The line-of-sight component (if it exists), p_1, arrives at the receiver antenna first, just like in the open field case. However, the rest of the energy does not simply dissipate. Rather, the energy is reflected by the walls and other objects in the room. Some of these reflections, such as p_2 and p_3, will make their way to the receiver antenna, although not with the same phase or amplitude. All these received components are functions of several parameters, including their overall distance between the transmitter and receiver antennas as well as the number of reflections.

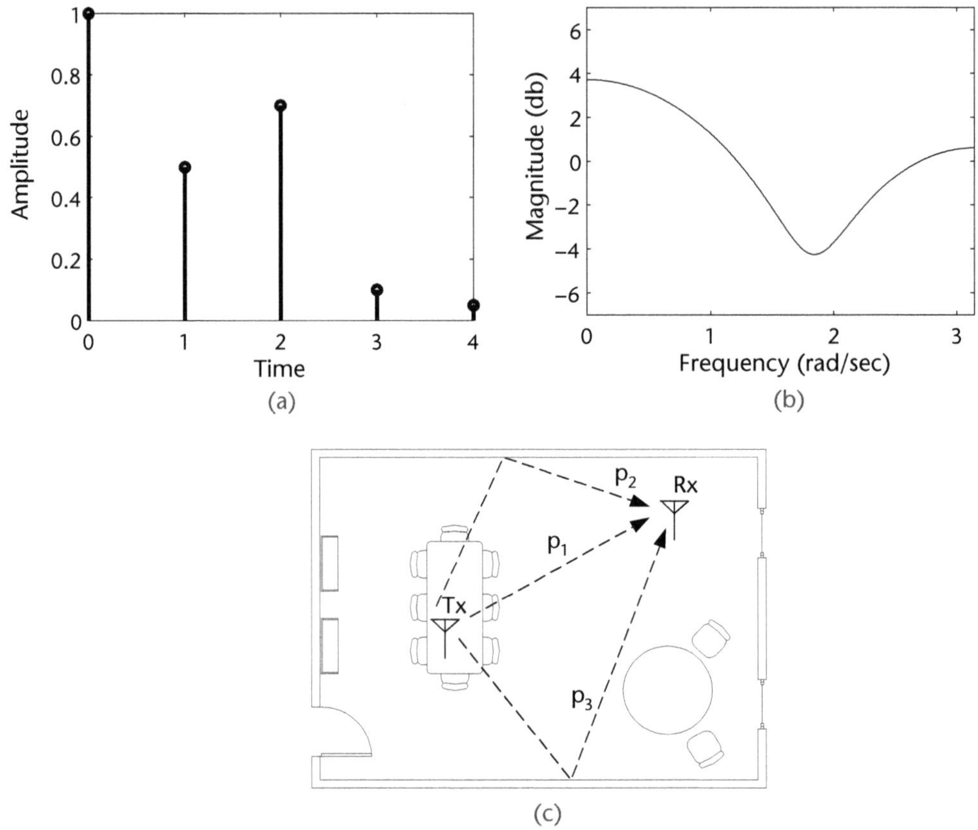

Figure 10.1 Example of a channel response due to dispersive propagation. Notice the three distinctive propagate paths, p_1, p_2, and p_3, that start at the transmitter Tx and are intercepted at the receiver Rx. (a) Impulse response, (b) frequency response, and (c) the process by which dispersive propagation arises.

At the receiver, these components are just copies of the same transmitted signal, but with different arrival times, amplitudes, and phases. Therefore, one can view the channel as an impulse response that is being convolved with the transmitted signal. In the open field case, the *channel impulse response* (CIR) would be a delta, since no other copies would be received by the receiver antenna. On the other hand, an indoor environment would have a several copies intercepted at the receiver antenna, and thus its CIR would be similar to the example in Figure 10.1(a). The corresponding frequency response of the example CIR is shown in Figure 10.1(b).

When observing Figure 10.1(b) more closely for the case of an operating environment with significant multipath characteristics, we can observe that the spectrum of the CIR appears to vary across the frequency domain. This is very problematic for any signal being transmitted across this type of channel since these signals will experience nonuniform attenuation that varies across frequency, which is relatively difficult to mitigate within the context of a single-carrier communication system since they will require the design and implementation of complex equalizer filters. On the other hand, MCM communication systems are well suited to handle this type of distortion since their multiple sub carriers are designed to divide-and-conquer the channel and treat each frequency-dependent attenuation individually, thus resulting in an implementation that has relatively low complexity.

10.2 General OFDM Model

OFDM is an efficient form of MCM, which employs the DFT and inverse DFT (IDFT) to modulate and demodulate multiple parallel data streams. As shown in Figure 10.2, the general structure of an OFDM communication system consists of a $2N$-point IDFT and a $2N$-point DFT at the transmitter and the receiver. The OFDM transceiver in Figure 10.2 operates as follows: A high-speed digital input, $d[m]$, is demultiplexed into N subcarriers using a commutator. The data on each subcarrier is then modulated into an M-QAM symbol, which maps a group of $\log_2(M)$ bits at a time. For subcarrier k, we will rearrange $a_k[\ell]$ and $b_k[\ell]$ into real and imaginary components such that the output of the modulator block is $p_k[\ell] = a_k[\ell] + jb_k[\ell]$. In order for the output of the IDFT block to be real, given N subcarriers we must use a $2N$-point IDFT, where terminals $k = 0$ and $k = N$ are don't-care inputs. For the subcarriers $1 \leq k \leq N-1$, the inputs are $p_k[\ell] = a_k[\ell] + jb_k[\ell]$, while for the subcarriers $N+1 \leq k \leq 2N-1$, the inputs are $p_k[\ell] = a_{2N-k}[\ell] + jb_{2N-k}[\ell]$.

The IDFT is then performed, yielding

$$s[2\ell N + n] = \frac{1}{2N} \sum_{k=0}^{2N-1} p_k[\ell] e^{j(2\pi nk/2N)}, \tag{10.1}$$

where this time $2N$ consecutive samples of $s[n]$ constitute an OFDM symbol, which is a sum of N different QAM symbols. This results in the data being modulated on several subchannels, which is achieved by multiplying each data stream by a $\sin(Nx)/\sin(x)$, several of which are shown in Figure 10.3.

The subcarriers are then multiplexed together using a commutator, forming the signal $s[n]$, and transmitted to the receiver. Once at the receiver, the signal is demultiplexed into $2N$ subcarriers of data, $\hat{s}[n]$, using a commutator and a $2N$-point DFT, defined as

$$\bar{p}_k[\ell] = \sum_{n=0}^{2N-1} \hat{s}[2\ell N + n] e^{-j(2\pi nk/2N)}, \tag{10.2}$$

is applied to the inputs, yielding the estimates of $p_k[\ell]$, $\bar{p}_k[\ell]$. The output of the equalizer, $\hat{p}_k[\ell]$, is then passed through a demodulator and the result multiplexed together using a commutator, yielding the reconstructed high-speed bit stream, $\hat{d}[m]$.

10.2.1 Cyclic Extensions

With the CIR be modeled as a finite impulse response filter that is convolved with a sampled version of the transmitted signal, this results in the CIR smearing past samples onto current samples, which are smeared onto future samples. The effect of this smearing causes distortion of the transmitted signal, thus increasing the aggregate BER of the system and resulting in a loss in performance.

Although equalizers can be designed to undo the effects of the channel, there is a trade-off between complexity and distortion minimization that is associated with the choice of an equalizer. In particular, the distortion due to the smearing of a previous OFDM symbol onto a successive symbol is a difficult problem. One simple solution is to put a few dummy samples between the symbols in order to capture the

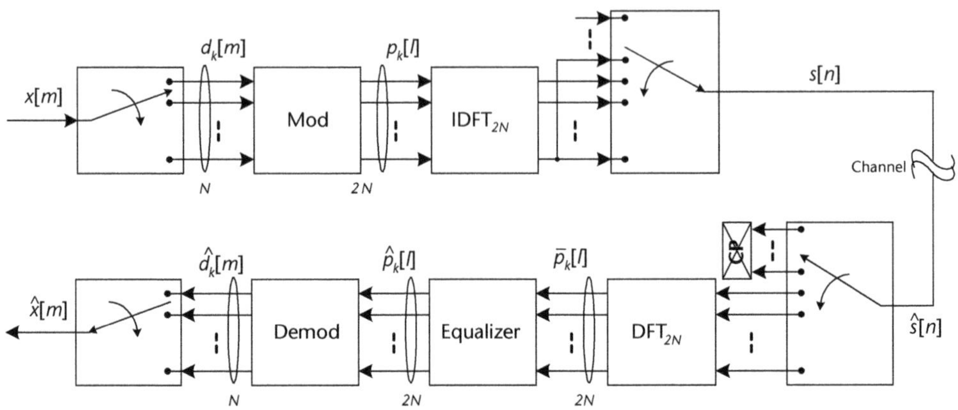

Figure 10.2 Overall schematic of an orthogonal frequency division multiplexing system, where the DFT and IDFT are employed to modulate and demodulate the data streams.

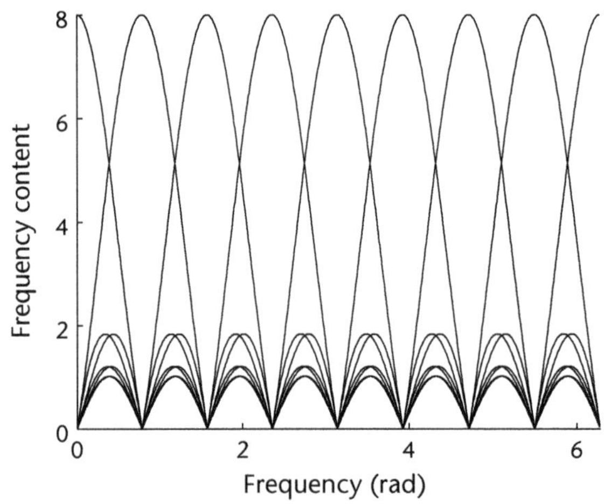

Figure 10.3 Characteristics of orthogonal frequency division multiplexing: frequency response of OFDM subcarriers.

intersymbol smearing effect. The most popular choice for these K dummy samples are the last K samples of the current OFDM symbol. The dummy samples in this case are known as a *cyclic prefix* (CP), as shown in Figure 10.4(a).

Therefore, when the OFDM symbols with cyclic prefixes are passed through the channel, the smearing from the previous symbols are captured by the cyclic prefixes, as shown in Figure 10.4(b). As a result, the symbols only experience smearing of samples from within their own symbol. At the receiver, the cyclic prefix is removed, as shown in Figure 10.4(c), and the OFDM symbols proceed with demodulation and equalization.

Despite the usefulness of the cyclic prefix, there are several disadvantages. First, the length of the cyclic prefix must be sufficient to capture the effects of the CIR. If not, the cyclic prefix fail to prevent distortion introduced from other symbols. The second disadvantage is the amount of overhead introduced by the cyclic prefix. By adding more samples to buffer the symbols, we must send more information across the channel to the receiver. This means to get the same throughput as a system without the cyclic prefix, we must transmit at a higher data rate.

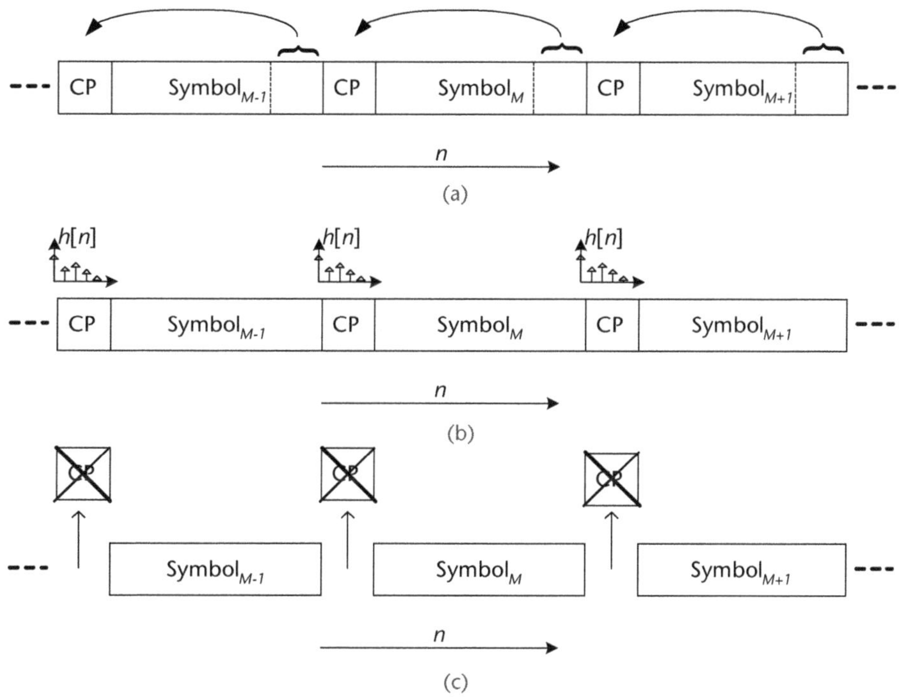

Figure 10.4 The process of adding, smearing capturing, and removal of a cyclic prefix. (a) Adding cyclic prefix to an OFDM symbol, (b) smearing by channel h(n) from previous symbol into cyclic prefix, and (c) removal of cyclic prefix.

10.3 Common OFDM Waveform Structure

In Section 10.2, we discussed a general OFDM transmission model and the CP used to limit ISI. Next we will examine a common structure within the OFDM symbol itself used among many standards such as IEEE 802.11. First, the number of subcarriers of an OFDM symbol N_s will typically be a base two number since the FFT and IFFT, which are efficient implementations of the DFT and IDFT used to modulate and demodulate the data are most efficient at these lengths. However, in a typical implementation N_g guard carriers will be utilized that occupy the outer frequency positions of the OFDM symbol. For example, in Figure 10.5, where $N_s = 64$ and $N_g = 14$, the first seven subcarriers and the last seven subcarriers are unoccupied. This is done to limit the out-of-band interference impacting the surrounding signals occupying neighboring channels. Nonetheless, the downside of using guard carriers is that it reduces the overall data rate of the transmission. Note that this is the OFDM symbol before the CP is added, which is of length N_{cp}. Furthermore, select subcarriers are chosen as training or pilot tones, which are used by equalizers at the receiver to correct for phase and frequency offsets, along with the channel effects. Similar to Figure 10.5, the selected subcarriers are often uniformly selected across the symbol in order to provide characterization across the entire OFDM symbol. These pilots tones will be known at the receiver.

Once OFDM modulation and CPs have been applied to the modulated symbols, additional structure can be added to the frame prior to analog transmission. These additions are based on the transmission type. For example, for continuously

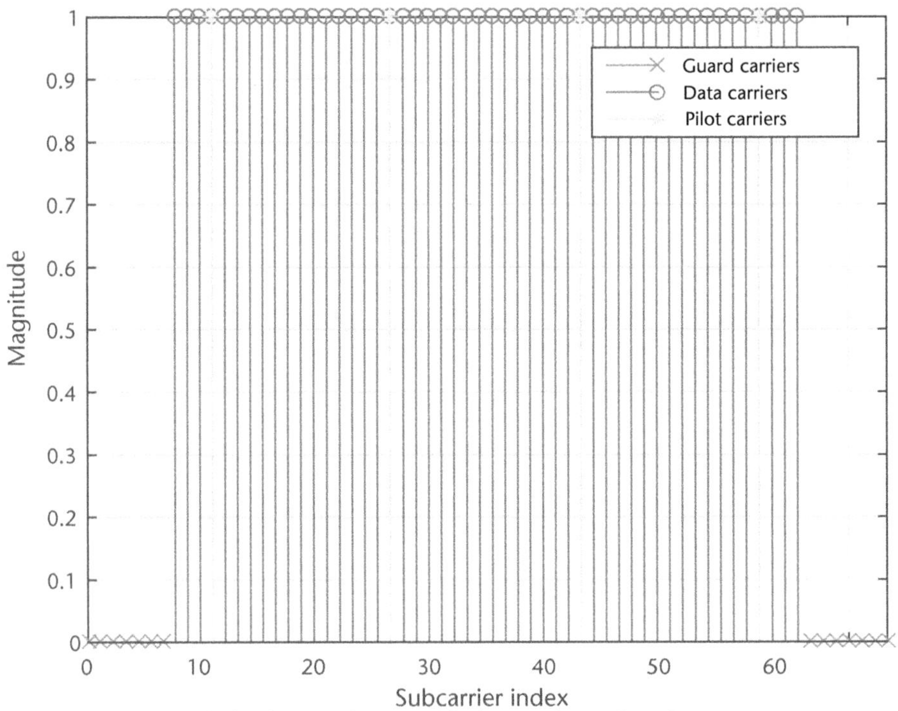

Figure 10.5 OFDM symbol subcarrier layout with data and guard carriers.

transmitting systems, such as LTE or broadcast signals, small insertions are made into the individual frames at a periodic rate in order to enable the synchronization of the receivers. These training or synchronization symbols can be minimal since the receiver can attempt to synchronize multiple times with the transmitter. However, in burst-based systems, such as wireless local area networks (WLANs), there will be large preamble structures that provide the receivers with more time to synchronize. This additional time allows recovery of frames with high probability without the need for retransmission. Retransmission can be expensive in terms of performance costs with wireless systems that have *carrier sense multiple access* (CSMA)/*collision avoidance* (CA) MAC schemes.

We will first examine the frame structure of WLAN since they are a common implementation of an OFDM system. Digital video broadcast terrestrial (DVB-T2) uses a similar structure to WLAN. However, the preamble data is modulated in a different way than WLAN [10]. We have chosen to examine OFDM transmission and reception from a standard's perspective for two reasons. First, OFDM implementations can be very unique based on their system design but generally follow two patterns, which are discussed here. Second, the standards discussed in this section are used extensively in industry and understanding how they work can be invaluable since engineers will most likely interact with WLAN or LTE at some point in their careers.

From Figure 10.6, we observe that the WLAN frame has a preamble that is made of up four full OFDM symbols and will always remain the same regardless of the mode or MCS link settings. The first section identified as the short portion of the preamble, called the Legacy Short Training Field (LSTF) in the IEEE 802.11

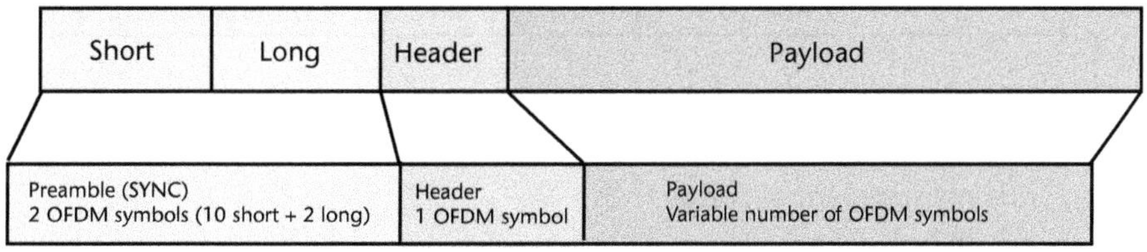

Figure 10.6 WLAN 802.11a frame outlining preamble, header, and payload sections.

standard [11], contains ten repeated copies of a short sequence. The purpose of this sequence is to allow the automatic gain control (AGC) of the receiver to stabilize over the 8 μs it is given and to perform carrier frequency offset (CFO) estimation. Generally, this is considered the packet detection phase of the receiver.

Once the packet has been detected and CFO corrected, the receiver will then utilize the long field, called the Legacy Long Training Field (LLTF) in the IEEE 802.11 standard [11], containing two repeated sequences and a CP. This portion of the preamble is designed to be used for channel estimation. Overall, an OFDM receiver will first identify the start of the packet, correct for frequency offsets, correct for channel effects, and remove residual phase distortions. This processing may be considered somewhat backward compared with the previous chapters that considered signal carrier transmissions.

Once the preambles have been utilized to correct the received frame, the header symbol will be utilized to determine the length and MCS of the remaining payload. Finally, pilot tones in the payload will be used to correct any remaining offsets and channel distortions. A similar frame can be generated from Code 10.1.

To provide an alternative perspective regarding standards that utilize OFDM, LTE is a great example to consider. Unlike WLANs, which focus all synchronization within a few symbols of the preamble, LTE uses primary synchronization signals (PSS) and secondary synchronization signals (SSS), as shown in Figure 10.7. The receiver detects via cross correlation, and utilizes both the frame boundaries along with the pilot tones sprinkled throughout the resource blocks (RBS) in order to correct at the receiver. However, since the downlink signals are constant at small intervals, if a PSS or SSS is missed it can be simply located on the next sequence provided. Once synchronized, the receiver will continually decode information from the base stations, unlike WLAN, which is more burst-based.

10.4 Packet Detection

The reasoning behind the structure of the LSTF in the preamble is based on the work by Schmidl and Cox [12]. In their paper, they outlined a symbol timing recovery stategy that relies on searching for a training symbol with two identical halves in the time domain, which will remain identical after passing through the channel, except that there will be a phase difference between them caused by the carrier frequency offset. The two halves of the training symbol are made identical (after the IFFT) by transmitting a pseudonoise (PN) sequence on the even frequencies while zeros are used on the odd frequencies. This means that at each even frequency one of the points of a source constellation (rotated BPSK for WLAN) is transmitted. Given this structure we can accomplish two tasks: first, estimate the start of the preamble

Code 10.1 Generate OFDM Packet: **genOFDMPacket.m**

```
NumFrames = 1;
%% Build OFDM Modulator
FFTLength              = 64;
NumGuardBandCarriers = [6; 5];
NumDataCarriers        = 48;
CyclicPrefixLength     = 16;
PilotCarrierIndices    = [12;26;40;54];
NumOFDMSymInPreamble = 5;
NumBitsPerCharacter    = 7;
% Convert message to bits
msgInBits = repmat(randi([0 1], NumDataCarriers, 1),10, 1);
PayloadBits = msgInBits(:);
% Calculate number of OFDM symbols per frame
NumOFDMSymbols = ceil(length(PayloadBits)/NumDataCarriers);
% Calculate number of bits padded in each frame
NumPadBits = NumDataCarriers * NumOFDMSymbols - length(PayloadBits);
% Get preamble for each frame
Preamble = double(getOFDMPreambleAndPilot('Preamble', ...
    FFTLength, NumGuardBandCarriers));
% Get pilot for each frame
Pilots = double(getOFDMPreambleAndPilot('Pilot', NumOFDMSymbols));
% BPSK modulator
BPSKMod = comm.BPSKModulator;
% OFDM modulator
DataOFDMMod = comm.OFDMModulator(...
    'FFTLength' ,            FFTLength, ...
    'NumGuardBandCarriers', NumGuardBandCarriers, ...
    'InsertDCNull',          true, ...
    'PilotInputPort',        true, ...
    'PilotCarrierIndices',   PilotCarrierIndices, ...
    'CyclicPrefixLength',    CyclicPrefixLength, ...
    'NumSymbols',            NumOFDMSymbols);

%% Modulate Data
symPostBPSK = BPSKMod.step([PayloadBits; randi([0 1], NumPadBits, 1)]);
% OFDM modulation for one frame
symPostOFDM = DataOFDMMod.step(reshape(symPostBPSK, ...
    NumDataCarriers, NumOFDMSymbols), Pilots);
% Repeat the frame
y = repmat([Preamble; symPostOFDM], NumFrames, 1);
```

using correlation in the time domain due to the known repeating sequences, and second, estimate the frequency base on the phase difference between the halves of the training symbols. There are many extensions to [12], such as Minn [13] and [14], but all maintain the principle of symmetric preambles across frequency and time.

Mathematically, the packet detector's timing metric from [12] is expressed as

$$M(k) = \frac{\sum_{m=0}^{L-1} r^*(k+m)r(k+m+L)}{\sum_{m=0}^{L-1} |r(k+m+L)|^2}, \tag{10.3}$$

where r is the received signal and L is the length of halve a training symbol. In WLAN, the length is equal t0 16 samples or 0.8 μs long. Equation (10.3) takes advantage of the the repeated sequences in time of length L present in the preamble.

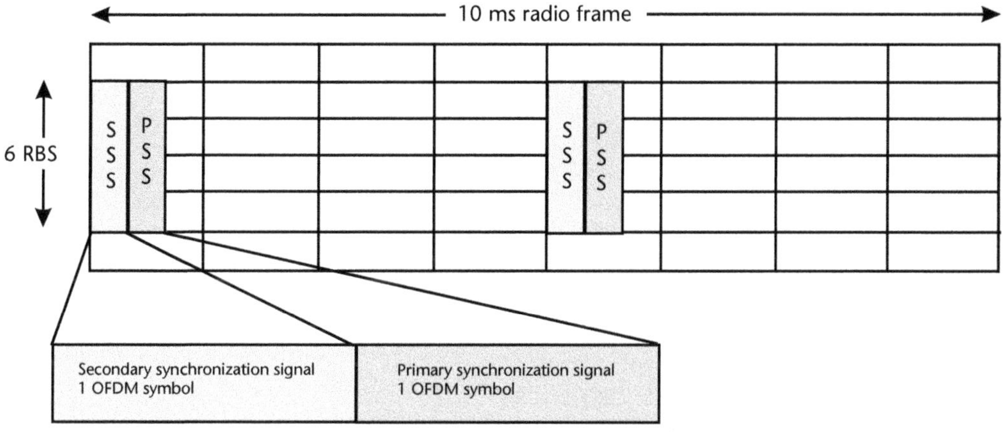

Figure 10.7 Simplified outline of LTE downlink frame with a focus on primary and secondary synchronization signals.

In Code 10.2, a simple example is provided that performs the operation shown in (10.3), which relies on our previous Code 10.1 to generate the necessary OFDM packet based on WLAN 802.11a specifications. The resulting metric is plotted in Figure 10.8, where we can observe an obvious plateau starting at the true position of the preamble start for different AWGN SNR values. At low SNR the plateau from the LSTF sequence becomes more difficult to observe, but still and be identified. Nonetheless, the purpose of the metric M is to determine in a binary degree if a packet was transmitted. The exact start of the preamble is not required, but still a rough estimate where the LSTF is located in the data is only required during this phase of the receiver.

In practice, this method becomes a threshold detection problem and we must build the receive to deal with different conditions. Equation (10.3) is an autocorrelation implementation since it utilizes only the received signal. This implementation is useful since it self-normalizes the input, trying to force the output to be between 0 and 1. This is not always possible, as observed in Figure 10.8, but the preamble does extend above the payload data for the output M even at low SNR. Therefore, it is important to select the correct value when thresholding the metric M in order to determine if a packet exists in the signal space. This parameter is actually tunable in some commercial hardware from Cisco Systems, which is defined as receiver start of packet detection Threshold (Rx SOP) [15]. Conceptually, changing this parameter will change change your wireless cell size and inversely the received packet error rate.

10.5 CFO Estimation

One problem associated with OFDM is the effect of frequency offsets. When the received symbols experience an offset greater than half the subcarrier bandwidth, the signal becomes nonorthogonal, which breaks the recovery assumptions associated with the signal. Therefore, before the OFDM signal is demodulated (i.e., processed via FFT), frequency correction must be considered. As mentioned in Section 10.4, the phase difference of sequence halves of the preamble are directly

Code 10.2 Detect Packet Start: **packetDetect.m**

```
%% Generate OFDM Waveform
genOFDMPacket;
% Add random offset
offset = randi([0 1e2]);
y = [zeros(offset,1);y];

%% Schmidl and Cox: Coarse Packet Detection
L = 16;  % Short sync field length
m = L;   % Distance between fields
N = 300; % Autocorrelation samples
M = zeros(N,1);
SNR = [0,5,10];
for SNRindx = 1:length(SNR)
    r = awgn(y,SNR(SNRindx),'measured');
    % Determine timing metric
    for k=1:N
        P = r(k:k+m)' *  r(k+L:k+m+L);
        a = abs(y(k+L:k+m+L));
        R = a'*a;
        M(k) = abs(P)^2/(R^2);
    end
    % Plot
    subplot(length(SNR),1,SNRindx);stem(M);
    hold on; stem(offset+1,M(offset+1),'r*'); hold off;
    grid on;xlabel('k');ylabel('M');
    legend('Autocorrelation','True Start');
    title(['SNR: ',num2str(SNR(SNRindx)),'dB']);
end
```

related to the frequency offset experience by receive waveform. This can be easily proven given a frequency offset Δ_f and the transmitted waveform $s(t)$:

$$r(t) = s(t)e^{j2\pi\Delta_f t/f_s} \tag{10.4}$$

Now utilizing the symbol halves in the LSTF, each of length L, we can write

$$\begin{aligned}
p(k) &= \sum_{m=0}^{L-1} r^*(k+m)r(k+m+L) \\
p(k) &= \sum_{m=0}^{L-1} s(k+m)^* e^{-j2\pi\Delta_f(k+m)/f_s} s(k+m+L)e^{j2\pi\Delta_f(k+m+L)/f_s} \\
p(k) &= e^{j2\pi\Delta_f(L)/f_s} \sum_{m=0}^{L-1} s(k+m)^* s(k+m+L) \\
p(k) &= e^{j2\pi\Delta_f(L)/f_s} \sum_{m=0}^{L-1} |s(k+m)|^2.
\end{aligned} \tag{10.5}$$

Therefore, we know the frequency offset is a direct result of the phase difference between the preamble halves. For a frequency offset of Δ_f, the phase difference ϕ

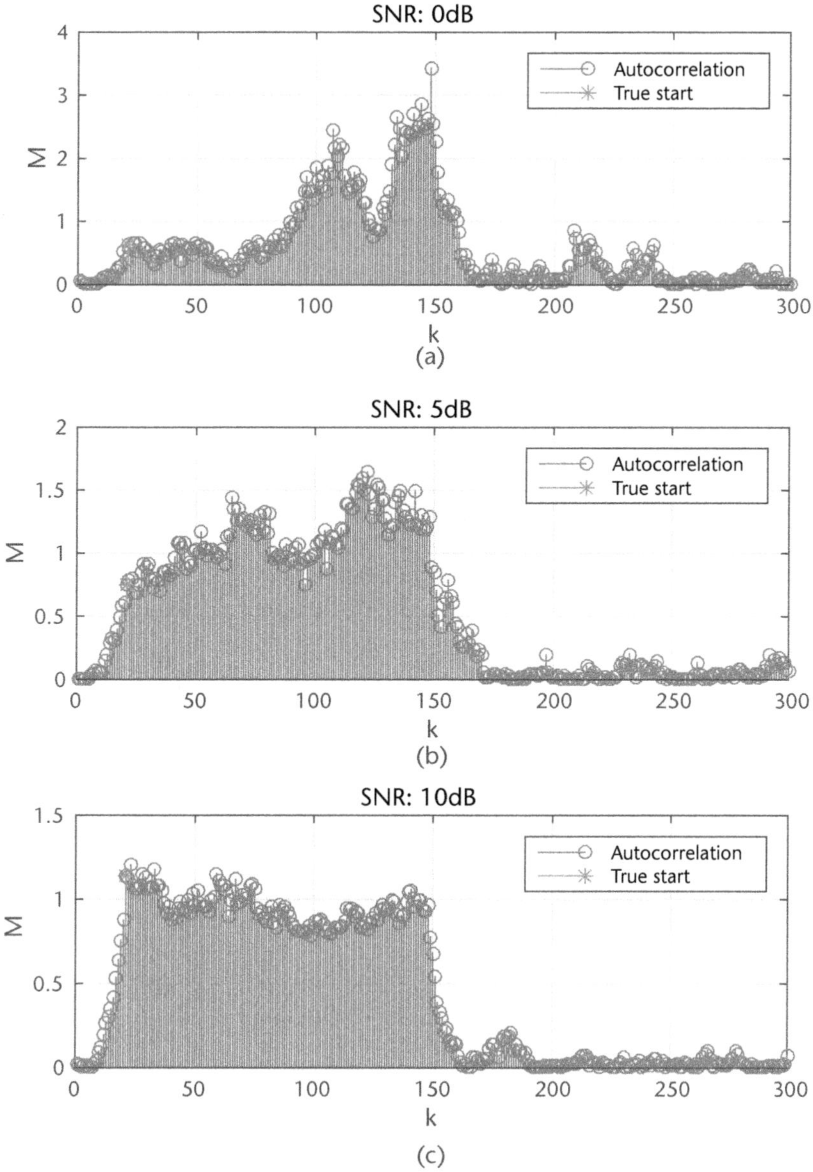

Figure 10.8 Timing metric M using in Schmidl and Cox algorithm at SNRs of 0, 5, and 10 dB. These provide rough estimates for the start of a receive frame in AWGN.

between halves of the preamble symbols can be directly related by

$$\phi = \frac{2\pi L \Delta_f}{f_s}, \tag{10.6}$$

where f_s is the sample rate. Therefore, ϕ can be estimated as

$$\hat{\phi} = tan^{-1}\left(\frac{\Im(P(k))}{\Re(P(k))}\right), \tag{10.7}$$

where

$$P(k) = \sum_{m=0}^{L-1} r^{*}(k+m)r(k+m+L), \tag{10.8}$$

and k in (10.7) is the within the first nine symbols of the short preamble sequence. Since ϕ can be at most π we can directly determine the largest offset:

$$\Delta_{f,max} = \frac{f_s}{2LN}. \tag{10.9}$$

In the case of WLAN 802.11a, $\Delta_{f,max} = 625$ kHz. However, in the IEEE standard a receiver must maintain a 20 PPM oscillator resulting in at most a 212-kHz difference between transmitter and receiver at 5.3 GHz [11].

Code 10.3 implements this technique from (10.5) to (10.8) over varying frequency offsets. The results are analyzed with respect to a normalized offsets. This is common since frequency offset will be dependent on the sampling rate of the system, which can be used to compensate for large offsets. Simply running a radio at a faster rate will reduce the normalized offset for the recovery algorithms.

Code 10.3 Determine CFO: `freqEstOFDM.m`

```
%% Generate OFDM Waveform
genOFDMPacket;
% Add random offset
offset = randi([0 1e2]); y = [zeros(offset,1);y];
SNR = 20;
%% CFO Estimation
L = 16;  % Short sync field length
m = L;   % Distance between fields
N = 300; % Autocorrelation samples
Fs = 1e6; % Sample rate
% CFO estimation over multiple frequencies
subchannelSpacing = Fs/FFTLength;
CFOs = subchannelSpacing.*(0.01:0.01:0.5);
cfoError = zeros(size(CFOs));
for cfoIndx = 1:length(CFOs)
    % Add noise
    r = awgn(y,SNR,'measured');
    % Frequency offset data
    pfOffset = comm.PhaseFrequencyOffset('SampleRate',Fs,...
        'FrequencyOffset',CFOs(cfoIndx));
    r = pfOffset(r);
    % Determine frequency offsets
    freqEst = zeros(N,1);
    for k=1:N
        P = r(k:k+m)' *  r(k+L:k+m+L);
        freqEst(k) = Fs/L*(angle(P)/(2*pi));
    end
    % Select estimate at offset estimated position
    freqEstimate = freqEst(offset);
    % Calculate CFO error
    cfoError(cfoIndx) = abs(freqEstimate-CFOs(cfoIndx))/subchannelSpacing;
end
```

Since the short section of the preamble contains ten copies of the same sequence, at most nine estimations can be made. However, it is unlikely that the early symbols of the short preamble will be detected or available due to AGC convergence. This is the reasoning behind providing additional copies of short preamble sequences. If the short preamble is recovered early, meaning a large number of the symbols were detected, then more frequency estimates can be made and eventually averaged together.

The long portion of the preamble maintains the same symmetry as the short portion, but there are only two halves to compare. Utilizing only the LLTF reduces the estimation range since L increases. Nonetheless, if the AGC is unable to lock in time the LLTF may need to be used for frequency estimation only. In the desirable case, where both the short and long preamble sequences can be used, a staged compensation design can be implemented. Similar to single carrier implementation discussed in Chapter 7, the short sequence can be used to provide a coarse offset estimate, and the long preamble can be used for fine frequency estimation. The estimation methods are segregated in this way due to the amount of data that can be used during estimation. Since the long preamble can generally use more data for a given calculation of P it can provide more reliable estimates. This analysis comes directly from [12], which estimates the variance of the estimate of Δ_f as

$$\sigma^2 = \frac{1}{L \times SNR}. \tag{10.10}$$

10.6 Symbol Timing Estimation

After packet detection and frequency correction, symbol timing can be performed. Symbol timing is similar to packet detection except that it provides symbol-level precision estimation of the preamble sequences in the time domain. To provide this additional precision a cross correlation is performed using the known transmitted preamble sequence. Using cross correlation provides better SNR of the correlation, but is not directly normalized to any range without additional scaling. This variable range of values make cross correlation not ideal for initial packet detection.

The cross-correlation technique applied in OFDM symbol timing estimation determines the boundary between the short and long training sequences with sample-level precision since at this point future usage of the LSTF is not required. An example implementation would be to utilize the first 80 samples of the LLTF sequence defined as $d_{LLTF}(t)$. The metric c of d_{LLTF} and the received signal r is

$$c(k) = \left| \sum_{m=0}^{L-1} d^*_{LLTF}(m) r(m+k) \right|^2. \tag{10.11}$$

In Code 10.4, two implementations for the calculation of c and are also presented in Figure 10.9. Both should provide identical results, but computationally the filter implementation will be more efficient.

Plotting c from either implementation in Code 10.4 should provide a plot similar to Figure 10.9, which clearly show two peaks 64 samples away from one another. These peaks represent strong correlations of the LLTF halves and will be pronounce

Code 10.4 Determine Fine Symbol Offset Metric: **fineSymEst.m**

```
genOFDMPacket;
% Add random offset with noise
offset = randi([0 1e2]); y = [zeros(offset,1);y];
SNR = 0; r = awgn(y,SNR,'measured');
%% Estimate fine sample offset
LSTF = Preamble(1:160);
LLTF = Preamble(161:161+160-1);
symLen = 80; % FFTLength + CPLen
known = LLTF(1:symLen,1);
% Filter
coeff = conj(flipud(known));
c_filt = abs(filter(coeff, 1, r)).^2;
% Correlation
m = abs(xcorr(r,known)).^2;
padding = length(r)-symLen+1;% Remove padding added to known sequence
c_cor = m(padding:end);
```

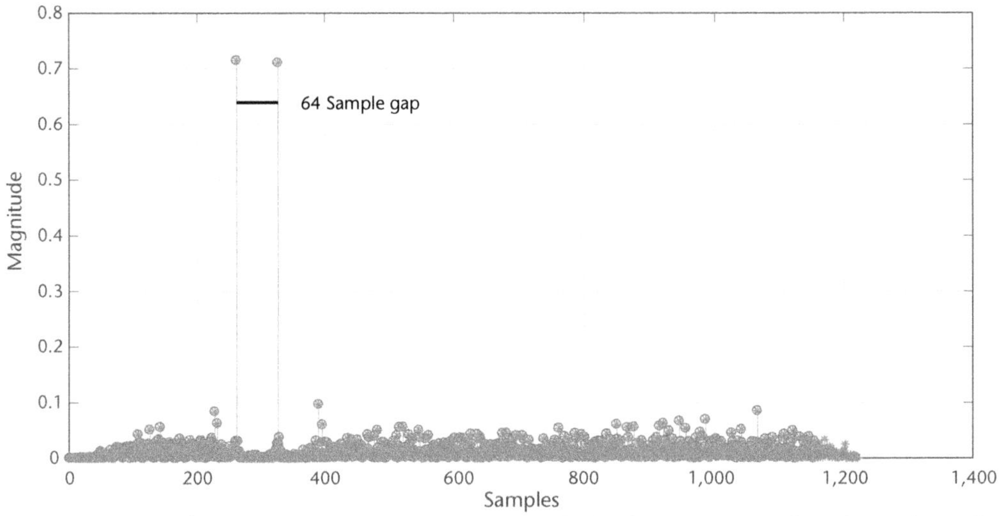

Figure 10.9 Symbol timing metric c using cross correlation for r at a SNR of 10 dB. Peaks at the positions of the LLTF sequences are clearly visible above the reset of the correlation samples.

in c even under low SNR conditions. From these sample positions OFDM symbol boundaries can be directly determined and OFDM demodulation can be performed with a FFT.

The actual offset can be calculated easily as follows in Code 10.5, since the first peak position will be shifted by half the length of the LLTF (1 OFDM symbol) due to the correlation operation.

10.7 Equalization

The last piece to the receiver is the equalizer, which is responsible for reducing channel effects and removing any residual phase or frequency offsets remaining the in the received signal. The technique discussed here is performed after OFDM

Code 10.5 Calculate Offset Position: **fineSymEst.m**

```
[v1,peak1] = max(c_cor);
c_cor(peak1) = 0;
[v2,peak2] = max(c_cor);
% Get numerical offset
if abs(peak2-peak1)==FFTLength
    % Adjust position from start of LLTF
    p = min([peak1 peak2]);
    LLTF_Start_est = p-symLen;
    LSTF_Start_est = LLTF_Start_est - length(LSTF);
    % Display
    disp([offset LSTF_Start_est]);
end
```

demodulation, although other methods do exist that can be implemented before the FFT operation.

One of the primary advantages of the cyclic prefix is that it helps transform the linear convolution between the transmitted signal $s[n]$ and the channel impulse response $h[n]$ into a symbol-by-symbol circular convolution. To clearly see this, let us take a closer look at a given OFDM symbol with cyclic prefix, the symbol starting at time $n = 0$. Denoting by $s[0], \ldots, s[2N-1]$ the $2N$ samples of output of the transmitter IDFT for the first OFDM symbol, the addition of the cyclic prefix gives rise to a new signal; namely,

$$\tilde{s}[n] = \begin{cases} s[n+2N-K] & 0 \leq n \leq K-1 \\ s[n-K] & K \leq n \leq 2N-1 \end{cases}$$

Denoted by $\tilde{r}[n]$, the result of the convolution of the signal $\tilde{s}[n]$ with the length-L channel impulse response $h[n]$ yields

$$\begin{aligned} \tilde{r}[n] &= \sum_{k=0}^{L-1} h[k]\tilde{s}[n-k] \\ &= \begin{cases} \displaystyle\sum_{k=0}^{n-K} h[k]s[n-K-k] + \sum_{k=n-K+1}^{L-1} s[n-k+2N-K] & K \leq n \leq K+L-1 \\ \displaystyle\sum_{k=0}^{L-1} h[k]s[n-K-k] & K+L \leq n \leq 2N-1 \end{cases} \end{aligned}$$

From the above equation, it is readily observed that after the removal of the cyclic prefix, the received sequence $r[n] = \tilde{r}[n+K]$ is equal to

$$r[n] = \sum_{k=0}^{2N-1} h[k]s[((n-k))_{2N}] = h[n] \textcircled{\scriptsize 2N} s[n]. \tag{10.12}$$

Thus, the received samples resulting from the removal of the cyclic prefix are made up of the circular convolution of the sent signal (i.e., $2N$ samples per symbol) with the channel impulse response $h[n]$. Observing (10.12) in the frequency domain, we see the following:

$$R[k] = H[k].S[k],$$

where capital letters represent $2N$-point DFTs of the corresponding sequences. Referring back to Figure 10.2, the $2N$-point DFT $R[k]$ of the received samples is already computed and is denoted by $\bar{p}_k[\ell]$.

Referring to Figures 10.3 and 10.1(b), if we consider the multiplication of the corresponding frequency samples together, we notice that each of the subcarriers experiences a different channel gain $H[k]$. Therefore, what must be done is to multiply each subcarrier with a gain that is an inverse to the channel frequency response acting on that subcarrier. This is the principle behind per tone equalization. Knowing what the channel frequency gains are at the different subcarriers, one can use them to reverse the distortion caused by the channel by dividing the subcarriers with them. For instance, if the system has 64 subcarriers centered at frequencies $\omega_k = 2\pi k/64$, $k = 0, \ldots, 63$, then one would take the CIR $h[n]$ and take its 64-point FFT, resulting with the frequency response $H[k]$, $k = 0, \ldots, 63$. Then, to reverse the effect of the channel on each subcarrier, simply take the inverse of the channel frequency response point corresponding to that subcarrier,

$$W[k] = \frac{1}{H[k]}, \tag{10.13}$$

and multiply the subcarrier with it.

Looking back at the WLAN-based frame structure, the LLTF can be best used for channel estimation since it will always been known at the receiver and has subcarriers that span the entire OFDM symbol. The LLTF has two complete OFDM symbols that are both used for channel estimation, averaging their estimates for more reliable results. To perform OFDM demodulation the `comm.OFDMDemodulator` will be utilize, which can be conveniently generated from a modulator object matching the necessary settings (see Code 10.8).

Code 10.6 OFDM Modulator/Demodulator: **`exampleMod.m`**

```
% Modulator
Mod = comm.OFDMModulator(...
        'FFTLength' ,            FFTLength,...
        'NumGuardBandCarriers', [6;5],...
        'CyclicPrefixLength',    0,...
        'NumSymbols', 2,...
        'InsertDCNull', true);
% Demodulator
Demod = comm.OFDMDemodulator(Mod);
```

Using this demodulator object the LLTF can be demodulated and the channel estimated using the least squares (zero-forcing) method described. This demodulation of the LLTF, channel estimation, and equalization of the data symbols with the LLTF is provided in Code 10.7. However, even with this additional correction frequency offsets can still exist in the data. Luckily, there are pilots still embedded within the data symbols to correct for these problems. Code 10.7 provides equalization with these inserted pilots as well.

The pilots are utilized in a very simple way to correct for the additive phase rotation that can be experienced from symbol to symbol in the data. This rotation again is a complex gain and will be estimated in a similar way as the LLTF was used

Code 10.7 OFDM Equalization: **eqOFDM.m**

```
genOFDMPacket;
% Add random offset
offset = randi([0 1e2]); y = [zeros(offset,1);y];
%% Equalization Example
r = awgn(y(offset+1:end),15,'measured'); Fs = 1e6;
pfOffset = comm.PhaseFrequencyOffset('SampleRate',Fs,...
    'FrequencyOffset',Fs*0.001);
r = pfOffset(r);
%% Channel estimation
preambleOFDMMod = comm.OFDMModulator(...
    'FFTLength' ,           FFTLength,...
    'NumGuardBandCarriers', NumGuardBandCarriers,...
    'CyclicPrefixLength',    0,'NumSymbols', 2,'InsertDCNull', true);
od = comm.OFDMDemodulator(preambleOFDMMod);
od.PilotOutputPort = true;
% OFDM Demodulate LLTF
LLTF = Preamble(161:161+160-1); rLLTF = r(161+32:161+160-1);
[rLLTFFreq,rp] = od(rLLTF); [LLTFFreq,p] = od(LLTF(33:end));% remove CP
% Estimate channel
ls = rLLTFFreq./LLTFFreq; % Least-square estimate
chanEst = mean(ls,2); % Average over both symbols
CSI = real(chanEst.*conj(chanEst));
ls = rp./p; % Least-square estimate
chanEstPilots = mean(ls,2); % Average over both symbols
CSIPilots = real(chanEstPilots.*conj(chanEstPilots));
%% Perform Equalization
data = r(2*length(LLTF)+1:end);
odd = comm.OFDMDemodulator(DataOFDMMod);
[dataFreq,pilots] = odd(data);
% Apply LLTF's estimate to data symbols and data pilots
postLLTFEqData = bsxfun(@times, dataFreq, conj(chanEst(:))./CSI(:));
postLLTFEqPilots = ...
    bsxfun(@times, pilots, conj(chanEstPilots(:))./CSIPilots(:));
% Visualization objects
tt1 = comm.ConstellationDiagram;tt2 = comm.ConstellationDiagram;
tt2.Position = tt2.Position + [500 0 0 0];
% Estimate remaining offsets with pilots
correctedSymbols = zeros(size(postLLTFEqData));
for symbol = 1:size(postLLTFEqData,2)
    % Estimate rotation across pilots
    p = postLLTFEqPilots(:,symbol);
    e = conj(mean(p.*conj(Pilots(:,symbol))));
    % Equalize
    sig = postLLTFEqData(:,symbol).*e;
    correctedSymbols(:,symbol) = sig;
    % Visualize
    tt1(sig);tt2(postLLTFEqData(:,symbol));pause(0.1);
end
```

for equalization [16]. Assuming that the pilots in the data were also corrected by the LLTF estimates, the N_p pilots from the data OFDM $\hat{p}(k)$ symbols are evaluated against the known pilots $p(k)$ to produce a single gain:

$$e(k) = \frac{1}{N_p} \sum_{n=0}^{N_p-1} p_n(k)\hat{p}_n^*(k). \tag{10.14}$$

This complex gain $e(k)$ is then multiplied by each sample in the kth OFDM symbol. To be clear $p_n(k)$ is the nth known pilot of the kth OFDM symbol.

10.8 Bit and Power Allocation

Resource allocation is an important aspect in OFDM design, since it determines the BER performance of the OFDM system. Given a fixed bit rate and a transmitted power constraint, the BER can be minimized by properly allocating the bit and power levels over the subcarriers. In this section, we will introduce the theory and technique concerning the bit and power allocation.

Most OFDM systems use the same signal constellation across all subcarriers, as shown in Figure 10.10(a), where the commutator allocates bit groupings of the same size to each subcarrier. However, their overall error probability is dominated by the subcarriers with the worst performance. To improve performance, adaptive bit allocation can be employed, where the signal constellation size distribution across the subcarriers varies according to the measured SNR values, as shown in Figure 10.10(b), where the commutator allocates bit groupings of different sizes. In extreme situations, some subcarriers can be turned off or nulled if the subcarrier SNR values are poor.

Assuming that both the transmitter and receiver possess knowledge of the subcarrier SNR levels, one is able to determine the subcarrier BER values. Since the subcarrier BER values are dependent on the choice of modulation used for a given SNR value, we can vary the modulation scheme used in each subcarrier in order to change the subcarrier BER value. Larger signal constellations (e.g., 64-QAM) require larger SNR values in order to reach the same BER values as smaller constellations (e.g., 4-QAM) which have smaller SNR values.

We will simply allocate the number of bits b_i on subcarrier i according to

$$b_i = \log_2\left(1 + \frac{\gamma_i}{\Gamma}\right), \tag{10.15}$$

where γ_i is the SNR of subcarrier i (not in dB). Of course, (10.15) gives rise to noninteger numbers of bits around the resulting b_i's, appropriately.

Channel responses are not always linear. In particular, certain frequencies can be far more attenuated than others. It can be shown that the optimum power distribution for the subcarriers should be a constant K. Determining what this level should be is done through water pouring, or choosing the power at a frequency, $P(f)$, such that all $P(f_1) = P(f_n)$.

$P(f)$ is given by

$$P(f) = \begin{cases} \lambda - \frac{N(f)}{|H(F)|^2}, & f \epsilon F \\ 0, & f \epsilon F' \end{cases} \tag{10.16}$$

where $N(f)$ is the PDF of Gaussian noise, $H(f)$ is the transfer function representing a linear channel, and λ is the value for which

$$\int_F P(f)df = P, \tag{10.17}$$

or the area under the curve in Figure 10.11. This water filling is done so that the probability of a bit error is the same for all subcarriers.

10.9 Putting It All Together

So far, all the necessary receiver algorithms to recover OFDM symbols transmitted over the air have been provided. Obviously the overall design could be enhanced by utilizing channel code and scramblers to reduce received BER, but the basic building blocks are provided. In review a flowchart of the receiver process is provided in Figure 10.12, which provides the basic comments on how the preamble and pilots are used. The goal is the maximally utilize the preamble symbols to provide the majority of the synchronization corrections, which differs in many ways from the single carrier implementations that heavily rely on just the modulation scheme itself.

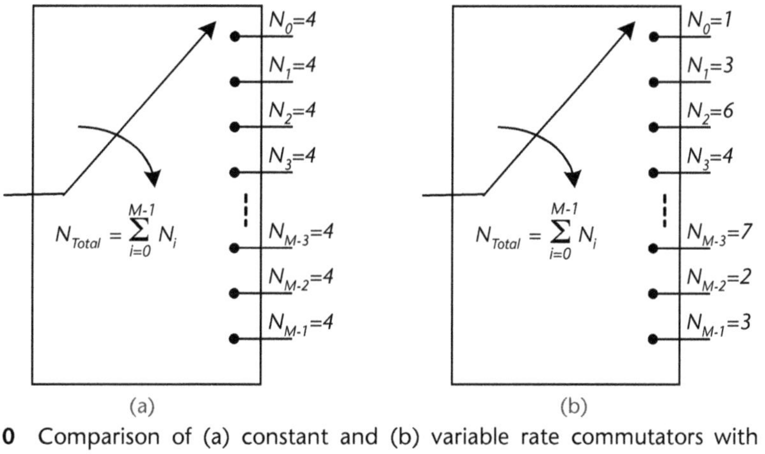

Figure 10.10 Comparison of (a) constant and (b) variable rate commutators with equivalent total rate.

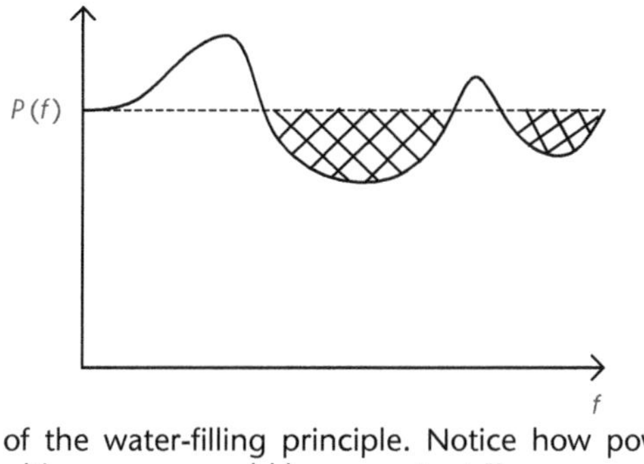

Figure 10.11 Illustration of the water-filling principle. Notice how power is allocated to each subcarrier, such that the resulting power would be a constant K.

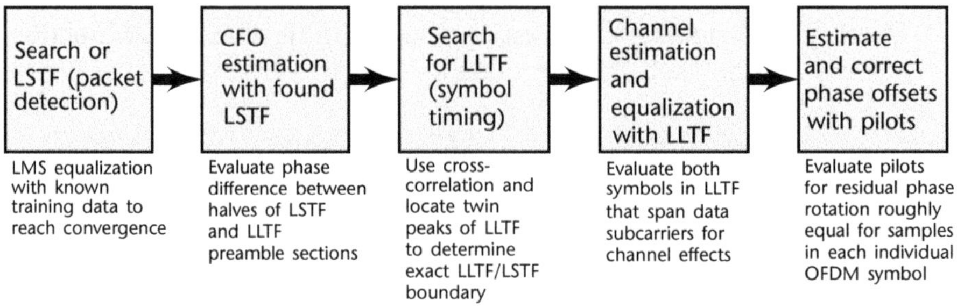

Figure 10.12 System overview for OFDM receiver for full frame recovery.

10.10 Chapter Summary

This chapter focused on the principle and implementation of multicarrier modulation, based on the 802.11a WLAN standard. Details were provided on the benefits of OFDM, including the simplistic receiver design. OFDM is the modern mechanism for high-speed transmission and reception, ranging from wired internet to even some satellite links. Although the implementation of the receiver is basically reversed from the previous succession of chapters, it can be considered much simpler than our single-carrier implementations.

References

[1] Cioffi, J. M., Chapter 34, "Asymmetric Digital Subscriber Lines," in *Communications Handbook*, CRC Press in Cooperation with IEEE Press, 1997.

[2] Golden, P., H. Dedieu, and K. Jacobsen, *Fundamentals of DSL Technology*, Boca Raton, FL: Auerbach Publications, 2004.

[3] Golden, P., and H. Dedieu, and K. Jacobsen, *Implementation and Applications of DSL Technology*, Boca Raton, FL: Auerbach Publications, 2007.

[4] Hanzo, L. L., Y. Akhtman, L. Wang, and M. Jiang, *MIMO-OFDM for LTE, WiFi and WiMAX: Coherent versus Non-coherent and Cooperative Turbo Transceivers*, Chichester, UK: Wiley-IEEE Press, 2010.

[5] Lee, B. G., and S. Choi, *Broadband Wireless Access and Local Networks: Mobile WiMaxand WiFi*, Norwood, MA: Artech House, 2008.

[6] Pareek, D., *WiMAX: Taking Wireless to the Max*, Auerbach Publications, 2006.

[7] Nuaymi, L., WiMAX: *Technology for Broadband Wireless Access*, Chichester, UK: Wiley, 2007.

[8] Sesia, S., I. Toufik, and M. Baker, *LTE—The UMTS Long Term Evolution: From Theory to Practice*, Second Edition, Chichester, UK: Wiley, 2011.

[9] Cox, C., *An Introduction to LTE: LTE, LTE-Advanced, SAE and 4G Mobile Communications*, Second Edition, Chichester, UK: Wiley, 2012.

[10] Digital Video Broadcasting (DVB), Implementation Guidelines for a Second Generation Digital Terrestrial Television Broadcasting System (DVB-T2), ETSI TS 102 831 V1.2.1, ETSI, 2012.

[11] IEEE Standard for Information Technology–Telecommunications and Information Exchange between Systems, Local and Metropolitan Area Networks–Specific Requirements–Part 11: Wireless LAN Medium Access Control (MAC) and Physical Layer (PHY) Specifications–Amendment 4: Enhancements for Very High Throughput for Operation in Bands below 6 GHz, IEEE Std 802.11ac-2013 (Amendment to IEEE Std 802.11-2012, as amended by IEEE Std 802.11ae-2012, IEEE Std802.11aa-2012, and IEEE Std 802.11ad-2012), December 2013, pp. 1–425.

[12] Schmidl, T. M., and D. C. Cox, "Robust Frequency and Timing Synchronization for OFDM," *IEEE Transactions on Communications*, Vol. 45, No. 12, December 1997, pp. 1613–1621.

[13] Minn, H., M. Zeng, and V. K. Bhargava, "On Timing Offset Estimation for OFDM Systems," *IEEE Communications Letters*, Vol. 4, No. 7, 2000, pp. 242–244.

[14] Kang, K. W., J. Ann, and H. S. Lee, "Decision-Directed Maximum-Likelihood Estimation of OFDM Frame Synchronisation Offset," *IEEE Electronics Letters*, Vol. 30, No. 25, December 1994, pp. 2153–2154.

[15] Cisco Inc., "Configuring Receiver Start of Packet Detection Threshold', Cisci Wireless Controller Guide, Release 8, April 2015, pp. 517–523

[16] van Zelst, A., and T. C. W. Schenk, "Implementation of a MIMO OFDM-Based Wireless LAN System," *IEEE Transactions on Signal Processing*, Vol. 52, No. 2, February 2004, pp. 483–494.

CHAPTER 11

Applications for Software-Defined Radio

Until now, the focus of this book was on understanding and mastering the tools of building a successful communication system using software-defined radio technology. Both theoretical concepts regarding the various building blocks of a communication system and practical insights on how to implement them have been extensively covered. The question one might be asking at this point is: What can I do with SDR? Indeed, SDR is a very powerful tool for designing, exploring, and experimenting with communication systems, but how can one wield this tool to innovate and create? In this chapter, two applications are discussed that significantly benefit from the versatility and performance of SDR: *cognitive radio* and *vehicular networking*. In particular, two approaches for implementing the intelligence and learning in cognitive radio will be discussed; namely, *bumblebee behavioral modeling* and *reinforcement learning*. As for vehicular networking, we will focus on the IEEE 802.11p and IEEE 1609 standards that define *vehicle-to-vehicle* and *vehicle-to-infrastructure* within vehicular ad hoc networks (VANETs). The goal of this chapter is to provide the reader with insights on how SDRs can be employed in these advanced applications.

11.1 Cognitive Radio

The concept of cognitive radio, whose term was coined in 2000 by Joseph Mitola [1], is a powerful methodology for performing communications where each radio within the network has the capability to sense its environment, adapt its operating behavior, and learn about new situations on-the-fly as they are encountered (see Figure 11.1). As a result of cognitive radio's ability to sense, adapt, and learn, it requires the communication system it is operating on to be highly versatile. Consequently, SDR technology is very well suited for implementing cognitive radio-based communication systems.

Although SDR seems to be a great fit for cognitive radio applications, there are several design and implementation considerations that need to be addressed. Referring to Figure 11.2, we see how the cognitive radio engine forms a layer on top of the baseband processing portion of the SDR platform. The baseband processing can be one of several computing technologies, such as general purpose microprocessor systems, FPGAs, DSPs, GPUs, ARM, and other embedded computing technologies. In fact, it might even be possible to have a SDR with several types of baseband processing technologies co-existing on the same system. Given a computing technology for a specific SDR system, one needs to be mindful that not all SDRs are built the same and that each computing technology has its advantages and disadvantages. For instance, FPGAs are not easily reprogrammable,

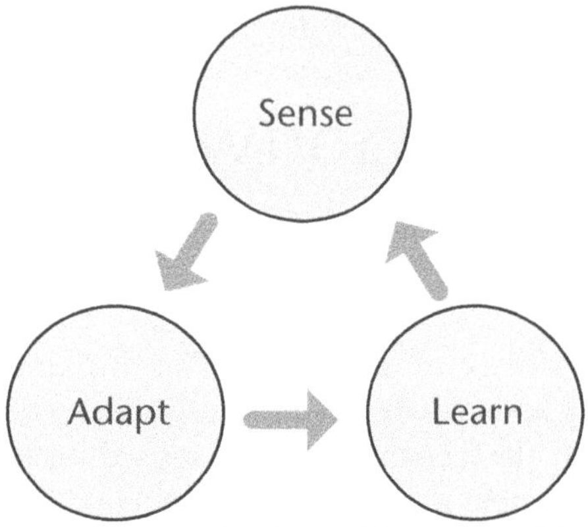

Figure 11.1 The sense, adapt, learn cycle of cognitive radio. This cycle is what differentiates cognitive radio from those that are purely automated or adaptive because of the presence of learning in this functional cycle.

which means they are not well suited for communication system operations that frequently change. On the other hand, they are very suitable for those applications requiring raw computational speed. Choosing the right computing hardware can significantly affect the performance of your cognitive radio.

If we study Figure 11.2 more closely, we can see that the cognitive radio engine consists of several inputs (sensed environment, expected performance metrics, available radio configurations), an output (radio configuration to be implemented), and a feedback loop. At the heart of the cognitive radio engine is the decision-making process, which determines the best possible radio configuration to be implemented based on past experiences, available radio configurations, sensed environmental parameters, and desired performance metrics. The decision-making process is often implemented using *machine learning*, of which there is a plethora of choices to select. To understand the design considerations for a cognitive radio engine, let us look at each of these elements more closely.

Machine learning techniques have been extensively studied to either partially or entirely automate the (re)configuration process (see [2–6] and references therein). However, the solutions produced by these techniques often require some knowledge of the wireless device and the target networking behavior (e.g., high data rates, high error robustness, and low network latency [7]). Nevertheless, machine learning techniques are well suited to handle scenarios possessing a very large device configuration solution space [4–6].

One major issue affecting cognitive radio systems is the accuracy of their decisions, which are based on the quality and quantity of input information to the cognitive radio engine. Thus, with more information available to the system, this enables the cognitive radio engine to make better decisions such that it achieves the desired user experience more precisely. Referring back to Figure 11.2, three types of parameters employed by a cognitive radio system exist:

1. *Device Configurations*: A collection of parameters that can be altered to change the current operating state of the device. Note that several potential configurations may not be possible to implement, and are thus disallowed by the adaptation algorithm.

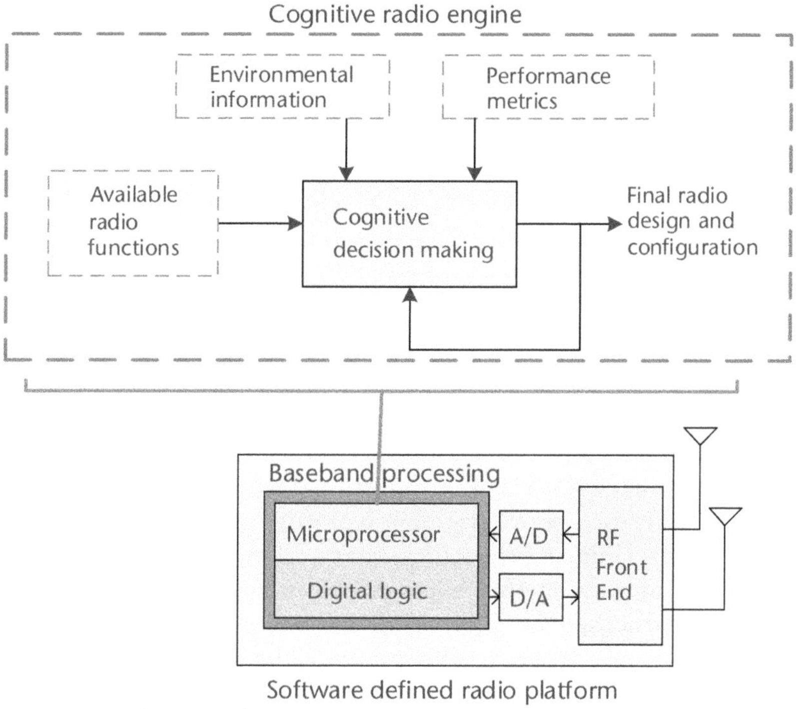

Figure 11.2 Concept diagram of a cognitive radio engine operating on a software-defined radio platform.

2. *Environmental Parameters*: These parameters represent the information about the current status of the device as well as its sensed wireless environment using external sensors.
3. *Target Networking Experience*: These metrics approximately describe the average human user's experience when operating the wireless networking device. The goal of the any cognitive radio is to achieve the best-possible value for a given metric.

Since all applications operate in different environments and possess different requires, a solution produced by the cognitive radio engine for one application that achieves superior performance might yield unacceptable performance when that same solution is applied to a different application.

The definition of an optimal decision is a combination of device configuration and environmental parameters that achieve the target networking experience better than any other combination available. Defining a proper list of parameters constituting a device configuration to be employed by a cognitive radio system is of prime importance. A well-constructed list consists of common wireless parameters that each possess a large impact on the target networking behavior. Table 11.1 shows a list of nine transmission parameters commonly used in wireless networking devices.

Environmental parameters inform the system of the surrounding environment characteristics. These characteristics include internal information about the device operating state and external information representing the wireless channel environment. Both types of information can be used to assist the cognitive radio

Table 11.1 Several Common Wireless Networking Device Configuration Parameters*

Parameter	*Description*
Transmit power	Raw transmission power
Modulation type	Type of modulation
Modulation index	Number of symbols for a given modulation scheme
Carrier frequency	Center frequency of the carrier
Bandwidth	Bandwidth of transmission signal
Channel coding rate	Specific rate of coding
Frame size	Size of transmission frame
Time division duplexing	Percentage of transmit time
Symbol rate	Number of symbols per second

* From [5, 8].

in making decisions. These variables along with the target networking experience are used as inputs to the algorithm. Table 11.2 shows a list of six environmental parameters that can affect the operational state of a cognitive radio device.

The purpose of a machine learning-based cognitive radio system is to autonomously improve the wireless networking experience for the human operator. However, improving this experience can mean several different things. Much research is focused on improving the accommodation of many wireless users within the same network. Other important aspects include providing error-free communications, high data rates, limiting interference between users, and even the actual power consumption of the wireless networking device, which is extremely important in mobile applications. As shown in Table 11.3, we have defined five common target networking experiences that guide the cognitive radio to a specific optimal solution for the cognitive radio system.

The target experiences presented in Table 11.2 represent the means for guiding the state of the cognitive radio-based wireless system. The cognitive radio makes use of these experiences through relationships that describe how modifying the device parameters achieve these objectives. To facilitate the decision making process, each target networking experience must be represented by a mathematical relationship that relates a device configuration and environmental parameters to a scalar value that describes how well this set achieve the specific goal [5, 8]. These functions will provide a way for the cognitive radio to rank combinations of configurations and environmental parameters, ultimately leading to a final decision.

Note that it is possible to specify several target networking experiences simultaneously, with the final score being represented by a numerical value. In this case, the individual scores of the target experiences are weighted according to their importance in the specific application and summed together, forming the final overall score [5].

11.1.1 Bumblebee Behavioral Model

So far we have focused on the framework surrounding the decision making process of a cognitive engine, but we have not really explored the different approaches for decision making on the radio itself. Consequently, let us explore two potential approaches for performing the decision making operation. The first approach is a *biologically inspired* method based on the behavior of bumblebees [9].

When people talk about cognitive radio, they hear the word cognitive and associate it with some sort of human intelligence that is driving the decision making

Table 11.2 Several Common Wireless Networking Environmental Parameters*

Parameter	*Description*
Signal power	Signal power as seen by the receiver
Noise power	Noise power density for a given channel
Delay spread	Variance of the path delays and their amplitudes for a channel
Battery life	Estimated energy left in batteries
Power consumption	Power consumption of current configuration
Spectrum information	Spectrum occupancy information

* From [5, 8].

Table 11.3 Several Common Wireless Networking Target Experiences*

Objective	*Description*
Minimize bit error rate	Improve overall BER of the transmission environment
Maximize data throughput	Increase overall data throughput transmitted by radio
Minimize power consumption	Decrease amount of power consumed by system
Minimize interference	Reduce the radio interference contributions
Maximize spectral efficiency	Maximize the efficient use of the frequency spectrum

* From [5, 8].

process. However, this might actually be overkill in terms of the performance we are seeking and the significant cost associated with the computational complexity. As a result, several researchers have instead focused on lifeforms with simpler cognitive capabilities as the basis for their cognitive radio engines as well as decision-making processes employed in other applications. Over the past several decades, researchers have investigated the behavior of ant colonies and honeybees as the basis for intelligent and efficient decision-making. These lifeforms possess the characteristic of being social animals, which means they exchange information with each other and perform an action that yields the best possible reward. However, ant colonies and honeybees suffer from being socially dependent lifeforms, which means that the actions of one entity is completely dependent on those of the collective. When translating these biologically inspired decision-making processes to cognitive radio and SDR, this yields a very challenging operating environment. Suppose that each radio operates a cognitive radio engine that collects information on its environment as well as information from nearby radios. As a result of this extensive information, we would expect that the radio would make excellent decisions on its own configuration. However, if these socially dependent models are used, this also constrains how these decisions are made on a per-radio basis.

Bumblebees are also social lifeforms that operate within a collective. However, unlike honeybees and ant colonies, bumblebees are socially independent by nature since they collect information from their environment and share information with each other, but they make their own decisions without control from the collective. It is this sort of flexility that makes bumblebees ideal for operating environments that could potentially change quickly over time. As for employing the bumblebee behavioral model in cognitive radio, it is well suited for operating environments where the network topology changes frequently, the channel conditions and spectral availability changes as a function of time, and the number of radios that are part of the network changes. Consequently, having each radio running a cognitive engine based on a bumblebee behavioral model is great since they gather information about the environment and then act to enhance their own performance.

In order to implement a bumblebee behavioral model for a cognitive radio engine, we need to leverage *foraging theory* on which bumblebees and many other lifeforms employ when gathering resources [9]. Essentially, foraging theory is a form of resource optimization employed by lifeforms to gather food, hunt prey, and seek shelter, along with various other operations. In the case of bumblebees, it is possible to map their activities and interpretations of the world that surrounds them to a wireless communication environment employing software-defined radio. For example, Table 11.4 presents the mapping between bumblebee behavior and perceptions to that of a vehicular ad hoc network that is dynamically accessing spectrum. It can be observed that many of the actions described for bumblebees possess some degree of similarity with those of a cognitive radio-based vehicle network.

11.1.2 Reinforcement Learning

Another decision-making process that has been receiving significant attention lately is reinforcement learning, which is a form of machine learning. As shown in Figure 11.3, reinforcement learning employs an agent that takes as inputs the reward of the previous action and the associated state and determines a new action. This action could be anything, but for the purposes of building cognitive radio engines the actions would mostly be specific radio configurations such that the performance of the communication system will be acceptable for the prevailing operating conditions, such as a dispersive wireless channel. At the receiver, the resulting reward associated with the action is calculated, which defines how well or how poorly the action was chosen. It is sent back via overhead channel to the agent in order to close the feedback loop such that the agent can adjust future actions (recall the feedback loop in the cognitive radio engine, as seen in Figure 11.2).

Although the structure described in Figure 11.3 appears to be straightforward, the one concern is about the overhead channel that is needed to close the loop. If there was some way of minimizing the impact of the overhead channel in this framework, this reinforcement learning approach could be made to be more robust. As a result, one approach for minimizing the overhead channel impact while still maintaining decent performance is to employ a neural network. The neural network is essentially a black box that can be trained using a large amount of data in order to create a complex input-output relationship in lieu of a closed-form mathematical expression. Referring to Figure 11.4, we can see an example of how a neural network

Table 11.4 Several Definitions in Connected Vehicular Communications and Their Equivalent Definitions in Bumblebees*

Vehicles	*Bumblebees*
In-band interference	Bees foraging on the same floral species
Out-of-band interference	Bees foraging on alternative floral species
Minimum channel energy level	Maximum nectar level per floral species
Computation/process time	Handling/searching time
Latency vs. reliability	Sampling frequency vs. choice accuracy
Switching cost/ time between channels	Switching cost/time between floral species
Channel activity over time	Floral species occupancy over time
Channel-user distribution	Bee distribution across floral species

* From: [9]

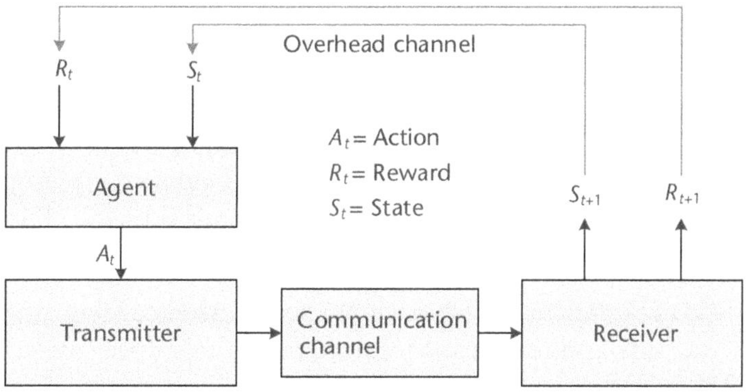

Figure 11.3 Concept diagram of a reinforcement learning approach for intelligently adapting a communication system to its operating environment [10].

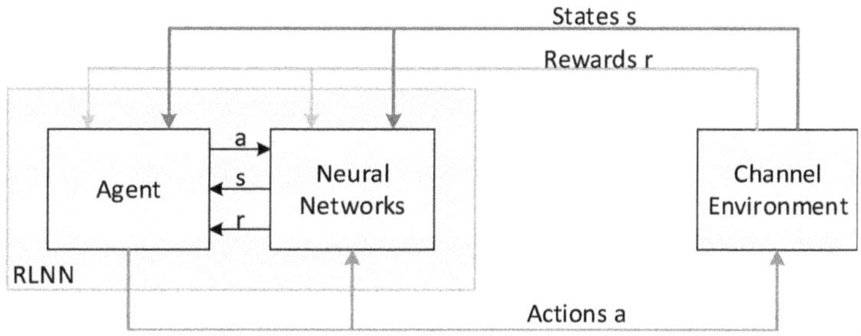

Figure 11.4 Hybrid reinforcement learning framework for communication systems [11].

can be employed within the reinforcement learning framework in order to assist the agent in deciding actions based on past rewards and states. It turns out that if the neural network is sufficiently trained such that it mathematically models associated rewards of the communication channel, it can be used to run the reinforcement learning agent until the channel conditions significantly change such that the neural network needs to be retrained.

11.2 Vehicular Networking

With some insight regarding cognitive radio, let us now proceed with exploring an application where cognitive radio combined with SDR technology can truly be a game-changer: *vehicular networking*.

Vehicular networking has been extensively researched over the past several decades [12], especially with respect to vehicle-to-vehicle (V2V) and vehicle-to-infrastructure (V2I) communications [13–16]. Given the complex nature of the operating environment, including a rapidly changing network topology [17], time-varying physical characteristics of the propagation medium [18, 19], and the need for a robust medium access control (MAC) protocol [20], vehicular networking is a challenging research area being addressed by both academia and industry.

IEEE 802.11p (*Dedicated Short Range Communications* or DSRC) and IEEE 1609 (*Wireless Access in Vehicular Environments* or WAVE) are ratified standards for the implementation of V2V and V2I network architectures [13, 16, 20–23]. Given that these standards are relatively simple extensions of the popular IEEE

802.11 family of wireless networking architectures, the ability to deploy compliant wireless devices is relatively inexpensive. However, unlike indoor environments employing Wi-Fi, vehicular networking environments are much more complex, introducing problems not experienced previously by the Wi-Fi community.

VANETs are one type of *mobile ad hoc networks* (MANETs) that specifically addresses scenarios involving moving ground vehicles. Three types of VANET applications include [16]

- *Road safety applications*: Warning applications and emergency vehicle warning applications. Messages from these applications possess top priority.
- *Traffic management applications*: Local and map information.
- *Infotainment*: Multimedia content based on the traditional IPv6 based internet.

In a VANET architecture, both V2V and V2I links may exist in order to support the communications within the network. In V2V, each vehicle is equipped with an *onboard unit* (OBU) where V2V communications is conducted between the OBUs of each vehicle mainly for road safety applications and traffic management applications [24]. The measurements for V2V DSRC are available from [15]. In V2I applications, roadside infrastructure might be equipped with a *road side unit* (RSU). In order to support these V2V and V2I communications within a VANET, two standardized protocols exist for VANETs: IEEE 802.11p and IEEE 1609. Figure 11.5 provides a graphical representation of the protocol stack of a vehicular radio unit employing IEEE 802.11p and IEEE 1609.

Referring to Figure 11.5, IEEE 802.11p [13] specifies the PHY and MAC layers, while the upper layers are defined by the IEEE 1609.x protocols. IEEE 802.11p possesses many similar characteristics relative to the IEEE 802.11-2012 standard [23]. However, to reduce the communications latency in a highly dynamic vehicular communications environment, the MAC layer needs to be defined in such a way that it can support rapid changes in the networking topology and the need for low-latency communications. Consequently, both IEEE 802.11p and IEEE 1609.4 define new characteristics for the MAC layer. For instance, in IEEE802.11 the wireless nodes could form a service set (SS) such that the nodes possess the same *service ID* (SSID) and share communications. The network possessing an

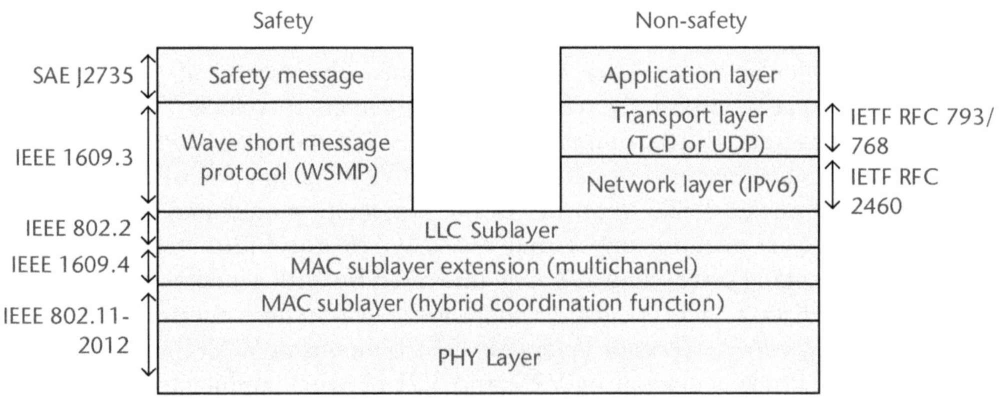

Figure 11.5 Protocol stack for a vehicular communication system.

access point (AP) is referred to as a *basic service set* (BSS) while a network with no AP (i.e., an ad hoc network), is referred to as an *independent BSS* (IBSS). Several BSS could connect together to form an *extended service set* (ESS), while all the BSSs in one ESS can share the same *extended SSID* (ESSID). The problem with respect to a VANET is that the formation of a SS before the start of any communications may potentially be time consuming, which is not suitable for rapidly changing vehicular networking environments. To handle this issue, the IEEE 802.11p standard proposed an *out-of-context BSS* (OCB), where the OCB mode applies to multiple devices within the coverage area of a single radio link. In OCB mode, the vehicle can send or receive data any time without forming or being a member of an SS. Additionally, the IEEE 802.11p standard removed the authentication, association, and data confidentiality mechanisms from the MAC layer and moved them to an independent higher layer defined in IEEE 1609.2 [21]. Conversely, IEEE 802.11p still keeps the BSS mode, which is mainly used for infotainment applications via V2I communication.

For MAC, conventional IEEE 802.11 uses carrier sense multiple access (CSMA)/collision avoidance (CA). IEEE 802.11p still employs the CSMA mechanism, but it also employs a *hybrid coordination function* (HCF), which ensures the *quality of service* (QoS) via an *enhance defense cooperation agreement* (EDCA) defined in IEEE 802.11e. Data from the different services have different priorities depending on its importance. For instance, the performance of CSMA/CA and the proposed *self-organizing time division multiple access* (STDMA) mechanism demonstrates the lower occurance of dropped packets relative to CSMA/CA [25]. However, this paper does not fully consider the HCF mechanism (i.e., the QoS based EDCA and contention-free period (CPF)-based HCCA are not evaluated). A priority-based TDMA MAC mechanism designed to decrease the packet drop rate in a transmission was also proposed for WAVE [26].

The PHY layer of a VANET based on IEEE 802.11p is derived from the IEEE 802.11a standard with three different channel width options: 5 MHz, 10 MHz, and 20 MHz, among which 10 MHz is recommended. As with IEEE 802.11a, IEEE 802.11p uses OFDM including 52 carriers, which consists of 48 data carriers and 4 pilots, and 8-μs symbol intervals. The physical channel supports BPSK, SPSK, 16-QAM, and 64-QAM. In addition to IEEE 802.11p, IEEE 1609.4 defines multichannel behavior in the MAC layer [20]. Given that the PHY layer consists of seven channels, IEEE 1609.4 defines the channel switching mechanism among the CCH and SCHs. IEEE 1609.3 defines two types of messages in VANET: *Wave Short Message Protocol* (WSMP) and IPv6 stack [22]. IPv6 is usually for infotainment applications while the safety applications are transmitted via *WAVE Short Messages* (WSM). Additionally, SAE J2945 specifies the minimum communication performance requirements of the SAE J2735 DSRC message sets and associated data frames and data elements. In order to ensure interoperability between vehicles, SAE J2945 further defines BSMs sending rate, transmit power control, and adaptive message rate control.

SAE standards have been extensively used by the automotive industry with respect to safety message implementation. In particular, SAE J2735 defines 15 types of safety messages such as the *basic safety message* (BSM), *signal phase time* (SPT) message, and MAP message [16].

BSM is broadcast to surrounding vehicles periodically at a frequency of 10 Hz, announcing the state information of the vehicle such as position, speed, acceleration, and heading direction [27]. Selective broadcasting is used, where other cars at the edge of the DSRC transmit range will rebroadcast a message sent by another vehicle. When the orginal message sender receives the rebroadcasted message, it will cancel its own broadcast. The BSM message feature is mandatory in DSRC. Note that selective broadcasting for VANETs has been implemented in NS-3 [28]. In SAEJ2735, the BSM message consists of two sections: the basic section and the optional section [29]. The basic section includes position, motion, time, and general status of the vehicle information, which are always sent using a combination of the DER encoding and some octet binary large-object encoding [27]. The optional section is only sent when it is necessary. This section provides information to assist the receiving devices in further processing.

Vehicles within the DSRC range can share situational awareness information among each other via BSM, including scenarios such as

- *Lane Change Warning*: Vehicles periodically share situational information including position, heading, direction, and speed via V2V communication within the DSRC range. When a driver signals a lane change intention, the OBU is able to determine if other vehicles are located in blind spots. The driver will be warned if other vehicles do exist in the blind spot; this is referred to as *blind spot warning*. On the other hand, if no vehicles exist in the blind spot, the OBU will predict whether or not there is enough of a gap for a safe lane change based on the traffic information via BSMs. If the gap in the adjacent lane is not sufficient, a lane change warning is provided to the driver.
- *Collision Warning*: The vehicle dynamically receives the traffic info from BSMs and compares that information with its own position, velocity, heading, and roadway information. Based on the results of the comparison algorithm, the vehicle will determine whether a potential collision is likely to happen and a collision warning is provided to the driver.
- *Emergency Vehicle Warning*: Emergency vehicles transmit a signal to inform nearby vehicles that an emergency vehicle is approaching.

In addition to the regular safety messages, BSM messages can be also be used to transmit control messages. It can help in a cooperative collision warning environment [30], in a safety message routing application [17], or improve the power control [31]. For the emergency channels (i.e., Channel 172 and Channel 184), BSM can convey power control information to coordinate the transmission power on each channel. Conversely, the BSM can be used as inputs to the vehicle's control algorithms. The control messages are transmitted among the vehicles within the range.

Given these specifications and standards regarding VANET communications, it is possible for an individual to implement their own radios capable of V2V and V2I communications. Although the complexity of the radio design is significant since the entire protocol stack is extensive, the information is sufficient to create a radio compliant with IEEE 802.11p and IEEE 1609. The primary issues to be considered when implementing IEEE 802.11p and IEEE 1609 on a SDR platform include the computing performance of the radio itself, the bandwidth limitations in terms of achievable throughput, and the real-time functionality of every function across the protocol stack. Despite these challenges, the opportunity exists to construct these vehicular communication SDR systems that can network on the road in real time.

11.3 Chapter Summary

In this chapter, we briefly examined two real-world applications that can extensively leverage SDR technology: cognitive radio and vehicular networking. With respect to cognitive radio, we explored how to set up the cognitive radio engine on a SDR platform and presented at least two ways to construct the decision-making process using either a bumblebee behavioral model or a reinforcement learning approach. Regarding vehicular network, we presented a short introduction to the IEEE 802.11p and IEEE 1609 standards that can enable us to construct our own vehicular networks from scratch using SDR technology.

In this book, we have delved into the theoretical foundations of signals, systems, and communications, and then explored the real-world issues associated with communication systems and the solutions needed to mitigate these impairments, and finally presented several advanced topics in equalization and OFDM implementations before introducing real-world applications such as cognitive radio and vehicular networking. Of course, this book only scratches the surface of the entire communication systems domain, but it is hoped that this book will serve as a starting point for mastering this very important topic.

References

[1] Mitola, J., *Cognitive Radio—An Integrated Agent Architecture for Software Defined Radio*, Ph.D. dissertation, Royal Institute of Technology, Stockholm, Sweden, 2000.

[2] Barker, B., A. Agah, and A. M. Wyglinski, "Mission-Oriented Communications Properties for Software Defined Radio Configuration," in *Cognitive Radio Networks*, Y. Xiao and F. Hu (eds.), Boca Raton, FL: CRC Press, 2008.

[3] Newman, T., A. M. Wyglinski, and J. B. Evans, "Cognitive Radio Implementation for Efficient Wireless Communication," in *Encyclopedia of Wireless and Mobile Communications*, B. Furht, (ed.), Boca Raton, FL: CRC Press, 2007.

[4] Newman, T., R. Rajbanshi, A. M. Wyglinski, J. B. Evans, and G. J. Minden. "Population Adaptation for Genetic Algorithm-Based Cognitive Radios," *Mobile Networks and Applications*, Vol. 13, No. 5, 2008, pp. 442–451.

[5] Newman, T. R., B. A. Barker, A. M. Wyglinski, A. Agah, J. B. Evans, and G. J. Minden, "Cognitive Engine Implementation for Wireless Multicarrier Transceivers" *Journal on Wireless Communications and Mobile Computing*, Vol. 7, No. 9, November 2007, pp. 1129–1142.

[6] Rieser, C. J., Biologically Inspired Cognitive Radio Engine Model Utilizing Distributed Genetic Algorithms for Secure and Robust Wireless Communications and Networking, Ph.D. Thesis, Virginia Polytechnic Institute and State University, Blacksburg, VA, 2005.

[7] Troxel, G. D., E. Blossom, and S. Boswell, et al., "Adaptive Dynamic Radio Open-source Intelligent Team (ADROIT): Cognitively Controlled Collaboration among SDR Nodes,"

in Proceedings of the IEEE Communications Society Conference on Sensors, Mesh and Ad Hoc Communications and Networks (SECON)—Workshop on Networking Technologies for Software-Defined Radio Networks. Reston, VA, August 2006.

[8] Wyglinski, A. M., M. Nekovee, and T. Hou, *Cognitive Radio Communications and Networks: Principles and Practice*, Burlington, MA: Academic Press, 2009, http://www.elsevierdirect.com/ISBN/9780123747150/Cognitive-Radio-Communications-and-Networks.

[9] Aygun, B., R. J. Gegear, E. F. Ryder, and A. M. Wyglinski, "Adaptive Behavioral Responses for Dynamic Spectrum Access-Based Connected Vehicle Networks," IEEE Comsoc Technical Committeeon Cognitive Networks, Vol. 1, No. 1, December 2015.

[10] Ferreira, P. V. R., R. Paffenroth, A. M. Wyglinski, T. M. Hackett, S. G. Bilen, R. C. Reinhart, and D. J. Mortensen., "Multi-Objective Reinforcement Learning for Cognitive Radio-Based Satellite Communications," in 34th AIAA International Communications Satellite Systems Conference, Cleveland, OH, October 2016.

[11] Ferreira, P. V. R., R. Paffenroth, A. M. Wyglinski, T. M. Hackett, S. G. Bilen, R. C. Reinhart, and D. J. Mortensen, "Multi-Objective Reinforcement Learning-Based Deep Neural Networks for Cognitive Space Communications," in Cognitive Communications for Aerospace Applications Workshop (CCAA), Cleveland, OH, June 2017.

[12] Hartenstein, H., and L. Laberteaux, "A Tutorial Survey on Vehicular Ad Hoc Networks," *IEEE Communications Magazine*, Vol. 46, No. 6, 2008, pp. 164–171.

[13] IEEE Standard for Information Technology–Local and Metropolitan Area Networks–Specific Requirements–Part 11: Wireless LAN medium access control (MAC) and physical layer (PHY) Specifications Amendment 6: Wireless Access in Vehicular Environments, Technical Report, July 2010.

[14] Kenney, J. B., "Dedicated Short-Range Communications (DSRC) Standards in the United States," *Proceedings of the IEEE*, Vol. 99, No. 7, 2011, pp. 1162–1182.

[15] Muller, M., "WLAN 802.11p Measurements for Vehicle to Vehicle (V2V) DSRC," Application Note Rohde & Schwarz, Vol. 1, 2009, pp. 1–25.

[16] IEEE Guide for Wireless Access in Vehicular Environments (WAVE)—Architecture, IEEE Std1609.0-2013, March 2014, pp. 1–78.

[17] Tonguz, O., N. Wisitpongphan, F. Bai, P. Mudalige, and V. Sadekar, "Broadcasting in VANET," 2007 Mobile Networking for Vehicular Environments, MOVE, June 2007, pp. 7–12.

[18] Akhtar, N., S. C. Ergen, and O. Ozkasap, "Vehicle Mobility and Communication Channel Models for Realistic and Efficient Highway VANET Simulation," *IEEE Transactions on Vehicular Technology*, Vol. 64, No. 1, January 2015, pp. 248–262.

[19] Akhtar, N., O. Ozkasap, and S. C. Ergen, "VANET Topology Characteristics under Realistic Mobility and Channel Models," in 2013 IEEE Wireless Communications and Networking Conference (WCNC), April 2013, pp. 1774–1779.

[20] IEEE Standard for Wireless Access in Vehicular Environments (WAVE)–Multi-Channel Operation Corrigendum 1: Miscellaneous Corrections, IEEE P1609.4-2010/Cor1/D4, August 2014, pp. 1–24.

[21] IEEE Draft Trial-Use Standard for Wireless Access in Vehicular Environments–Security Services for Applications and Management Messages," IEEE Std P1609.2/D6, 2006.

[22] IEEE Approved Draft Standard for Wireless Access in Vehicular Environments (WAVE)–Networking Services, IEEE P1609.3v3/D6, November 2015, pp. 1–162.

[23] IEEE Standard for Information Technology–Telecommunications and Information Exchange between Systems Local and Metropolitan Area Networks–Specific Requirements Part 11: Wireless LAN Medium Access Control (MAC) and Physical Layer (PHY) Specifications, Technical Report, March 2012.

[24] Uzcátegui, R. A., A. J. De Sucre, and G. Acosta-Marum, "WAVE: A Tutorial," *IEEE Communications Magazine*, Vol. 47, No. 5, 2009, pp. 126–133.

[25] Bilstrup, K., E. Uhlemann, E. G. Strom, and U. Bilstrup, "Evaluation of the IEEE 802.11p MAC Method for Vehicle-to-Vehicle Communication," in Vehicular Technology Conference, VTC2008-Fall, IEEE 68th IEEE, 2008, pp. 1–5.

[26] Li, B., M. S. Mirhashemi, X. Laurent, and J. Gao, "Wireless Access for Vehicular Environments." *Journal on Wireless Communications and Networking*, 2009: 576217, https://doi.org/10.1155/2009/576217.

[27] *DSRC Implementation Guide: A Guide to Users of SAE J2735 Message Sets over DSRC*, SAE International, February 2010.

[28] Bür,K., and M. Kihl, "Selective Broadcast for VANET Safety Applications," in SNOW–the 2nd Nordic Workshop on System and Network Optimization for Wireless, Salen, Sweden, 2011.

[29] *Vehicle Information Exchange Needs for Mobility Applications*, RITA Intelligent Transportation Systems Joint Program Office, February 2012 [online], available athttp://www.its.dot.gov/.

[30] ElBatt, T., S. K. Goel, G. Holland, H. Krishnan, and J. Parikh, "Cooperative Collision Warning Using Dedicated Short Range Wireless Communications," Proceedings of the 3rd International Workshop on Vehicular Ad Hoc Networks–VANET '06, p. 1, 2006 [online], available at http://portal.acm.org/citation.cfm?doid=1161064.1161066.

[31] Yoon, Y., and H. Kim, "Resolving Distributed Power Control Anomaly in IEEE 802.lip WAVE," *IEICE Transactions on Communications*, Vol. E94-B, No. 1, 2011, pp. 290–292.

APPENDIX A

A Longer History of Communications

A.1 History Overview

This section is included not to encourage mindless memorization of names or dates, or to overwhelm the reader, but to help understand how communications systems have changed, how adoption of new concepts can be lengthy, and how the technology we use every day in our modern communications infrastructure came to be. Some may wonder if this section is necessary at all; however, as Carl Sagan famously said, "If you wish to make an apple pie from scratch, you must first invent the universe" [1]. The more the history of communications is understood, the more of a modern marvel the current communications are. It was only 260 short years ago that an an anonymous letter in a magazine was published theorizing instantaneous communication by routing 26 wires (one for each letter of the alphabet; coding schemes were not commonplace), sending static electric pulses down the wire, and watching pith balls move at the other end (Volta had not developed the battery yet). This was an early trigger for the multicentury communications revolution.

It goes without saying that communications systems are limited to the transmission media that are available at the time. Aboriginal Australians in the Western Desert, indigenous peoples of North America, ancient Chinese soldiers stationed along the Great Wall, and the ancient Greeks all devised complex systems of how to transmit messages based on fire or smoke signals. Polybius, a Greek historian, developed a an early coding system of alphabetical smoke signals around 150 BCE that converted Greek alphabetic characters into numeric characters [2]. Each of these approaches possesses the same sort of engineering trade-offs that modern communication systems struggle with today, including *transmission distance* and *data rate*.

Some of these communication schemes are still used today; for example the semaphores shown in Figure A.1 that are used when stealth outside of the line of sight is required (flags don't emit RF, which could give away location).

It wasn't until the early 1990s when everything was beginning to come together. It was the unique combination of RF capability, software, algorithms, and computer hardware that are necessary to actually implement things. In fact it was only in 1993 that the term software-defined radio was mentioned.

This brief history will ignore the improvements of Johannes Gutenberg's movable type printing press, developed in around 1440, which spawned an entire printing and publishing economy. By 1700, the number of various books and scholarly journals published in Europe was an estimated 150 to 200 million copies [3]. This ability to share ideas and concepts with contemporary scientists,

Figure A.1 Persian Gulf (Novmber 29, 2005) Quartermaster Seaman Ryan Ruona signals with semaphore flags during a replenishment at sea aboard the Nimitz-class aircraft carrier USS Theodore Roosevelt (CVN 71). Roosevelt and embarked Carrier Air Wing Eight (CVW-8) are currently underway on a regularly scheduled deployment conducting maritime security operations. (U.S. Navy photo by Photographer's Mate Airman Javier Capella (RELEASED).)

inventors, and engineers of the day and of future helped to usher in the continuing scientific revolution.

People like Leonardo da Vinci are not noted in this list. While he made substantial discoveries in anatomy, civil engineering, geology, optics, and hydrodynamics, he did not publish his findings. Future generations of inventors, scientists, and engineers had no formal documentation to study or expand upon [4].

A.2 1750–1850: Industrial Revolution

By many accounts, 1750 kicks off the Industrial Revolution, which occurred from approximately 1750 to 1850. This transition created change for many:

- Hand production giving way to to mechanization;
- New processes for chemical and iron production;
- The increased adoption of steam power;
- The rise of the factory system;
- People moving from cottage-based industries to factory jobs.

Innovations developed late in the period, such as adoption of locomotives, steamboats and steamships, hot blast iron smelting and other new technologies, such as the electrical telegraph, continued to change daily life for many [5].

1748: Leonhard Euler publishes *Introductio in Analysin Infinitorum* ("Introduction to the analysis of the infinite") [6]. Euler was not only one of the most important and influential mathematicians, but also the most prolific. He wrote more than 500 books and papers during his lifetime [7].

1752: Benjamin Franklin shows that lightning is caused by electricity during his well-known kite experiment in Philadelphia [8].

1753: An anonymous writer in the *Scots Magazine* suggests an electrostatic telegraph. Using multiple wires, a message would be transmitted by observing the deflection of pith balls [9].

1769: James Watt develops a reciprocating steam engine, capable of powering a fly wheel [10].

1792: Claude Chappe constructs a semaphore network across France during the French Revolution to provide a swift and reliable communication system. It brought news over a distance of 230 kilometers in under one hour, faster than any other communications method at the time [11].

1800: Alessandro Volta develops the voltaic pile (now known as a battery). With a constant source of DC voltage, more practical experiments could be made and new apparatus invented [12].

1804: Improving on James Watt's steam engine, Richard Trevithicks build the first full-scale working railway steam locomotive [13].

1820: Johann Schweigger builds the first sensitive galvanometer, an electromechanical instrument for detecting and indicating electric current [14].

1827: Georg Ohm publishes the book *Die galvanische Kette, mathematisch bearbeitet* [15] ("The Galvanic Circuit Investigated Mathematically"), where the fundamental law of electric circuits, $V = I \times R$—what becomes known as Ohm's law—is shared. At the time, Ohm's employor, the Dreikönigsgymnasium in Cologne, did not appreciate his work and Ohm resigned from his teaching position [16].

1827: André-Marie Ampère publishes the book *Mémoire sur la théorie mathématique des phénomènes électrodynamiques uniquement déduite de lexperience* ("Memoir on the Mathematical Theory of Electrodynamic Phenomena, Uniquely Deduced from Experience") [17]. This is the text that created electrodynamics as a field of study.

1831: Michael Faraday produces current from a moving magnet, called a disc dynamo, later to be known as an electrical generator [18].

1833: Carl Friedrich Gauss installs a 1,200m long wire above Göttingen's roofs to experiment with the telegraph [19].

1837: Charles Babbage designs the mechanical analytical engine, the first a general-purpose computer [20].

1837: Samuel Morse patents a recording electric telegraph. In order to sell his new equipment, an alphabet encoding scheme referred to as *Morse code* was used in order to communicate strings of characters [21]. Simple to use, electric telegraphs were extensively employed during most of the 1800s, with data rates dependent on the skill of the telegraph operator and the capabilities of the equipment.

A.3 1850–1945: Technological Revolution

The Technological Revolution, or what many call the Second Industrial Revolution, was punctuated by advancements in manufacturing that enabled the widespread adoption of preexisting technological systems such as telegraph and railroad networks, gas and water supply, and sewage systems. The enormous expansion of rail and telegraph lines after 1870 allowed unprecedented movement of people and ideas. In the same time period, new technological systems were introduced, most significantly electrical power and telephones.

1860: Giovanni Caselli demonstrates his pantelegraph, an early form of facsimile machine, which transmitted a likeness of a signature over 140 km over normal telegraph lines [22].

1864: James Maxwell publishes a paper, *A Dynamical Theory of the Electromagnetic Field*, which tells the world about his findings about electromagnetic waves and the math behind them. A collection of 20 equations become known as *Maxwell's equations* [23].

1865: At the International Telegraph Convention in Paris, the *Union Tèlègraphique Internationale (International Telegraph Union or ITU)* was formed to codify and ensure interoperability within the growing international telegraphic traffic within Europe [24].

1869: The First Transcontinental Railroad is completed. It was a 3,077 km continuous railroad line that connected the eastern U.S. rail network with the Pacific coast in San Francisco Bay [25]

1873: Maxwell publishes his text *A Treatise on Electricity and Magnetism* [26]. This would be read by many, including Hermann von Helmholtz, a faculty member at the University of Berlin.

1874: Karl Braun develops the cat's-whisker detector, later known as a point-contact rectifier. Consisting of a thin wire that lightly touches a crystal of semiconducting mineral (usually galena), this device was used as the detector in early crystal radios, from the early twentieth century through 1945 [27].

1876: Alexander Graham Bell successfully implementes the world's first telephone, ushering in a new age of voice-based electronic communications [28]. It was now possible for any individual to communicate with any other individual anywhere on the planet, constrained only by the wired medium connecting the two of them to each other.

1879: Thomas Edison files for a U.S. patent for an electric lamp using "a carbon filament or strip coiled and connected ... to platina contact wires" [29].

1879: Heinrich Hertz, a Ph.D. student, is given the challenge by his faculty advisor, Hermann von Helmholtz, to prove the practical uses of Maxwell's equations, which were published only a few years before. Although Hertz did some analysis, he thought this suggestion was too difficult and worked on electromagnetic induction instead [30].

1885: Karl Benz builds the first practical motorized automobile [31].

1886: Heinrich Hertz receives his professorship at the Karlsruhe Institute of Technology and is finally able to fulfill his advisor's challenge and builts an apparatus for generating and detecting radio waves [30]. This eliminated the needed wired medium for communications (no more cables), since radio waves would be generated by a transmitter and propagate throughout an area until they are intercepted by a receiver.

1894: Jagadish Bose uses a cat's-whisker detector to detect radio waves in his pioneering experiments with microwaves [32], finally applying for a patent on a galena detector in 1901 [33].

1897: Karl Braun builds the first cathode-ray tube (CRT) and cathode ray tube oscilloscope [34].

1901: Guglielmo Marconi demonstrates the very first trans-Atlantic wireless transmission between Poldhu, Cornwall, in England and Signal Hill, St. John's, in Newfoundland in 1901 [35], using messages encoded using Morse code [36]. The viability of wirelessly transmitting information over very large distances was proven. With the success of this experiment, wireless data transmission became an option for connecting people around the world without the need for any wired communication infrastructure.

1904: John Fleming develops the oscillation valve, later to be called the two-electrode vacuum-tube rectifier (now known as a diode) [37]. Fleming's diode was used in radio receivers and radars for many decades afterward as part of the ring mixer, as shown in Figure A.2.

1905: While accepting his 1909 Nobel Prize for physics, Karl Braun describes how he carefully arranged three antennas to transmit a directional signal [38]. This invention led to the development of radar, smart antennas, and MIMO.

1905: Reginald Fessenden describes a proposed method of producing an audible signal from the new Morse code continuous wave transmitters he calls the *heterodyne* principle [39]. It was not a practical device at the time, since the oscillating vacuum-tube had not been developed yet.

1906: The International Radiotelegraph Union was unofficially established at first International Radiotelegraph Convention in Berlin [40].

1907: Lee De Forest develops grid audions, the first vacuum tube based amplifier, the precursor to the triode [41].

1912: Henry Ford develops and manufactures the first affordable automobile, turning the automobile into a practical conveyance [42].

1913: While an undergraduate student at Columbia, Edwin Armstrong prepares a series of comprehensive demonstrations and papers that carefully document his research on De Forest's grid audions, employing regeneration (later known as positive feedback) producing amplification hundreds of times greater than previously thought possible [43].

1915: Alexander Graham Bell and Thomas A. Watson talk to each other over a 3,400-km wire between New York and San Francisco [44].

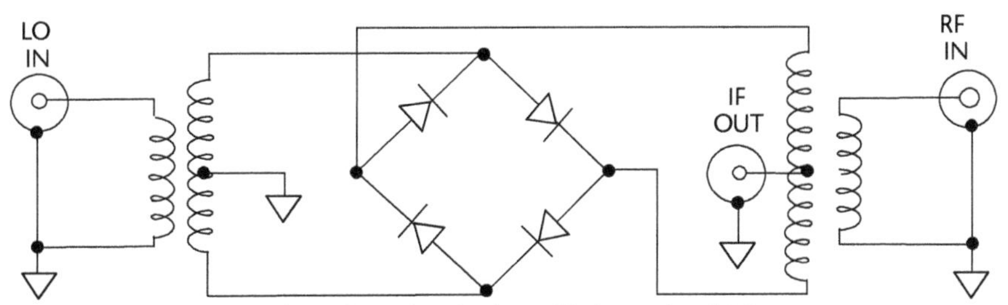

Figure A.2 Diode ring mixer based on Fleming's oscillation valve [45].

1917: Eric Tigerstedt files a patent for a pocket-size folding telephone with a very thin carbon microphone, a suggestion of the mobile phones we know today [46].

1917: The superheterodyne is patented. With three different patent applications being filed by three different people, Lucien Lèvy, Edwin Armstrong, and Walter Schottky, it is unknown who actually was first. Eventually, the U.S. patent office recognizes both Armstrong and Lèvy as the inventors [47]. The superheterodyne is the design used in almost all modern radio receivers. Figure A.3 shows a block diagram of a single-conversion superheterodyne radio receiver used as the fundamental architecture for many modern radios.

1924: Harry Nyquist, an engineer in AT&T's Department of Development and Research (later to be known as Bell Telephone Laboratories) studies telegraph signaling with the objective of finding the maximum signaling rate that could be used over a channel with a given bandwidth, and publishes "Certain Factors Affecting Telegraph Speed" [48].

1927: Charles Ducas patents a method of electroplating circuit patterns, the fundamental technology used in the manufacturing of printed circuit boards [49].

1928: Harold Black of Bell Labs patents negative feedback in amplifiers, where he defines negative feedback as a type of coupling that reduces the gain of the amplifier, in the process greatly increasing its stability and bandwidth [50].

1928: Nyquist also publishes futher findings in his article, "Certain Topics in Telegraph Transmission Theory" [51]. Both his papers provide fundamental contributions to a quantitative understanding of data transmission.

1932: The International Telegraph Convention and International Radiotelegraph Convention merge into the International Telecommunication Union (ITU) in Madrid. This new convention covers the three fields of communication: telegraphy, telephony, and radio [40].

1932: Since superheterodyne architecture was normally implemented in multiple stages, and required large components, a team of British scientists develop what they call the homodyne, later renamed as synchrodyne, what we now

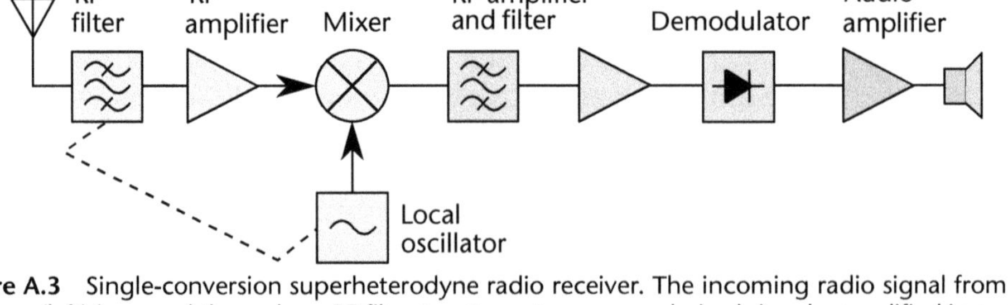

Figure A.3 Single-conversion superheterodyne radio receiver. The incoming radio signal from the antenna (left) is passed through an RF filter to attenuate some undesired signals, amplified in a radio frequency (RF) amplifier, and mixed with an unmodulated sine wave from a local oscillator. The result is a beat frequency or heterodyne at the difference between the input signal and local oscillator frequencies, a lower frequency called the IF. The IF signal is selected and strengthened by several IF stages that bandpass filter and amplify the signal. The IF signal is then applied to a demodulator that extracts the modulated audio signal. An audio amplifier further amplifies the signal, and the speaker makes it audible.

call the zero-IF radio architecture [52]. While the homodyne had shown performance improvements from the superheterodyne, it was difficult to implement due to component tolerances at the time, which must be of small variation for this type of circuit to function successfully and the requirement of a PLL (the superhet only requires a oscillator).

1940: The term RADAR was used by the United States Navy as an acronym for radio detection and ranging [53].

1945: Arthur C. Clarke, science fiction writer, pens a letter to the editor of *Wireless World* describing how geostationary satellites would be ideal telecommunications relays [54].

A.4 1946–1960: Jet Age and Space Age

While research was being done and patents filed about jet engines since the 1920s, it was not until Hans von Ohain and Max Hahn designed and built what would be the first gasoline-fueled jet engine, known as HeS 3. Providing 5 kN of force, it was fitted to simple and compact He 178 airframe and in 1939, the Jet Age began [55]. Commercial aviation introduced jets with the first scheduled flight of the de Havilland Comet, the world's first commercial jetliner, in 1952 [56, 57].

With people hurling through the sky at hundreds of kilometers per hour, the need for sophisticated communications grew, and knowing your location, and where other people are (collision avoidance), becomes more critical.

1946: Electronic Numerical Integrator and Computer (ENIAC) is formally dedicated at the University of Pennsylvania, becoming one of the earliest electronic general-purpose computers made [58].

1947: Engineers at Bell Labs observe that when two gold contacts are applied to a crystal of germanium, a signal is produced with the output power greater than the input, and this device is soon to be known as the transistor. Group leader William Shockley sees the potential in this, and over the next few months works to greatly expand the knowledge of semiconductors [59].

1948: Claude Shannon publishes "A Mathematical Theory of Communication" [60] in the *Bell System Technical Journal*. Shannon shows that all communication (voice, radio, data, etc.) is fundamentally the same, and furthermore, that any source can be represented by digital data. He was able to quantify that communication channels had a speed limit, measured in binary digits per second, something that was revolutionary for its time. Shannon gave a mathematical proof that there had to exist codes that would approach the limit without losing any information at all, and this proof still holds to this day [61].

The concepts of digital communication took off in the late 1950s, since Shannon showed *perfect information transfer* is possible no matter the noise, distortion, or signal amplitude. The concept of *relay stations* became important, as data could be transfered from the source at A, to B (the relay) to C (the destination) without the loss of information.

1950: The *Auto-Sembly* process is patented by the United States Army [62]. This was the solder dipped circuit process of assembling electrical circuits and was widely adopted by both military and industrial designers [63].

1950: William Shockley publishes *Electrons and Holes in Semiconductors* [64], becoming the reference text for other scientists working to develop and improve new variants of the transistor.

1954: Epsco, a company founded by former UNIVAC engineer Bernie Gordon, releases the first commercially offered analog-to-digital (ADC) to utilize the shift-programmable successive approximation architecture and include sample-and-hold function. It was called the 11-bit, 50-kSPS, 150 lbs, Datrac converter. Implemented with vacuum tubes, it dissipated 500 watts, was designed for rack mounting (19" × 15" × 26"), and sold for $8,000 to $9,000. It was the first commercial ADC suitable for digitizing AC waveforms, such as speech [65]. Because of vacuum tube technology, the converters were very expensive, bulky, and dissipated lots of power. There was practically no volume commercial usage of these devices.

1956: William Shockley moves to Mountain View, California, to start Shockley Semiconductor Laboratory, starting the silicon valley revolution [66].

1957: Because of Shockley's management style, a eight unhappy employees leave Shockley Semiconductor Laboratory and form Fairchild Semiconductor [66]. This is the start of a trend in silicon valley that continues to this day.

1957: The Soviet Union launches Elementary Satellite 1, or Sputnik 1, into elliptical low earth orbit [67].

1958: Jack Kilby, a junior engineer at Texas Instruments (TI), comes to the conclusion that the manufacturing of circuit components could be done on a single piece of semiconductor material. He then files for the first integrated circuit patent, *miniaturized electronic circuits*, the next year [68].

1962: The first communications satellite, *Telstar 1* is launched [69]. Telstar 1 relayed data, a single television channel, or multiplexed telephone circuits between Andover, Maine, Goonhilly Downs in southwestern England, and at Pleumeur-Bodou, in northwestern France. Its use for transatlantic signals was limited to 20 minutes in each 2.5 hour when the non geosynchronous orbit passed the satellite over the Atlantic Ocean [70].

1965: Ray Stata and Matthew Lorber, two MIT graduates, found Analog Devices Inc. [71]. The same year, the company releases its first product, the model 101 op-amp, which was a hockey-puck-sized module used in test and measurement equipment [72].

1965: Gordon Moore, director of research and development at Fairchild Semiconductor, observes that the number of components (transistors, resistors, diodes, or capacitors) in a dense integrated circuit had doubled approximately every year, and speculated that it would continue to do so for at least the next ten years. He later revised the forecast rate to approximately every two years [73]. This has become known as Moore's law.

1966: Analog-to-digital converters (ADC) are now not enjoying the same amount of scaling that their pure digital counterparts were. Computer Labs (acquired by Analog Devices in 1978) were selling an 8-bit, 10-MSPS converter that was rack mounted, contained its own linear power supply, dissipating nearly 150 watts, and selling for approximately $10,000 [65].

1967: Barrie Gilbert develops a ring mixer based on transistors, which becomes known as the *Gilbert cell* or the four quadrant multiplier [74]. Building on the diode based ring mixer shown in Figure A.2, the diodes were replaced by four transistors, performing the same switching function. This formed the basis of the now-classical bipolar circuit shown in Figure A.4, which is a minimal configuration for the fully balanced version. Millions of such mixers have been made, now including variants in CMOS, BiCMOS, and GaAs.

The Gilbert cell is attractive as an active mixer for the following reasons:

- It can be monolithically integrated with other signal processing circuitry.
- It can provide conversion gain, whereas a diode-ring mixer always has an insertion loss.
- It requires much less power to drive the LO port.
- It provides improved isolation between the signal ports.
- Is far less sensitive to load-matching, requiring neither diplexer nor broadband termination.

1967: Andrew Viterbi, a faculty of electrical engineering at UCLA and UCSD, publishes an algorithm to decode convolutionally encoded data [75], known as the *Viterbi algorithm*. It is still used widely in cellular phones for error correcting codes, as well as many other applications. Viterbi did not patent the algorithm, a decision he does not regret [76].

1969: The first successful message on the ARPANET, which laid the foundations of the internet is sent by a UCLA student programmer [77].

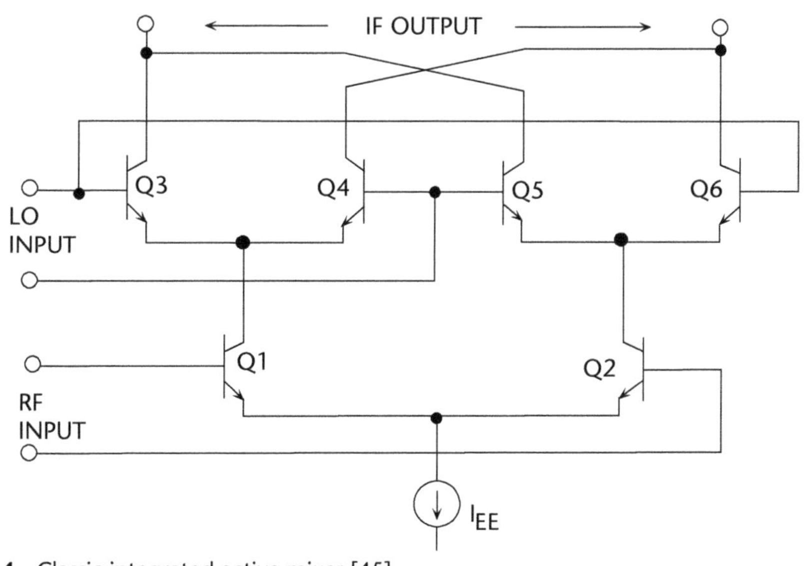

Figure A.4 Classic integrated active mixer [45].

A.5 1970–1979: Information Age

The Information Age is characterized by the shift from traditional industry that the Industrial Revolution, to an economy based on information computerization. The 1970s were driven by a large number of emerging applications, including high-volume calculators, high-resolution digital voltmeters, industrial process control, digital video, military phased array radar, medical imaging, vector scan displays, and raster scan displays. These systems had formerly utilized conventional (at the time) analog signal processing techniques, or ASICs, but the increased availability of low-cost computing technology generated a desire to take advantage of the increased performance and flexibility offered by digital signal processing, as well as the need for compatible data converters. Central to this age is the mass production and widespread use of digital logic circuits and their derived technologies, including the computer and cellular phones.

1970: The term digital receiver is coined by a researcher at a United States Department of Defense laboratory, where a software baseband analysis tool called Midas (multiuser interactive digital analysis system) was created, which had its operation defined in software running on a mainframe of the time [78].

1971: The Intel Corporation releases, the Intel 4004, the first commercially available 4-bit central processing unit (CPU) [79].

1971: Texas Instruments invents and releases the first microcontroller (MCU), the TMS1802NC [80].

1971: ALOHAnet connects the Hawaiian islands with a UHF wireless packet network. The goal was to use low-cost commercial radio equipment to connect users on Oahu and the other Hawaiian islands with a central time-sharing computer on the main University of Hawaii, Oahu campus [81].

1972: Signetics (later acquired by Philips Semiconductors, and now NXP, and, at the time of this writing is being acquired by Qualcomm) introduces the NE555V timer. As of 2003, it was estimated that 1 billion units were manufactured every year [82].

1972: Pulse code modulation (PCM) of voice frequencies, the ITU-T standard for audio companding known as G.711 is released. It is a narrowband (300–3400 Hz) audio codec that provides voice quality similar to analog signal transmission over copper lines, known as plain old telephone service (POTS) [83].

1973: Martin Cooper of Motorola makes the first publicized handheld mobile phone call on a prototype DynaTAC, a 4.4-lb (2-kg) phone.

1973: The U.S. Department of Defense kicks off planning for the Defense Navigation Satellite System (DNSS), soon to be renamed Navigation System Using Timing and Ranging (Navstar), and eventually Global Positioning System (GPS). Ten prototype satellites are launched between 1978 and 1985.

1975: Paul Allen and Bill Gates officially establish Microsoft and license Altair BASIC [84].

1975: Analog Devices releases the AD7570, a 10-bit, 50-kHz CMOS SAR ADC on a monolithic device. Due to the difficulty of designing good comparartors, amplifiers, and references on the early CMOS processes, the AD7570 required an external comparator as well as an external voltage reference [65].

1975: Barrie Gilbert joins Analog Devices and releases the AD534 [85] Precision IC Multiplier. This is the first device to utilize laser trimming and the Gilbert cell, to provide less than ±0.25 % 4-quadrant error [86].

1975: Steven Sasson develops the first digital camera [87]. At 3.6 kg, using a 100,100-pixel (0.01 megapixels) sensor, the resulting black-and-white digital image is recorded onto a cassette tape, taking 23 seconds [88].

1976: Steve Jobs, Steve Wozniak, and Ronald Wayne found Apple Computer [89]. The company's first product is the Apple I, a computer designed and hand-built by Wozniak [90].

1977: What *Byte magazine* later refers to the 1977 Trinity of personal computing [91]:

- Apple releases the Apple, based on 6502 running at 1.023 MHz
- Commodore releases the Commodore Personal Electronic Transactor (PET), based on 6502 running at 1 MHz
- Tandy releases the TRS-80 based on a Zilog Z80 at 1.77 MHz

1978: The EEPROM (electrically erasable programmable read-only memory) is developed by Hughes Microelectronics Division, which offers a huge improvement from EPROM in that it is now not necessary to shine UV light on the memory to erase it [92].

1978: Texas Instrument produces the first Speak & Spell, with the technological centerpiece being the TMS5100, the industry's first digital signal processor (DSP). It also sets other milestones, being the first chip to use linear predictive coding to perform speech synthesis [93].

1978: The AD574 is released by Analog Devices, the most significant SAR ADC ever produced [94]. It is the first device representing a complete solution, including Zener reference, timing circuits, and three-state output buffers for direct interfacing to an 8-, 12- or 16-bit microcontroller or microprocessor bus [65]. This becomes the industry standard 12-bit ADC for its time, but is based on two die in the same package.

1978: Fred Harris, professor of electrical engineering, San Diego State University, publishes his most cited paper, "On the use of windows for harmonic analysis with the Discrete Fourier Transform" [95], receiving over 10,000 peer citations for the papers he has published and information he has shared [96]. Dr Harris started publishing papers in 1962, and continues to this day.

A.6 1980–1989: Digital Revolution

The Digital Revolution is the continued change from mechanical and analog electronic technology to digital electronics which began, depending on the market segment, anywhere from the late 1950s to the late 1970s with the adoption of digital signal processing and desktop computing that continues to the present day.

1981: IBM Personal Computer is introduced, based on a 4.77 MHz Intel 8088 16 KB RAM, color graphics adapter, and no disk drives [97].

1982: The Internet Protocol Suite (TCP/IP) is standardized by the U.S. Department of Defense, which permits worldwide proliferation of interconnected networks [98].

1983: The Motorola DynaTAC 8000X commercial portable cellular phone is released. A full charge takes roughly 10 hours and offers 30 minutes of talk time. It was priced at \$3, 995 in 1984, and used an analog modulation scheme, known as advanced mobile phone system, or AMPS.

1983: John G. Proakis [99] publishes his first of ten books on digital communications and signal processing, including *Digital Communications* (McGraw Hill) [100]. He goes on write many books on the subject: *Digital Signal Processing* [101], *Digital Signal Processing Laboratory* [102], *Advanced Digital Signal Processing* [101], *Digital Processing of Speech Signals* [103], and *Communication Systems Engineering* [104]. His books have educated a generation of students and engineers about the fundamentals and theory associated with SDR.

1983: Altera is founded and delivers the industry's first reprogrammable logic device in 1984, the EP300, which features a quartz window in the package that requires users to shine an ultraviolet lamp on the die to erase the EPROM cells that hold the device configuration [105].

1983: President Ronald Reagan issues a directive making GPS freely available for civilian use, once it is sufficiently developed, as a common good. The first production satellite was launched on February 14, 1989, and the 24th satellite was launched in 1994.

1983: Work begins to develop a European standard for digital cellular voice telecommunications, which eventually becomes GSM.

1984: Apple Computer introduces the first Macintosh computer. It is the first commercially successful personal computer to use a graphical user interface (GUI) and mouse, which become standard features in computers [106]. It features a 7.83 MHz 32-bit Motorola 68000 CPU, built-in 9-inch monochrome screen, 512 × 342 graphics, 400 kB 3.5-inch Sony floppy disk drive, mouse, 128 kB RAM. Weight is 20 pounds, size is 9.7 × 10.9 × 13.5 inches, and price ranges from \$1,995 to \$2,495.

1984: The term software radio is coined by a team at the Garland, Texas Division of E-Systems Inc. (now Raytheon) to refer to a digital baseband receiver and the team publishes in their E-Team company newsletter [107].

1984: MathWorks is founded, and MATLAB made its public debut at the IEEE Conference on Decision and Control in Las Vegas, Nevada [108]. Used in industry and education, in particular the teaching of linear algebra, numerical analysis, it is popular with scientists involved in signal processing.

1984: The ITU releases the V.32 recommendation for a dial-up modem, allowing bidirectional data transfer at either 9.6 kbit/s at a symbol rate of 2,400 baud using QAM modulation [109].

1985: Analog Devices rereleases AD574 in single-chip monolithic form, as shown in Figure A.5. For the first time a low-cost commercial plastic converter is available in mass volume.

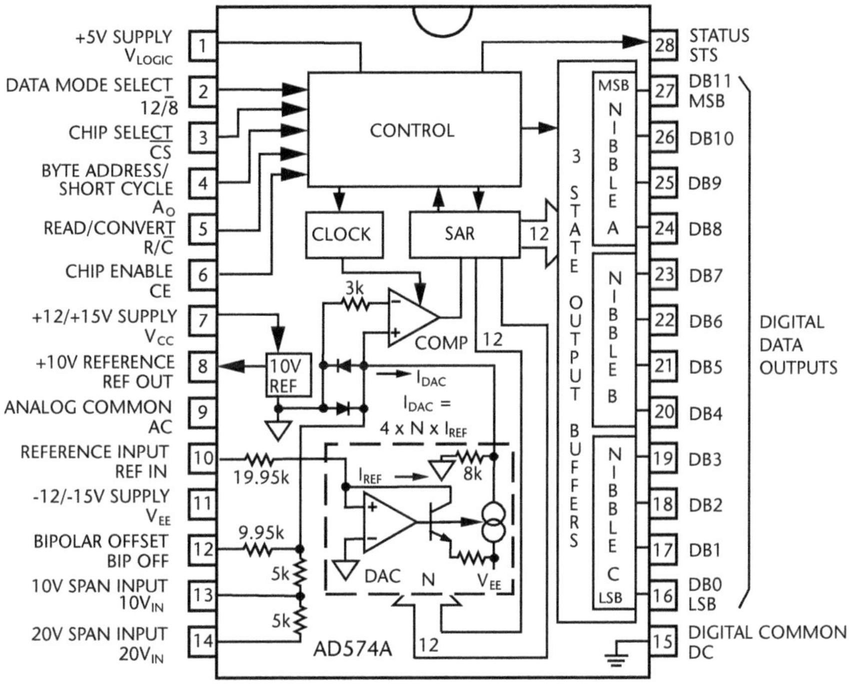

Figure A.5 The industry standard AD574 12 bit 28.5 kSPS ADC.

1985: A ruling by the U.S. Federal Communications Commission releases 902 to 928, 2400 to 2483.5 and 5725 to 5850 MHz bands for unlicensed use [110]. This creates the industrial, scientific, and medical (ISM) radio bands.

1987: Fifteen representatives from 13 European countries sign a memorandum of understanding in Copenhagen to develop and deploy a common cellular telephone system across Europe, and EU rules are passed to make GSM a mandatory standard.

1987: Flash memory (both NOR and NAND types) is developed by Toshiba and presented at the IEEE 1984 International Electron Devices Meeting (IEDM) held in San Francisco [111].

1987: Over 15 billion USD from venture capital is invested in semiconductor startups from 1980 through 1987 [112]. The semiconductor industry continues to grow, with startups such as Linear Technologies (founded 1987, acquired by Analog Devices 2016), Hittite Microwave (founded acquired by Analog Devices), Wolfson Microeltronics (founded acquired by Cirrus Logic), Atmel (founded 1984, acquired by Microchip 2016), Xilinx (founded 1984), Altera (founded 1983, acquired by Intel 2015), Maxim (1983), and Micron (1978).

1988: The ITU-T approves G.722, an wideband (507000 Hz) audio codec operating at 48, 56 and 64 kbit/s. Technology of the codec is based on subband ADPCM (SB-ADPCM). G722 provides improved speech quality due to a the wider bandwidth of G.711 released 16 years earlier [113].

1988: Crystal Semiconductor (later purchased by Cirrus Logic) releases the monolithic CSZ5316, the first commercial Σ-Δ ADC. With 16-bit resolution, and an effective throughput rate of 20 kSPS is suitable for voiceband digitization [65].

1989: Nils Rydbeck, CTO at Ericsson Mobile, proposes a short-link radio technology to connect wireless headsets to mobile phones, which eventually becomes known as Bluetooth.

1989: public commercial use of the internet begins with the connection of MCI Mail and Compuserve's email capabilities [114].

A.7 1990–1999: Age of the Public Internet (Web 1.0)

Digital signal processing and computing technology continue to revolutionize the 1990s. Between 1990 and 1997, individual PC ownership in the United States rose from 15% to 35%. Cell phones of the early 1990s and earlier were very large, lacked extra features, and were used by only a few percent of the population. Only a few million people used online services in 1990, and the World Wide Web had only just been invented. The first Web browser went online in 1993 and by 2001, more than 50% of some Western countries had internet access, and more than 25% had cell phone access. Advancements in computer modems, ISDN, cable modems, and DSL lead to faster connections to the internet.

Wireless transmission technology has gradually evolved over time; the superheterodyne architecture developed in 1917, shown in Figure A.3, is still the go-to radio architecture of the time. It has taken advantage of the current integrated circuits and morphed into multistage receiver architectures typically consisting of one or two mixing stages, which are fed into single-chip ADCs and DACs. A multistage superheterodyne transceiver architecture can be seen in Figure A.6.

The first conversion stage converters (up or down) the RF frequencies to a lower intermediate frequency (IF_1). The first IF_1 is then translated down to a lower frequency (IF_2) baseband frequency that the ADC or DAC can digitize. In addition to the mixers, there are filters, amplifiers, and step attenuators at each stage. The filtering is used to reject unwanted out-of-band (OOB) signals. If unchecked, these OOB signals can create spurious signals (spurs) that falls on top of a desired signal, making it difficult or impossible to demodulate. The actual IF_1 or IF_2 frequencies depends on the frequency and spur planning, as well as mixer performance and available filters for the RF front-end. These radios are mostly fixed frequency (not very tunable), as the desire for performance (sensitivity) has been traded off for flexibility.

1990: Advanced RISC Machines Ltd (ARM) is founded as a joint venture between Acorn Computers, Apple Computer (now Apple Inc.) and VLSI Technology. The company develops the Acorn RISC Machine (ARM) processor, which is now used in 98% of the more than 1 billion mobile phones sold each year [115].

1991: Algorithms for MPEG-1 Audio Layer I, II and III, which become known as international standard ISO/IEC 11172-3 [116] (MPEG-1 Audio or MPEG-1 Part 3), or mp3, are approved in 1991 and finalized in 1992 [117].

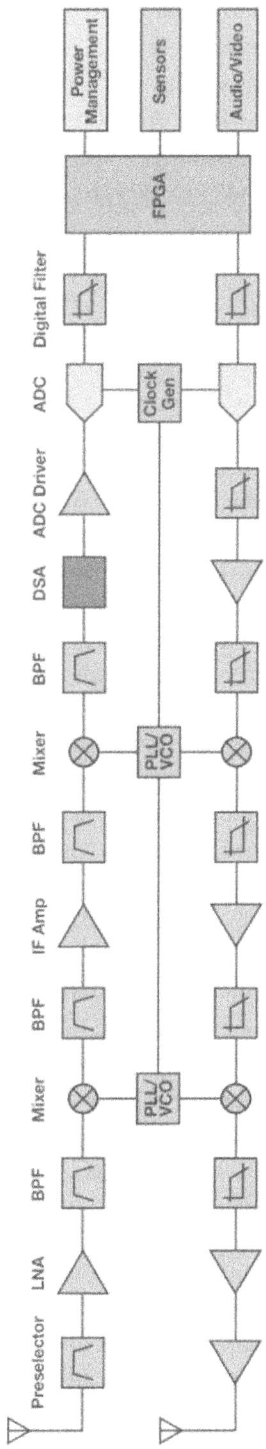

Figure A.6 Multistage superheterodyne receive and transmit signal chains [118].

1991: The ITU releases the V.32bis recommendation for a dial-up modem, allowing bidirectional data transfer at 14.4 kbit/s using trellis-modulated data using the Viterbi algorithm [109].

1991: Linus Torvalds, a computer science student, starts working on some simple ideas for an operating system, and 10,239 lines of code, version 0.01 of the Linux kernel, is released on the FTP server [119].

1991: Former Finnish prime minister Harri Holkeri makes the world's first GSM call, calling Kaarina Suonio (mayor of the city of Tampere, where Nokia was founded) [120].

1991: SpeakEasy I, a software-defined radio project, is kicked off by the U.S. military [121] Although different parts of SDR technology were available since the 1970s and the first demonstration as a SDR prototype (with a fixed RF implementation) was presented in 1988 [122], SpeakEasy was the first large-scale software radio.

When a communication device is designed for one unique unique purpose, only using one unique waveform, at one unique frequency, different devices made by different people are not interoperable with one another. The U.S. military had this exact problem—they had many different types of radios and waveforms for each branch and group of the armed services, and in times of conflict, they could not talk to each other in real time. This was not an academic exercise—lives were being lost due to the inability to communicate. They needed a single system that could communicate with over 10 other different types of military radios.

This was the key milestone for the advancement of SDR technology. The developers wanted to test the feasibility of a multiband (different RF frequencies), multimode (different waveforms) radio in practical settings.

1991: CERN, a European organization for particle research, publicizes the new World Wide Web project.

1992: The first short messaging service (SMS or text message) message is sent over the GSM network.

1993: Joseph Mitola coins the term software-defined radio [123].

1994: The SpeakEasy project is successfully demonstrated. It was not only a success, but a significant leap forward in SDR technology. It used four Texas Instruments TMS320C40 processor, which ran at 40 MHz and a SUN Sparc 10 workstation as the man-machine interface, implementing more than ten military communication standards, with transmission carrier frequencies ranging from 2 to 2000 MHz, which at that time was a major advancement in communication systems engineering. The SpeakEasy implementation allowed for software upgrades of new functional blocks, such as modulation schemes and coding schemes. Note that given the microprocessor technology at the time, the physical size of the SpeakEasy prototypes were large enough to fit in the back of a truck and required a significant amount of power.

Since the SpeakEasy program had taken 3 years to complete, (two cycles of Moore's law), many thought it would be easy to scale down to a faster, lower-power processor. However, since everything was hand-coded in assembly (to maximize performance), this was not possible. Development was stuck. The observation was that it had taken 3 years to write software for a platform that Moore's law made obsolete in 18 months.

1994: Linus Torvalds releases the Linux kernel 1.0, with 176,250 lines of code [124].

1995: Sony releases the Playstation worldwide. It is the first computer entertainment platform to ship 100 million units, which it reached 9 years and 6 months after its initial launch [125]

1996: John O'Sullivan takes a technique he developed in 1977 for sharpening and improving picture clarity in radio astronomy images using Fourier, and reapplies the technique for reducing multipath interference of radio signals as part of a research project to detect exploding mini black holes the size of an atomic particles. While his main experiment failed, his signal processing techniques were patented and applied to future 802.11 standards.

1996: Linus Torvalds releases the Linux kernel 2.0, with 632,062 lines of code [126].

1997: The first version of the 802.11 protocol is released and provides up to 2 Mbit/s link speeds.

1998: The Bluetooth Special Interest Group (SIG) is formed, and the first specification is released.

1998: The ITU releases the V.90 recommendation for a dial-up modem, allowing 56 kbit/s digital download and 33.6 kbit/s analog upload. using digital modulation [127].

1999: The IEEE ratifies an upgrade to the 802.11 specification with 802.11b to permit 11 Mbit/s link speeds.

2000: Bluetooth products begin to ship, including mice, PC cards, headsets, and phones [128].

2001: Jimmy Wales and Larry Sanger launch Wikipedia [129].

A.8 Post-2000: Everything comes together

Since 2000, the amount of research and development activities has exploded. New, faster-, or lower-power communications schemes are almost being announced on a yearly basis. In 2008, worldwide GSM subscribers exceeded three billion people; however in 2016, operators are decommissioning networks to make ready for LTE and other standards. The rate of change has increased substantially.

References

[1] Sagan, C., *Cosmos*, New York: Random House. 1980.

[2] Polybius, *The Histories of Polybius*, published in Vol. IV of the Loeb Classical Library edition, 1922 through 1927, http://penelope.uchicago.edu/Thayer/E/Roman/Texts/Polybius/10*.html#45.6.

[3] Febvre, L., and Martin, H.-J., *The Coming of the Book: The Impact of Printing 1450–1800*, London: New Left Books, 1976, quoted in Anderson, B., Comunidades Imaginadas. Reflexiones sobre el origen y la difusin del nacionalismo, Fondo de cultura econmica, Mexico, 1993, p. 58f.

[4] Capra, F., *The Science of Leonardo; Inside the Mind of the Genius of the Renaissance*, New York: Doubleday, 2007.

[5] Landes, D. S., *The Unbound Prometheus*, Cambridge, UK: Press Syndicate of the University of Cambridge, 1969.

[6] Euler, L., *E101–Introductio in analysin infinitorum*, Volume 1, 1748, The Euler Archive, in the original Latin, http://eulerarchive.maa.org/pages/E101.html, and a modern English translation by Ian Bruce, a retired physics professor, formerly of the University of Adelaide, South Australia, at http://www.17centurymaths.com/contents/introductiontoanalysisvol1.htm.

[7] Wilson, R., Read Euler, Read Euler, he Is the Master of Us All, March 1, 2007, https://plus.maths.org/content/os/issue42/features/wilson/index.

[8] Franklin, B., "The Kite Experiment," *The Pennsylvania Gazette*, October 19, 1752, in *The Papers of Benjamin Franklin*, The American Philosophical Society and Yale University, digital edition by The Packard Humanities Institute, Vol. 4, p. 360a. Retrieved November 24, 2017.

[9] Marland, E. A., *Early Electrical Communication*, London: Abelard-Schuman Ltd., 1964, pp. 17–19.

[10] Carnegie, A., *James Watt*, The Minerva Group, Inc, p. 215, http://www.jameswatt.info/andrew-carnegie/3-captured-by-steam.html.

[11] Grandstrand, O., "Semaphore Signalling," in R. W. Burns (ed.), *Communications: An International History of the Formative Years*, Herts, UK: The Institution of Engineering and Technology, 2004.

[12] Decker, F., "Volta and the 'Pile,'" *Electrochemistry Encyclopedia*, Case Western Reserve University, January 2005.

[13] Amgueddfa Cymru National Museum Wales, "Richard Trevithick's Steam Locomotive," 2008, Museumwales.ac.uk. https://museum.wales/articles/2008-12-15/Richard-Trevithicks-steam-locomotive/.

[14] Wickens, A. P., *A History of the Brain: From Stone Age Surgery to Modern Neuroscience*, London: Psychology Press, 2014, https://books.google.com/books?id=gSKcBQAAQBAJ&pg=PA124.

[15] Ohm, G., http://www2.ohm-hochschule.de/bib/textarchiv/Ohm.Die_galvanische_Kette.pdf.

[16] Chisholm, H. (ed.), "Ohm, Georg Simon," Encyclopedia Britannica, 20 (11th Edition), Cambridge, UK: Cambridge University Press, 1911, p. 34.

[17] Ampère, A,-M., *Mémoire sur la théorie mathématique des phénomènes électrodynamiques uniquement déduite de lexperience*, http://www.ampere.cnrs.fr/textes/theoriemathematique/pdf/theorie_mathematique.pdf.

[18] Alglave, M., and Boulard, J., *The Electric Light: Its History, Production, and Applications*, translated by T. O'Conor Sloan, New York: D. Appleton & Co., 1884, p. 224, also available on Google Books, http://books.google.com/books?id=zh5pbMMwARQC&pg=PA224).

[19] Vail, A., *The American Electro Magnetic Telegraph: With the Reports of Congress, and a Description of All Telegraphs Known, Employing Electricity Or Galvanism*, 1847, Lea & Blanchard.

[20] Graham-Cumming, J., *The 100-Year Leap* (2010-10-04), O'Reilly Radar, http://radar.oreilly.com/2010/10/the-100-year-leap.html. Retrieved August 1, 2012.

[21] Burns, R. W., *Communications: An International History of the Formative Years*, London: The Institution of Engineering and Technology, 2004.

[22] Huurdeman, A. A., *The Worldwide History of Telecommunications*, John Wiley & Sons, 2003.

[23] Maxwell, J., "A Dynamical Theory of the Electromagnetic Field," *Philosophical Transactions of the Royal Society of London*, 1865, http://rstl.royalsocietypublishing.org/content/155/459.full.pdf.

[24] International Telecommunications Union, *ITU History*, 1993, http://www.itu.int/itudoc/about/itu/history/history.txt.

[25] Vernon, E. (ed.), *Travelers' Official Railway Guide of the United States and Canada*, Philadelphia: The National General Ticket Agents' Association, June, 1870, Tables 215, 216, http://www.lancerlovers.com/Resources/Travel/Travellers%20guide%20to%20trains%201871.pdf.

[26] *Treatise on Electricity and Magnetism*, 1873, http://www.aproged.pt/biblioteca/MaxwellI.pdf.

[27] Braun, E., and S. MacDonald, *Revolution in Miniature: The History and Impact of Semiconductor Electronics*, Second Edition, Cambridge, UK: Cambridge University Press, 1982, pp. 11–12.

[28] Bell, A. G., *Improvement in Telegraphy*, U.S. Patent No. 174465, 2 1876, issued March 7, 1876.

[29] Levy, J., *Really Useful: The Origins of Everyday Things*, New York: Firefly Books, 2003, p. 124.

[30] Heilbron, J. L. (ed.), *The Oxford Guide to the History of Physics and Astronomy*, Oxford, UK: Oxford University Press, 2005, p. 148.

[31] Company History, *Benz Patent Motor Car: The First Automobile (1885–1886)*, https://www.daimler.com/company/tradition/company-history/1885-1886.html.

[32] Emerson, D. T., "The Work of Jagadish Chandra Bose: 100 Years of mm Wave Research," *IEEE Transactions on Microwave Theory and Techniques*, Vol. 45, No. 12, December 1997, pp. 2267–2273, http://ieeexplore.ieee.org/document/643830/.

[33] Bose, J. C., *Detector for Electrical Disturbances*, U.S. Patent No. 755840, published September 30, 1901, issued March 29, 1904.

[34] Braun F., *Ueber ein Verfahren zur Demonstration und zum Studium des zeitlichen Verlaufs variabler Strme* (On a Process for the Display and Study of the Course in Time of Variable Currents), 1897, Annalen der Physik und Chemie, 3rd series, http://onlinelibrary;wiley.com/doi/10.1002/andp.18972960313/abstract

[35] Marconi National Historic Site, http://www.pc.gc.ca/en/lhn-nhs/ns/marconi/index.

[36] Belrose, J. S., "Marconi and the History of Radio," *IEEE Antennas and Propagation Magazine*, Vol. 46, No. 2, 2004, pp. 130.

[37] Harr, C., Ambrose J. Fleming Biography. Pioneers of Computing. The History of Computing Project, June 23, 2003, http://www.thocp.net/biographies/fleming_ambrose.htm.

[38] Braun, K. F., Nobel Lecture: Electrical Oscillations and Wireless Telegraphy, p. 239, Nobel Media AB 2013, https://www.nobelprize.org/nobel_prizes/physics/laureates/1909/braun-lecture.pdf.

[39] Karwatka, D., "Reginald Fessenden and Radio Transmission," Tech Directions, Vol. 63, No. 8, March 2004, p. 12.

[40] ITU Constitution and Convention, http://www.itu.int/en/history/Pages/ConstitutionAndConvention.aspx.

[41] L., De Forest, *Device for Amplifying Feeble Electrical Currents*, U.S. Patent No. 841387, issued January 15, 1907.

[42] Henry Ford Museum & Greenfield Village, "The Life of Henry Ford." Retrieved November 28, 2013, https://web.archive.org/web/20011005164558/http://www.hfmgv.org/exhibits/hf/.

[43] Armstrong, E., U.S. Patent No. 1,113,149, *Wireless Receiving System*, published October 6, 1914.

[44] *The New York Times*, "Phone to Pacific from the Atlantic," January 26, 1915. Retrieved July 21, 2007, http://www.nytimes.com/learning/general/onthisday/big/0125.html.

[45] Mixers and Modulators, http://www.analog.com/media/en/training-seminars/tutorials/MT-080.pdf.

[46] Tigerstedt, E., Danish Patent 22091, https://ie.espacenet.com/publicationDetails/originalDocument?CC=DK&NR=22091C#ND=3&date=19170430 19170430.

[47] Klooster, J. W., *Icons of Invention: The Makers of the Modern World from Gutenberg to Gates*, Santa Barbara, CA: ABC-CLIO, 2009.

[48] Nyquist, H., "Certain Factors Affecting Telegraph Speed," *Bell System Technical Journal*, Vol. 3, April 1924, pp. 324–346.

[49] Harper, C. A., *Electronic Materials and Processes Handbook*, McGraw-Hill, pp. 7.3 and 7.4.

[50] Brittain, J. E., "Electrical Engineering Hall of Fame: Harold S Black" *Proceedings of the IEEE*, Vol. 99, No. 2, February 2011, pp. 351–353, doi:10.1109/jproc.2010.2090997. https://www.ieee.org/documents/proc_scanpast0211.pdf.

[51] Nyquist, H., "Certain Topics in Telegraph Transmission Theory," *A.I.E.E. Transactions*, Vol. 47, April 1928, pp. 617–644.

[52] Hayward, W., and Dick Bingham, "Direct Conversion–A Neglected Technique," November 1968, *QST*. ARRL: 1517, 156.

[53] The United States Army Signal Corps Officer Candidate School Association, "Part 4, America between the Wars," 2013, http://www.armysignalocs.com/index_oct_13.html.

[54] Clarke, A. C., "V2 for Ionosphere Research?" *Wireless World*, February 1945, p. 45.

[55] Warsitz, L., *The First Jet Pilot: The Story of German Test Pilot Erich Warsitz*, South Yorkshire, UK: Pen and Sword Books, 2009.

[56] "Great Airliners 11: de Havilland Comet," *Flight*, March 14, 1974. Retrieved 26 April 2012, https://www.flightglobal.com/pdfarchive/view/1974/1974%20-%200411.html.

[57] Walker, T., *The First Jet Airliner: The Story of the de Havilland Comet*, Newcastle upon Tyne, UK: Scoval Publishing Ltd., 2000.

[58] Weik, M. H., *The ENIAC Story. O R D N A N C E*, Washington, DC: American Ordnance Association, January/February 1961. Retrieved March 29, 2015, https://web.archive.org/web/20110814181522/http://ftp.arl.mil/~mike/comphist/eniac-story.html.

[59] American Physical Society, *This Month in Physics History, November 17–December 23, 1947: Invention of the First Transistor*, 2000, http://www.aps.org/publications/apsnews/200011/history.cfm.

[60] Shannon, C. E., "A Mathematical Theory of Communication," *Bell System Technical Journal*, Vol. 27, No. 3, July 1948,pp. 379–423, https://en.wikipedia.org/wiki/A_Mathematical_Theory_of_Communication.

[61] Waldrop, M. M., "Claude Shannon: Reluctant Father of the Digital Age," *MIT Technology Review*, July 1, 2001, https://www.technologyreview.com/s/401112/claude-shannon-reluctant-father-of-the-digital-age/.

[62] Abramson, M., and S. F. Danko, United States Patent 2,756,485, "Process of assembling electrical circuits," August 28, 1950.

[63] http://smithsonianchips.si.edu/danko/danko.htm.

[64] Shockley, W., *Electrons and Holes in Semiconductors with Applications to Transistor Electronics*, Bell Laboratory series, 1950, https://en.wikipedia.org/wiki/Electrons_and_Holes_in_Semiconductors_with_Applications_to_Transistor_Electronics.

[65] Kester, W., *The Data Conversion Handbook*, Analog Devices, 2005, http://www.analog.com/en/education/education-library/data-conversion-handbook.html.

[66] Brock, G. W., *The Second Information Revolution*, Harvard University Press, 2003.

[67] Zak, A., "Sputnik's Mission," 2015. Retrieved 27 December 2015, http://www.russianspaceweb.com/sputnik_mission.html.

[68] Kilby, J. S., *Miniaturized Electronic Circuits*, 1964, US Patent No. 3,138,743, https://www.google.com/patents/US3138743.

[69] U.S. Registry of Objects Launched into Outer Space, US Space Objects Registry 1962-ALPHA EPSILON 1, June 19, 2013. Retrieved 2013-10-02. https://web.archive.org/

web/20131005021146/https://usspaceobjectsregistry.state.gov/Lists/SpaceObjects/DispForm.aspx?ID=90

[70] Mann, A., "Telstar 1: The Little Satellite That Created the Modern World 50 Years Ago," *Wired Magazine*, July 2012, https://www.wired.com/2012/07/50th-anniversary-telstar-1/.

[71] Analog Devices, *Our 50-Year Journey*, 2015, http://www.analog.com/en/timeline/50th-timeline.html.

[72] Analog Devices, *10 Facts to Know About Analog Devices, Inc.*, 2015, http://www.analog.com/media/en/other/about-ADI/Ten-Facts-to-Know-About-Analog-Devices.pdf.

[73] Moore, G. E., "Lithography and the Future of Moore's Law," *SPIE*, Vol. 2438, 1995, http://www.lithoguru.com/scientist/CHE323/Moore1995.pdf.

[74] Analog Multipliers, http://www.analog.com/media/en/training-seminars/tutorials/MT-079.pdf.

[75] Viterbi, A., "Error Bounds for Convolutional Codes and an Asymptotically Optimum Decoding Algorithm," *IEEE Transactions on Information Theory*, Vol. 13, No. 2, April 1967, http://ieeexplore.ieee.org/document/1054010/.

[76] Morton, D., *Andrew Viterbi, Electrical Engineer, An Oral History*, San Diego: IEEE Global History Network, http://ethw.org/Oral-History:Andrew_Viterbi#Linkabit_and_M.2FA-COM.3B_development_and_consumers.

[77] Savio, J., "Browsing History: A Heritage Site Has Been Set Up in Boelter Hall 3420, the Room the First Internet Message Originated in," *Daily Bruin*, UCLA, April 1 2011.

[78] Grayver, E., *Implementing Software Defined Radio*, New York: Springer Science & Business Media.

[79] Intel Corporation, https://www.intel.com/content/www/us/en/history/museum-story-of-intel-4004.html.

[80] http://www.circuitstoday.com/microcontroller-invention-history.

[81] Abramson, N., "The ALOHA System–Another Alternative for Computer Communications," Proc. 1970 Fall Joint Computer Conference, AFIPS Press, 1970, https://robotics.eecs.berkeley.edu/~pister/290Q/Papers/MAC%20protocols/ALOHA%20abramson%201970.pdf.

[82] Ward, J., "The 555 Timer IC, An Interview with Hans Camenzind," The Semiconductor Museum, 2004, http://www.semiconductormuseum.com/Transistors/LectureHall/Camenzind/Camenzind_Index.htm.

[83] International Telecommunications Union, G.711: Pulse Code Modulation (PCM) of Voice Frequencies, http://www.itu.int/rec/T-REC-G.711/.

[84] Allan, R. A., *A History of the Personal Computer,* London, Ontario: Allan Publishing, 2001, https://books.google.com/books?id=FLabRYnGrOcC&hl=en.

[85] Analog Devices, AiD534 www.analog.com/AD534.

[86] Gilbert, B., "A New Technique for Analog Multiplication," *IEEE Journal Solid State Circuits*, Vol. 10, No. 6, December 1975, pp. 437–447.

[87] Jarvis, A., "How Kodak Invented the Digital Camera in 1975," May 9, 2008, techradar.com, https://web.archive.org/web/20120110030031/http://www.techradar.com/news/photography-video-capture/how-kodak-invented-the-digital-camera-in-1975-364822.

[88] Estrin, J., "Kodak's First Digital Moment," *The New York Times*, August 12, 2015, https://lens.blogs.nytimes.com/2015/08/12/kodaks-first-digital-moment/?smid=pl-share.

[89] Williams, R., "Apple Celebrates 39th Year on April 1," *The Telegraph*, April 1, 2015, Telegraph Media Group. Retrieved July 9, 2017, http://www.telegraph.co.uk/technology/apple/11507451/Apple-celebrates-39th-year-on-April-1.html.

[90] Wozniak, S., and G. Smith, *iWoz: Computer Geek to Cult Icon: How I Invented the Personal Computer, Co-Founded Apple, and Had Fun Doing It*, New York: W. W. Norton & Company, 2007.

[91] *Byte Magazine*, "Most Important Companies," September 1995. Archived from the original on June 18, 2008. Retrieved June 10, 2008, https://web.archive.org/web/20080618072507/http://www.byte.com/art/9509/sec7/art15.htm.

[92] http://ethw.org/Milestones:The_Floating_Gate_EEPROM,_1976_-_1978.

[93] Wiggins, R., "An Integrated Circuit for Speech Synthesis," conference paper, May 1980, DOI: 10.1109/ICASSP.1980.1170897. Source: IEEE Xplore Conference: Acoustics, Speech, and Signal Processing, IEEE International Conference on ICASSP '80, Vol. 5, http://ieeexplore.ieee.org/document/1170897/.

[94] Analog Devices, www.analog.com/AD574.

[95] Harris, F. J., "On the Use of Windows for Harmonic Analysis with the Discrete Fourier Transform," *Proceedings of the IEEE*, Vol. 66, No. 1, 1978, pp. 51–83.

[96] Harris, F., Google Scholar, https://scholar.google.com/citations?hl=en&user=6A5XoroAAAAJ&view_op=list_works.

[97] IBM Computers, Presenting the IBM of Personal Computers, *PC Magazine* (advertisement), February/March 1982, inside front cover. Retrieved October 20, 2013, https://books.google.com/books?id=w_OhaFDePS4C&lpg=RA2-PA18&pg=PP2#v=onepage&q&f=false.

[98] Hauben, R., "From the ARPANET to the Internet," *TCP Digest* (UUCP), January 2001. Retrieved July 5, 2007, http://www.columbia.edu/~rh120/other/tcpdigest_paper.txt.

[99] Proakis, J. G., http://www.cdsp.neu.edu/info/faculty/proakis/proakis.html https://www.jacobsschool.ucsd.edu/faculty/faculty_bios/index.sfe?fmp_recid=94 http://ethw.org/John_G._Proakis

[100] Proakis, J. G., *Digital Communications (Third Edition)*, New York: McGraw Hill, 1994.

[101] Proakis, J. G., C. M. Rader, and F. Ling, *Advanced Digital Signal Processing*, New York: Macmillan, 1992.

[102] Ingle, V. K., and J. G. Proakis, *Digital Signal Processing Laboratory: Using the ADSP-2101 Microcomputer*, Englewood Cliffs, NJ: Prentice Hall, 1991.

[103] Deller, J. R., J. G. Proakis, and J. H. L. Hansen, *Discrete-Time Processing of Speech Signals*, Wiley-IEEE Press, 1999.

[104] Proakis, J. G., and M. Salehi, *Communication Systems Engineering*, Upper Saddle River, NJ: Prentice Hall, 2002.

[105] https://www.altera.com/solutions/technology/system-design/articles/_2013/in-the-beginning.html.

[106] Polsson, K., *Chronology of Apple Computer Personal Computers*. Archived from the original on August 21, 2009. See May 3, 1984, https://web.archive.org/web/20090821105822/http://www.islandnet.com/~kpolsson/applehis/appl1984.htm.

[107] Johnson, P., "New Research Lab Leads to Unique Radio Receiver," *E-Systems Team*, Vol. 5, No. 4, May 1985, pp. 6–7, http://chordite.com/team.pdf.

[108] Schrader, C. B., and M. W. Spong, "The IEEE Conference on Decision and Control–Tracing CDC History," *IEEE Control Systems Magazine*, Vol. 24, No. 6, December 2004, p. 5666, doi:10.1109/MCS.2004.1368481.

[109] International Telecommunication Union, V.32, Series V: Data Communication over the Telephone Network, Interfaces and Voiceband Modems, 1998, https://www.itu.int/rec/dologin_pub.asp?lang=e&id=T-REC-V.34-199802-I!!PDF-E&type=items.

[110] Federal Communications Commission of the United States, Authorization of Spread Spectrum Systems under Parts 15 and 90 of the FCC Rules and Regulations June 18, 1985, https://web.archive.org/web/20070928054826/http://www.marcus-spectrum.com/documents/81413RO.txt.

[111] Masuoka, F., M. Momodomi, Y. Iwata, and R Shirota, New Ultra High Density EPROM and Flash EEPROM with NAND Structure Cell, Electron Devices Meeting, 1987 International. IEEE 1987, http://ieeexplore.ieee.org/document/1487443/?arnumber=1487443.

[112] DataQuest, *A Decade of Semiconductor Companies*, 1988 Edition, http://archive.computerhistory.org/resources/access/text/2013/04/102723194-05-01-acc.pdf.

[113] International Telecommunications Union, G.722: 7 kHz Audio-Coding within 64 kbit/s, https://www.itu.int/rec/T-REC-G.722.

[114] InfoWorld Media Group, InfoWorld (September 25, 1989). InfoWorld Media Group, Inc. Archived from the original on January 29, 2017, via Google Books.

[115] *The Telegraph*," History of ARM: From Acorn to Apple," January 6, 2011, http://www.telegraph.co.uk/finance/newsbysector/epic/arm/8243162/History-of-ARM-from-Acorn-to-Apple.html.

[116] International Organization for Standardization, Information Technology–Coding of Moving Pictures and Associated Audio for Digital Storage Media at Up to about 1,5 Mbit/s–Part 3: Audio, https://www.iso.org/standard/22412.html.

[117] International Organization for Standardization, MPEG press release, Kurihama, November 1991. Archived from the original on May 3, 2011, Retrieved July 17, 2010, https://web.archive.org/web/20110503174827/http://mpeg.chiariglione.org/meetings/kurihama91/kurihama_press.htm.

[118] Hall, B., and W. Taylor, "X- and Ku-Band Small Form Factor Radio Design," Analog Devices Inc., Wilmington, MA, 2017 http://www.analog.com/en/technical-articles/x-and-ku-band-small-form-factor-radio-design.html.

[119] Torvalds, L. B., Free Minix-Like Kernel Sources for 386-AT, October 5, 1991, newsgroup, comp.os.minix https://groups.google.com/forum/#!msg/comp.os. minix/4995SivOl9o/GwqLJlPSlCEJ

[120] Rediff, "GSM is 20! Transcript of World's First Such Call," July 1, 2011, http://www.rediff.com/business/slide-show/slide-show-1-tech-transcript-of-the-worlds-first-gsm-call/20110701.htm.

[121] Lackey, R. J., and D. W. Upmal, "Speakeasy: The Military Software Radio," *IEEE Communications Magazine*, May, 1995.

[122] Hoeher, P., and H. Lang. "Coded-8PSK Modem for Fixed and Mobile Satellite Services Based on DSP," in Proceedings of the First International Workshop on Digital Signal Processing Techniques Applied to Space Communications, Noordwijk, the Netherlands, 1988.

[123] Mitola, J., III, "Software Radios: Survey, Critical Evaluation and Future Directions" *IEEE Aerospace and Electronic Systems Magazine*, April 1993.

[124] Torvalds, L., Linux Kernel Source Code, Version 1.0 https://www.kernel.org/pub/linux/kernel/v1.0/.

[125] "Sony PlayStation 2 Breaks Record as the Fastest Computer Entertainment Platform to Reach Cumulative Shipment of 100 Million Units," press release, Sony Computer Entertainment, November 30, 2005. Archived from the original (PDF) on January 3, 2006; retrieved June 8, 2008, https://web.archive.org/web/20060103211119/http://www.scei.co.jp:80/corporate/release/pdf/051130e.pdf.

[126] Torvalds, L., Linux Kernel Source Code, Version 2.0, https://www.kernel.org/pub/linux/kernel/v2.0/.

[127] International Telecommunication Union, V.90 Series V: Data Communication over the Telephone Network, Simultaneous Transmission of Data and Other Signals, 1998, https://www.itu.int/rec/dologin_pub.asp?lang=e&id=T-REC-V90-199809-I!!PDF-E&type=items.

[128] Bluetooth SIG: Our History, https://www.bluetooth.com/about-us/our-history.

[129] https://en.wikipedia.org/wiki/History_of_Wikipedia.

APPENDIX B

Getting Started with MATLAB and Simulink

You will be using MATLAB and Simulink for the experiments and for the open-ended design projects in this book. This appendix serves as a brief refresher of MATLAB, since you should have used it before. However, if you don't have extensive experience with Simulink, then this appendix shows you how to get started with the tool. Please note the MATLAB portion of this appendix is mainly based on the MATLAB documentation presented in [1] and the Simulink portion is based on the Simulink Basics Tutorial presented in [2], but here we extract the most important and fundamental concept so that you can quickly get started after reading this appendix. For more information about these two products, you are encouraged to refer to [1] and [2].

B.1 MATLAB Introduction

MATLAB is widely used in all areas of applied mathematics, in education and research at universities, and in industry. MATLAB stands for Matrix Laboratory and the software is built up around vectors and matrices. Consequently, this makes the software particularly useful for solving problems in linear algebra, but also for solving algebraic and differential equations as well as numerical integration. MATLAB possesses a collection of graphic tools capable of producing advanced GUI and data plots in both 2-D and 3-D. MATLAB also has several toolboxes useful for performing communications, signal processing, image processing, optimization, and other specialized operations.

MathWorks has created an excellent online tutorial to review basic and advanced concepts, as well as provide instructor lead tutorials to show off the various capabilities of MATLAB. It can be found at https://matlabacademy.mathworks.com

B.2 Useful MATLAB Tools

This section introduces general techniques for finding errors, as well as using automatic code analysis functions in order to detect possible areas for improvement within the MATLAB code. In particular, the MATLAB debugger features located within the Editor, as well as equivalent Command Window debugging functions, will be covered.

Debugging is the process by which you isolate and fix problems with your code. Debugging helps to correct two kinds of errors:

- Syntax errors: For example, misspelling a function name or omitting a parenthesis.
- Run-time errors: These errors are usually algorithmic in nature. For example, you might modify the wrong variable or code a calculation incorrectly. Run-time errors are usually apparent when an M-file produces unexpected results. Run-time errors are difficult to track down because the function's local workspace is lost when the error forces a return to the MATLAB base workspace.

B.2.1 Code Analysis and M-Lint Messages

MATLAB can check your code for problems and recommend modifications to maximize the performance and maintainability through messages, sometimes referred to as M-Lint messages. The term lint is the name given to similar tools used with other programming languages such as C. In MATLAB, the M-Lint tool displays a message for each line of an M-file it determines possesses the potential to be improved. For example, a common M-Lint message is that a variable is defined but never used in the M-file.

You can check for coding problems using three different ways, all of which report the same messages:

- Continuously check code in the Editor while you work. View M-Lint messages and modify your file based on the messages. The messages update automatically and continuously so you can see if your changes addressed the issues noted in the messages. Some messages offer extended information, automatic code correction, or both.
- Run a report for an existing MATLAB code file: From a file in the Editor, select **Tools** > **Code Analyzer** > **Show Code Analyzer Report.**
- Run a report for all files in a folder: In the Current Folder browser, click the Actions button, then select **Reports** > **Code Analyzer Report.**

For each message, review the message and the associated code in order to make changes to the code itself based on the message via the following process:

- Click the line number to open the M-file in the Editor/Debugger at that line.
- Review the M-Lint message in the report and change the code in the M-file based on the message.
- Note that in some cases, you should not make any changes based on the M-Lint messages because the M-Lint messages do not apply to that specific situation. M-Lint does not provide perfect information about every situation.
- Save the M-file. Consider saving the file to a different name if you made significant changes that might introduce errors. Then you can refer to the original file as you resolve problems with the updated file.
- If you are not sure what a message means or what to change in the code as a result, use the Help browser to look for related topics.

You can also get M-Lint messages using the `mlint` function. For more information about this function, you can type `help mlint` in the Command Window. Read the online documentation [3] for more information about this tool.

B.2.2 Debugger

The MATLAB Editor, graphical debugger, and MATLAB debugging functions are useful for correcting run-time problems. They enable access to function workspaces and examine or change the values they contain. You can also set and clear *breakpoints*, which are indicators that temporarily halt execution in a file. While stopped at a breakpoint, you can change the workspace contexts, view the function call stack, and execute the lines in a file one by one.

There are two important techniques in debugging: one is the *breakpoint* while the other is the *step*. Setting breakpoints to pause the execution of a function enables you to examine values where you think the problem might be located. While debugging, you can also step through an M-file, pausing at points where you want to examine values.

There are three basic types of breakpoints that you can set in the M-files:

- A standard breakpoint, which stops at a specified line in an M-file.
- A conditional breakpoint, which stops at a specified line in an M-file only under specified conditions.
- An error breakpoint that stops in any M-file when it produces the specified type of warning, error, or NaN or infinite value.

You cannot set breakpoints while MATLAB is busy (e.g., running an M-file, unless that M-file is paused at a breakpoint). While the program is paused, you can view the value of any variable currently in the workspace, thus allowing you to examine values when you want to see whether a line of code has produced the expected result or not. If the result is as expected, continue running or step to the next line. If the result is not as expected, then that line, or a previous line, contains an error.

While debugging, you can change the value of a variable in the current workspace to see if the new value produces expected results. While the program is paused, assign a new value to the variable in the Command Window, Workspace browser, or Array Editor. Then continue running or stepping through the program. If the new value does not produce the expected results, the program has a different or another problem.

Besides using the Editor, which is a graphical user interface, you can also debug MATLAB files by using debugging functions from the Command Window, or you can use both methods interchangeably. Read the online documentation [4] for more information about this tool.

B.2.3 Profiler

Profiling is a way to measure the amount of time a program spends on performing various functions. Using the MATLAB Profiler, you can identify which functions in your code consume the most time. You can then determine why you are calling them and look for ways to minimize their use. It is often helpful to decide whether the number of times a particular function is called is reasonable. Since programs often have several layers, your code may not explicitly call the most time-consuming functions. Rather, functions within your code might be calling other time-consuming functions that can be several layers down into the code. In this case, it is important to determine which of your functions are responsible for such calls.

Profiling helps to uncover performance problems that you can solve by

- Avoiding unnecessary computation, which can arise from oversight.
- Changing your algorithm to avoid costly functions.
- Avoiding recomputation by storing results for future use.

When you reach the point where most of the time is spent on calls to a small number of built-in functions, you have probably optimized the code as much as you can expect. You can use any of the following methods to open the Profiler:

- Select **Desktop** → **Profiler** from the MATLAB desktop.
- Select **Tools** → **Open Profiler** from the menu in the MATLAB Editor/Debugger.
- Select one or more statements in the Command History window, right-click to view the context menu, and choose **Profile Code.**
- Enter the following function in the Command Window: `profile viewer`.

To profile an M-file or a line of code, follow these steps after you open the Profiler, as shown in Figure B.1:

1. In the **Run this code** field in the Profiler, type the statement you want to run.
2. Click **Start Profiling** (or press **Enter** after typing the statement).
3. When profiling is complete, the **Profile Summary** report appears in the Profiler window.

Read the online documentation [5] for more information about this tool.

B.3 System Objects

System objects are a specialization of a class in MATLAB, which define a specific set of methods that make initialization, runtime operation, and tear-down simple.

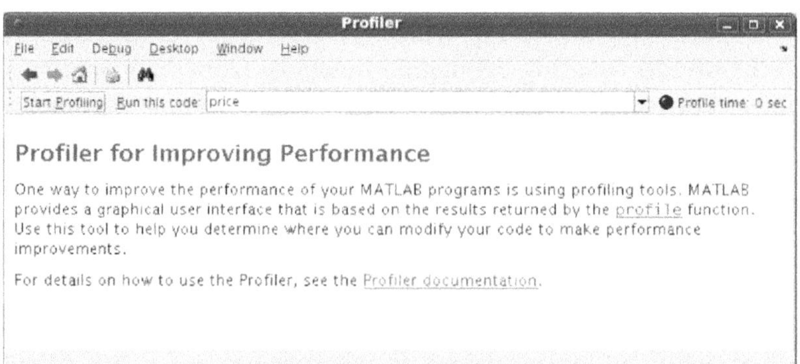

Figure B.1 The profiler window.

A class itself is basically a set of functions that share a set of variables called parameters. These parameters are defined within the class and have defined scopes. Although many methods are implemented by system objects, the three main methods a user should understand are `setupImpl`, `stepImpl`, and `releaseImpl`. They will be written as

```
methods (Access = protected)
    function setupImpl(obj)
        % Set parameter
        obj.x = 1;
    end
end
```

`setupImpl` is used to initial parameters and perform calculations that are needed for the life of the system object. `stepImpl` is the runtime function (method) that is called generally multiple times and will consume or produce inputs/outputs. Finally, `releaseImpl` is used to tear-down the object that will clear its memory or perform closing actions. For example, if the object is to write data to a file it will close the file when this method is called.

When a system object's operator or step method is called, the initial call will actually first call the `setupImpl` method. At this point the system object is considered locked. Then the `stepImpl` method will be called. Successive operator or step calls will only call the `stepImpl` method. To unlock the object the `releaseImpl` method must be called. Once unlock the `setupImpl` will again be called at the next operator or step call. We outline this process in a script here:

```
% Instantiate object
ss = dsp.SignalSource;
% setupImpl and stepImpl are called
data = ss();
% stepImpl is only called
data = ss();
% Object is unlocked
ss.release();
% setupImpl and stepImpl are called
data = ss.step();
```

System objects are extremely useful for encapsulating a large degree of functionality and when state needs to be maintained. For example, filters require state and are an obvious use for system objects. The toolboxes that make up MATLAB utilize extensions for their system objects that are related to their abbreviation. For example, system objects that are from the Communication Systems Toolbox will have the extension `comm`, resulting in objects with names such as `comm.AGC`, `comm.CarrierSynchronizer`, or `comm.EVM`. Examples in the DSP Systems Toolbox are: `dsp.FIRDecimator`, `dsp.SpectrumAnalyzer`, and `dsp.SignalSource`. More information about system objects can be found in the MathWorks documentation with extensive details on their implementation and use.

References

[1] The MathWorks, *MATLAB Documentation*, http://www.mathworks.com/help/techdoc/.

[2] University of Michigan, *Simulink Basics Tutorial*, http://www.engin.umich.edu/group/ctm/working/mac/simulink_basics/.

[3] The MathWorks, *Avoid Mistakes While Editing Code*, http://www.mathworks.com/help/techdoc/matlab_env/brqxeeu-151.html.

[4] The MathWorks, *Debugging Process and Features*, http://www.mathworks.com/help/techdoc/matlab_env/brqxeeu-175.html.

[5] The MathWorks, *Profiling for Improving Performance*, http://www.mathworks.com/help/techdoc/matlab_env/f9-17018.html

APPENDIX C

Equalizer Derivations

C.1 Linear Equalizers

Suppose we assume a transceiver model where the information source produces amplitude values I_n applied to an infinite impulse train; namely,

$$s(t) = \sum_{n=-\infty}^{\infty} I_n \delta(t - nT), \tag{C.1}$$

where $\delta(t)$ is the Dirac delta function. Applying a transmit pulse shaping filter $h_T(t)$ to the information signal $s(t)$, we obtain the transmitter's output signal:

$$v(t) = \sum_{n=-\infty}^{\infty} I_n h_T(t - nT), \tag{C.2}$$

which is then sent through a propagation channel that is characterized by a channel filter $h_C(t)$ and an additive white Gaussian noise signal $z(t)$. The output of the channel filter yields a signal:

$$p(t) = \sum_{n=-\infty}^{\infty} I_n h(t - nT), \tag{C.3}$$

where $h(t) = h_T(t) * h_C(t)$ is the channel impulse response. The signal intercepted by the receiver is expressed as

$$r(t) = p(t) + z(t) = \sum_{n=-\infty}^{\infty} I_n h(t - nT) + z(t). \tag{C.4}$$

At the receiver, we would like to find the expression for the mean squared error (MSE) and minimize it. To achieve this objective, we choose the receive filter $h_R(t)$ to be matched to $h(t)$, yielding the following optimal result:

$$h_R(t) = h^*(-t), \tag{C.5}$$

which results in the output of the receiver filter being equal to

$$y(t) = p(t) * h^*(-t) + z(t) * h^*(-t) = \sum_{n=-\infty}^{\infty} I_n g(t - nT) + z(t) * h^*(-t), \tag{C.6}$$

where $g(t) = h(t) * h_R(t)$, and

$$x(t) = \phi_h(t) = h(t) * h^*(-t) = \int_{-\infty}^{\infty} h(t+\tau)h^*(\tau)d\tau. \tag{C.7}$$

Supposed that $v(t) = z(t) * h^*(-t)$, then we get the following expression:

$$y(t) = \sum_{n=-\infty}^{\infty} I_n x(t - nT) + v(t). \tag{C.8}$$

Now, let us sample the output of the receive filter such that

$$y_k = y(kT) = \sum_{n=-\infty}^{\infty} I_n x((k-n)T) + v(kT), \tag{C.9}$$

$$= \sum_{n=-\infty}^{\infty} I_n x_{k-n} + v_k. \tag{C.10}$$

Note that x_k is referred to as the channel autocorrelation. Furthermore, the v_k samples are not white due to the filtering by $h^*(t)$, which means that $E\{v_k v_{k+n}\} \neq \delta(n)$. In these circumstances, we have a couple of options to deal with this colored noise. One approach involves finding $E\{v_k v_l^*\}$ for the noise sequence v_k. In this approach, we know that $v_k(t)$ is Gaussian since $z(t)$ is Gaussian. Furthermore, we know that $z(t)$ is white; that is, $S_z(f) = N_0$. Thus, if $E\{z^{ast}(s)z(t)\} = N_0\delta(t-s)$, we can then solve $E\{v_k v_l^*\}$ as follows:

$$E\{v_k v_l^*\} = \int_{-\infty}^{\infty} ds \int_{-\infty}^{\infty} dt h(t - lT)h^*(s - kT)E\{z^*(s)z(t)\}, \tag{C.11}$$

$$= N_0 \int_{-\infty}^{\infty} dt h(t - lT)h^*(t - kT), \tag{C.12}$$

$$= N_0 \int_{-\infty}^{\infty} dt h(t + |k - l|T)h^*(t). \tag{C.13}$$

Since we have:

$$x_k = \int_{-\infty}^{\infty} h^*(t)h(t + kT)dt, \tag{C.14}$$

which then gives us the expression for $E\{v_k v_l^*\}$ as:

$$E\{v_k v_l^*\} = N_0 x_{k-l}, \tag{C.15}$$

that can then be written as the autocorrelation function and the power spectral density of v_k; namely,

$$R_v(k) = E\{v_n v^*_{k+n}\} = N_0 x_k, \tag{C.16}$$

$$S_v(z) = \mathcal{Z}\{R_v(k)\} = N_0 X(z). \tag{C.17}$$

The second approach for dealing with colored noise is to implement something referred to as a *whitening filter*. In this case, we try to reverse the effects of the receiver filter on the noise signal $z(t)$. In other words, since we have the power spectral density of v_k equal to $S_v(z) = N_0 X(z)$, we ultimately would like to have the whitened noise power spectral density only equal to N_0. To achieve this, we assume that the z-transform of x_k, $X(z)$, can be represented by the following:

$$X(z) = F(z)F^*(1/z^*). \tag{C.18}$$

Thus, in we have a whitening filter whose transfer function is equal to $1/F^*(1/z^*)$, the resulting power spectral density at the output of this whitening filter should be equal to:

$$S_n(z) = \frac{1}{F(z)F^*(1/z^*)} S_v(z) = \frac{N_0 X(z)}{X(z)} = N_0, \tag{C.19}$$

which yields an output noise signal that is white.

C.2 Zero-Forcing Equalizers

Suppose we assume a discrete time model for the receiver that is equal to the following:

$$w_k = \sum_{n=-\infty}^{\infty} I_n f_{k-n} + n_k, \tag{C.20}$$

where w_k is the output signal of the whitening filter, f_k is the impulse response of the whitening filter, and n_k is the whitened noise signal with power spectral density equal to N_0.

In the *zero-forcing equalizer* (ZFE), we choose a transfer function $C(z)$ such that each ISI-compensated term $q_k = \delta_k$, which implies that there is no ISI present and the resulting equalized outputs $I_k + n'_k$ are subsequently quantized, yielding a probability of error equal to $P_e = Q(1/\sigma')$ for $I_k = \pm 1$. Although P_e is often used as a performance metric for a communication system, in the case of equalizer design we are going to use the minimum mean squared error (MMSE) as our metric; that is, $E\{|I_k - \hat{I}_k|^2\}$.

Suppose we take the z-transform of (C.20), thus obtaining:

$$W(z) = I(z)F(z) + N(z) \tag{C.21}$$

and once this has been filtered by the ZFE equalizer we get the output equal to

$$W(z)C(z) = I(z)F(z)C(z) + N(z)C(z), \tag{C.22}$$

with $Q(z) = F(z)C(z)$. The objective of our ZFE is to generate an output equal to

$$\mathcal{Z}^{-1}\{W(z)C(z)\} = I_k * q_k + n'_k, \tag{C.23}$$

where $q_k = \delta_k$; that is, no ISI present in the signal. Consequently, the ISI term in this expression is $F(z)C(z)$ and thus we want it to be equal to $Q(z) = F(z)C(z) = 1$ for $q_k = \delta_k$. Rearranging the terms, our ZFE should be equal to $C(z) = 1/F(z)$.

C.3 Decision Feedback Equalizers

Although ZFE filters are conceptually straightforward, they almost always possess an infinite number of taps, thus making them impossible to implement in real world applications. Furthermore, when the signal resulting at the output of the whitening filter is equalized using the ZFE, any noise contributions contained within the equalized signal will no longer be white. As a result, there exists other equalizer designs that can be more readily implemented in hardware. One of these is the *decision feedback equalizer* (DFE), which consists of two filters: a feedforward filter $a(z)$ and a feedback filter $b(z) - 1$.

The *feedforward filter* $a(z)$ is anti-causal and has the following form:

$$a(z) = a_{-1}z + a_{-2}z^2 + a_{-3}z^3 + a_{-4}z^4 + \dots, \tag{C.24}$$

while the *feedback filter* $b(z)$ is a causal filter possessing the form:

$$b(z) - 1 = b_1 z^{-1} + b_2 z^{-2} + b_3 z^{-3} + b_4 z^{-4} + \dots. \tag{C.25}$$

Consequently, we can design $a(z)$ and $b(z)$ such that

$$a(z) = CF^*(1/z^*) \tag{C.26}$$

$$b(z) = F(z) \tag{C.27}$$

$$F(z)F^*(1/z^*) = X(z) + N_0. \tag{C.28}$$

APPENDIX D

Trigonometric Identities

$$\exp(\pm j\theta) = \cos(\theta) \pm j\sin(\theta)$$

$$\cos(\theta) = \frac{1}{2}[\exp(j\theta) + exp(-j\theta)]$$

$$\sin(\theta) = \frac{1}{2j}[\exp(j\theta) - \exp(-j\theta)]$$

$$\sin^2(\theta) + \cos^2(\theta) = 1$$

$$\cos^2(\theta) - \sin^2(\theta) = \cos(2\theta)$$

$$\cos^2(\theta) = \frac{1}{2}[1 + \cos(2\theta)]$$

$$\sin^2(\theta) = \frac{1}{2}[1 - \cos(2\theta)]$$

$$2\sin(\theta)\cos(\theta) = \sin(2\theta)$$

$$\sin(\alpha \pm \beta) = \sin(\alpha)\cos(\beta) \pm \cos(\alpha)\sin(\beta)$$

$$\cos(\alpha \pm \beta) = \cos(\alpha)\cos(\beta) \mp \sin(\alpha)\sin(\beta)$$

$$\tan(\alpha \pm \beta) = \frac{\tan(\alpha) \pm \tan(\beta)}{1 \mp \tan(\alpha)\tan(\beta)}$$

$$\sin(\alpha)\sin(\beta) = \frac{1}{2}[\cos(\alpha - \beta) - \cos(\alpha + \beta)]$$

$$\cos(\alpha)\cos(\beta) = \frac{1}{2}[\cos(\alpha - \beta) + \cos(\alpha + \beta)]$$

$$\sin(\alpha)\cos(\beta) = \frac{1}{2}[\sin(\alpha - \beta) + \sin(\alpha + \beta)]$$

About the Authors

Dr. Travis F. Collins holds M.S. and Ph.D. degrees in electrical and computer engineering from Worcester Polytechnic Institute (WPI) in Worcester, MA. Dr. Collins' research is focused on small cell interference modeling, phased-array localization, and high performance computing for software-defined radio (SDR). He has extensive experience developing for SDR applications in many different software environments, hardware architectures, and remains active in several open-source-based SDR projects. Currently, Dr. Collins works as a development engineer for Analog Devices, Inc. (ADI) in the Systems Development Group. At ADI, Dr. Collins is responsible for transceiver applications and works heavily on hardware and software integration projects.

Robin Getz has spent 25 years in the semiconductor industry, and has held positions ranging from applications engineer to field applications engineer to director of engineering. He has worked on a variety of systems in the analog, digital, RF, and software domains, as well as with large direct customers and smaller customers. Robin has been with Analog Devices for 17 years and is the director of engineering for Analog Devices Inc. Systems Development Group, where he works creating HDL interfaces and device drivers for ADI's mixed-signal IC products. He holds four patents and a B.Sc. (EE) from the University of Saskatchewan.

Dr. Di Pu was born in Suzhou, Jiangsu, China. She received her M.S. and Ph.D. degrees from Worcester Polytechnic Institute (Worcester, MA) in 2009 and 2013. During her time at WPI, she was a member of the Wireless Innovation Laboratory, where she conducted research into cognitive radio system implementations. Dr. Pu is a recipient of the 2007 Institute Fellowship and is a 2013 Sigma Xi Research Award winner in doctoral dissertation. Dr. Pu was an applications engineer at Analog Devices, Inc., in Wilmington, MA after she graduated from WPI. She is now a software engineer at Adobe Systems Inc. in Seattle, WA.

Professor Alexander M. Wyglinski is internationally recognized as an expert in the field of wireless communications, cognitive radio, connected vehicles, software-defined radio, dynamic spectrum access, electromagnetic security, vehicular technology, wireless system optimization and adaptation, autonomous vehicles, and cyber-physical systems. Dr. Wyglinski is an associate professor of electrical and computer engineering at Worcester Polytechnic Institute, (Worcester, MA) as well as the director of the Wireless Innovation Laboratory (WI Lab). Dr. Wyglinski is very active in the technical community, serving on the organizing committees of numerous technical conferences and several journal editorial boards. These activities include serving as the general cochair for both the 2013 IEEE Vehicular Networking Conference and the 82nd IEEE Vehicular Technology Conference in the Fall of 2015. Dr. Wyglinski's editorial board commitments include the IEEE Communications Magazine, IEEE Transactions on Wireless Communications, and IEEE Transactions on Communications. In January 2018, Dr. Wyglinski will be the president of the IEEE Vehicular Technology Society, an applications-oriented society of approximately 4,200 members that focuses on the theoretical, experimental, and operational aspects of electrical and electronics engineering in mobile radio, motor vehicles,and land transportation. Throughout his academic career, Dr. Wyglinski has published approximately 40 journal papers, over 80 conference papers, 9 book chapters, and 3 textbooks. He is currently being or has been sponsored by organizations such as the Defense Advanced Research Projects Agency, the Naval Research Laboratory, the MITRE Corporation, the Office of Naval Research, the Air Force Research Laboratory Space Vehicles Directorate, The MathWorks Inc., Toyota InfoTechnology Center U.S.A., and the National Science Foundation. Dr. Wyglinski is a senior member of the IEEE, as well as a member of Sigma Xi, Eta Kappa Nu, and the ASEE.

Index

Q

R

Recent Titles in the Artech House Mobile Communications Series

William Webb, Series Editor

3G CDMA2000 Wireless System Engineering, Samuel C. Yang

3G Multimedia Network Services, Accounting, and User Profiles, Freddy Ghys, Marcel Mampaey, Michel Smouts, and Arto Vaaraniemi

5G Spectrum and Standards, Geoff Varrall

802.11 WLANs and IP Networking: Security, QoS, and Mobility, Anand R. Prasad and Neeli R. Prasad

Achieving Interoperability in Critical IT and Communications Systems, Robert I. Desourdis, Peter J. Rosamilia, Christopher P. Jacobson, James E. Sinclair, and James R. McClure

Advances in 3G Enhanced Technologies for Wireless Communications, Jiangzhou Wang and Tung-Sang Ng, editors

Advances in Mobile Information Systems, John Walker, editor

Advances in Mobile Radio Access Networks, Y. Jay Guo

Applied Satellite Navigation Using GPS, GALILEO, and Augmentation Systems, Ramjee Prasad and Marina Ruggieri

Artificial Intelligence in Wireless Communications, Thomas W. Rondeau and Charles W. Bostian

Broadband Wireless Access and Local Network: Mobile WiMax and WiFi, Byeong Gi Lee and Sunghyun Choi

CDMA for Wireless Personal Communications, Ramjee Prasad

CDMA Mobile Radio Design, John B. Groe and Lawrence E. Larson

CDMA RF System Engineering, Samuel C. Yang

CDMA Systems Capacity Engineering, Kiseon Kim and Insoo Koo

CDMA Systems Engineering Handbook, Jhong S. Lee and Leonard E. Miller

Cell Planning for Wireless Communications, Manuel F. Cátedra and Jesús Pérez-Arriaga

Cellular Communications: Worldwide Market Development, Garry A. Garrard

Cellular Mobile Systems Engineering, Saleh Faruque

Cognitive Radio Interoperability through Waveform Reconfiguration, Leszek Lechowicz and Mieczyslaw M. Kokar

Cognitive Radio Techniques: Spectrum Sensing, Interference Mitigation, and Localization, Kandeepan Sithamparanathan and Andrea Giorgetti

The Complete Wireless Communications Professional: A Guide for Engineers and Managers, William Webb

Digital Communication Systems Engineering with Software-Defined Radio, Di Pu and Alexander M. Wyglinski

EDGE for Mobile Internet, Emmanuel Seurre, Patrick Savelli, and Pierre-Jean Pietri

Emerging Public Safety Wireless Communication Systems, Robert I. Desourdis, Jr., et al.

From LTE to LTE-Advanced Pro and 5G, Moe Rahnema and Marcin Dryjanski

The Future of Wireless Communications, William Webb

Geographic Information Systems Demystified, Stephen R. Galati

Geospatial Computing in Mobile Devices, Ruizhi Chen and Robert Guinness

GPRS for Mobile Internet, Emmanuel Seurre, Patrick Savelli, and Pierre-Jean Pietri

GPRS: Gateway to Third Generation Mobile Networks, Gunnar Heine and Holger Sagkob

GSM and Personal Communications Handbook, Siegmund M. Redl, Matthias K. Weber, and Malcolm W. Oliphant

GSM Networks: Protocols, Terminology, and Implementation, Gunnar Heine

GSM System Engineering, Asha Mehrotra

Handbook of Land-Mobile Radio System Coverage, Garry C. Hess

Handbook of Mobile Radio Networks, Sami Tabbane

High-Speed Wireless ATM and LANs, Benny Bing

Inside Bluetooth Low Energy, Second Edition, Naresh Gupta

Interference Analysis and Reduction for Wireless Systems, Peter Stavroulakis

Interference and Resource Management in Heterogeneous Wireless Networks, Jiandong Li, Min Sheng, Xijun Wang, and Hongguang Sun

Internet Technologies for Fixed and Mobile Networks, Toni Janevski

Introduction to 3G Mobile Communications, Second Edition, Juha Korhonen

Introduction to 4G Mobile Communications, Juha Korhonen

Introduction to Communication Systems Simulation, Maurice Schiff

Introduction to Digital Professional Mobile Radio, Hans-Peter A. Ketterling

Introduction to GPS: The Global Positioning System, Ahmed El-Rabbany

An Introduction to GSM, Siegmund M. Redl, Matthias K. Weber, and Malcolm W. Oliphant

Introduction to Mobile Communications Engineering, José M. Hernando and F. Pérez-Fontán

Introduction to Radio Propagation for Fixed and Mobile Communications, John Doble

Introduction to Wireless Local Loop, Broadband and Narrowband, Systems, Second Edition, William Webb

IS-136 TDMA Technology, Economics, and Services, Lawrence Harte, Adrian Smith, and Charles A. Jacobs

Location Management and Routing in Mobile Wireless Networks, Amitava Mukherjee, Somprakash Bandyopadhyay, and Debashis Saha

LTE Air Interface Protocols, Mohammad T. Kawser

Metro Ethernet Services for LTE Backhaul, Roman Krzanowski

Mobile Data Communications Systems, Peter Wong and David Britland

Mobile IP Technology for M-Business, Mark Norris

Mobile Satellite Communications, Shingo Ohmori, Hiromitsu Wakana, and Seiichiro Kawase

Mobile Telecommunications Standards: GSM, UMTS, TETRA, and ERMES, Rudi Bekkers

Mobile-to-Mobile Wireless Channels, Alenka Zajić

Mobile Telecommunications: Standards, Regulation, and Applications, Rudi Bekkers and Jan Smits

Multiantenna Digital Radio Transmission, Massimiliano Martone

Multiantenna Wireless Communications Systems, Sergio Barbarossa

Multi-Gigabit Microwave and Millimeter-Wave Wireless Communications, Jonathan Wells

Multipath Phenomena in Cellular Networks, Nathan Blaunstein and Jørgen Bach Andersen

Multiuser Detection in CDMA Mobile Terminals, Piero Castoldi

OFDMA for Broadband Wireless Access, Slawomir Pietrzyk

Personal Wireless Communication with DECT and PWT, John Phillips and Gerard MacNamee

Practical Wireless Data Modem Design, Jonathon Y. C. Cheah

Prime Codes with Applications to CDMA Optical and Wireless Networks, Guu-Chang Yang and Wing C. Kwong

Quantitative Analysis of Cognitive Radio and Network Performance, Preston Marshall

QoS in Integrated 3G Networks, Robert Lloyd-Evans

Radio Engineering for Wireless Communication and Sensor Applications, Antti V. Räisänen and Arto Lehto

Radio Propagation in Cellular Networks, Nathan Blaunstein

Radio Resource Management for Wireless Networks, Jens Zander and Seong-Lyun Kim

Radiowave Propagation and Antennas for Personal Communications, Third Edition, Kazimierz Siwiak and Yasaman Bahreini

RDS: The Radio Data System, Dietmar Kopitz and Bev Marks

Resource Allocation in Hierarchical Cellular Systems, Lauro Ortigoza-Guerrero and A. Hamid Aghvami

RF and Baseband Techniques for Software-Defined Radio, Peter B. Kenington

RF and Microwave Circuit Design for Wireless Communications, Lawrence E. Larson, editor

RF Positioning: Fundamentals, Applications, and Tools, Rafael Saraiva Campos and Lisandro Lovisolo

Sample Rate Conversion in Software Configurable Radios, Tim Hentschel

Signal Processing Applications in CDMA Communications, Hui Liu

Signal Processing for RF Circuit Impairment Mitigation, Xinping Huang, Zhiwen Zhu, and Henry Leung

Smart Antenna Engineering, Ahmed El Zooghby

Software-Defined Radio for Engineers, Travis F. Collins, Robin Getz, Di Pu, and Alexander M. Wyglinski

Software Defined Radio for 3G, Paul Burns

Spread Spectrum CDMA Systems for Wireless Communications, Savo G. Glisic and Branka Vucetic

Technical Foundations of the IoT, Boris Adryan, Dominik Obermaier, and Paul Fremantle

Technologies and Systems for Access and Transport Networks, Jan A. Audestad

Third-Generation and Wideband HF Radio Communications, Eric E. Johnson, Eric Koski, William N. Furman, Mark Jorgenson, and John Nieto

Third Generation Wireless Systems, Volume 1: Post-Shannon Signal Architectures, George M. Calhoun

Traffic Analysis and Design of Wireless IP Networks, Toni Janevski

Transmission Systems Design Handbook for Wireless Networks, Harvey Lehpamer

UMTS and Mobile Computing, Alexander Joseph Huber and Josef Franz Huber

Understanding Cellular Radio, William Webb

Understanding Digital PCS: The TDMA Standard, Cameron Kelly Coursey

Understanding GPS: Principles and Applications, Second Edition, Elliott D. Kaplan and Christopher J. Hegarty, editors

Understanding WAP: Wireless Applications, Devices, and Services, Marcel van der Heijden and Marcus Taylor, editors

Universal Wireless Personal Communications, Ramjee Prasad

WCDMA: Towards IP Mobility and Mobile Internet, Tero Ojanperä and Ramjee Prasad, editors

Wireless Communications in Developing Countries: Cellular and Satellite Systems, Rachael E. Schwartz

Wireless Communications Evolution to 3G and Beyond, Saad Z. Asif

Wireless Intelligent Networking, Gerry Christensen, Paul G. Florack and Robert Duncan

Wireless LAN Standards and Applications, Asunción Santamaría and Francisco J. López-Hernández, editors

Wireless Sensor and Ad Hoc Networks Under Diversified Network Scenarios, Subir Kumar Sarkar

Wireless Technician's Handbook, Second Edition, Andrew Miceli

For further information on these and other Artech House titles,including previously considered out-of-print books now available through our In-Print-Forever® (IPF®) program, contact:

Artech House
685 Canton Street
Norwood, MA 02062
Phone: 781-769-9750
Fax: 781-769-6334
e-mail: artech@artechhouse.com

Artech House
16 Sussex Street
London SW1V 4RW UK
Phone: +44 (0)20 7596-8750
Fax: +44 (0)20 7630-0166
e-mail: artech-uk@artechhouse.com

Find us on the World Wide Web at: www.artechhouse.com

www.ingramcontent.com/pod-product-compliance
Ingram Content Group UK Ltd.
Pitfield, Milton Keynes, MK11 3LW, UK
UKHW052015210726
473529UK00004B/17

9 781630 814571